MODERN SURVEY SAMPLING

MODERN SURVEY SAMPLING

ARIJIT CHAUDHURI

INDIAN STATISTICAL INSTITUTE,
KOLKATA, INDIA

CRC Press
Taylor & Francis Group
Boca Raton London New York

CRC Press is an imprint of the
Taylor & Francis Group, an **informa** business

A CHAPMAN & HALL BOOK

CRC Press
Taylor & Francis Group
6000 Broken Sound Parkway NW, Suite 300
Boca Raton, FL 33487-2742

© 2014 by Taylor & Francis Group, LLC
CRC Press is an imprint of Taylor & Francis Group, an Informa business

No claim to original U.S. Government works

Printed on acid-free paper
Version Date: 20140505

International Standard Book Number-13: 978-1-4665-7260-7 (Hardback)

This book contains information obtained from authentic and highly regarded sources. Reasonable efforts have been made to publish reliable data and information, but the author and publisher cannot assume responsibility for the validity of all materials or the consequences of their use. The authors and publishers have attempted to trace the copyright holders of all material reproduced in this publication and apologize to copyright holders if permission to publish in this form has not been obtained. If any copyright material has not been acknowledged please write and let us know so we may rectify in any future reprint.

Except as permitted under U.S. Copyright Law, no part of this book may be reprinted, reproduced, transmitted, or utilized in any form by any electronic, mechanical, or other means, now known or hereafter invented, including photocopying, microfilming, and recording, or in any information storage or retrieval system, without written permission from the publishers.

For permission to photocopy or use material electronically from this work, please access www.copyright.com (http://www.copyright.com/) or contact the Copyright Clearance Center, Inc. (CCC), 222 Rosewood Drive, Danvers, MA 01923, 978-750-8400. CCC is a not-for-profit organization that provides licenses and registration for a variety of users. For organizations that have been granted a photocopy license by the CCC, a separate system of payment has been arranged.

Trademark Notice: Product or corporate names may be trademarks or registered trademarks, and are used only for identification and explanation without intent to infringe.

Visit the Taylor & Francis Web site at
http://www.taylorandfrancis.com

and the CRC Press Web site at
http://www.crcpress.com

Dedication

*To
Bulu*

Contents

Chapter 1   Exposure to Sampling 1

    1.0   Abstract 1
    1.1   Introduction 1
    1.2   Concepts of Population, Sample, and Sampling 2

Chapter 2   Initial Ramifications 7

    2.0   Abstract 7
    2.1   Introduction 7
    2.2   Sampling Design, Sampling Scheme 7
    2.3   Random Numbers and Their Uses in Simple Random Sampling (SRS) 9
    2.4   Drawing Simple Random Samples with and without Replacement 10
    2.5   Estimation of Mean, Total, Ratio of Totals/Means: Variance and Variance Estimation 11
    2.6   Determination of Sample Sizes 17
    A.2   Appendix to Chapter 2 20
        A.2.1   More on Equal Probability Sampling 20
        A.2.2   Horvitz-Thompson Estimator 24
        A.2.3   Sufficiency 31
        A.2.4   Likelihood 33
        A.2.5   Non-Existence Theorem 34

Chapter 3   More Intricacies 41

    3.0   Abstract 41
    3.1   Introduction 41
    3.2   Unequal Probability Sampling Strategies 41
    3.3   PPS Sampling 43

Chapter 4   Exploring Improved Ways 69

    4.0   Abstract 69
    4.1   Introduction 69
    4.2   Stratified Sampling 70
    4.3   Cluster Sampling 78
    4.4   Multi-Stage Sampling 86
    4.5   Multi-Phase Sampling: Ratio and Regression Estimation 102

	4.6	Controlled Sampling 112
Chapter 5		Modeling 117
	5.1	Introduction 117
	5.2	Super-Population Modeling 118
	5.3	Prediction Approach 121
	5.4	Model-Assisted Approach 126
	5.5	Bayesian Methods 129
	5.6	Spatial Smoothing 133
	5.7	Sampling on Successive Occasions: Panel Rotation ... 134
	5.8	Non-Response and Not-at-Homes 140
	5.9	Weighting Adjustments and Imputation 145
	5.10	Time Series Approach in Repeated Sampling 147
Chapter 6		Stigmatizing Issues 149
	6.0	Abstract 149
	6.1	Introduction 149
	6.2	Early Growth of RR and the Current Status 150
		6.2.1 Warner (1965) 151
		6.2.2 Unrelated Question Model 153
		6.2.3 RRT with Quantitative Variables 154
	6.3	Optional Randomized Response Techniques 156
	6.4	Indirect Questioning 162
Chapter 7		Developing Small Domain Statistics 169
	7.0	Abstract 169
	7.1	Introduction 169
	7.2	Some Details 169
Chapter 8		Network and Adaptive Procedures 183
	8.0	Abstract 183
	8.1	Introduction 183
	8.2	Estimation by Network Sampling and Estimation by Adaptive Sampling 185
		8.2.1 Network Sampling and Estimation 185
		8.2.2 Adaptive Sampling and Estimation 187
	8.3	Constraining Network Sampling and Constraining Adaptive Sampling 187
		8.3.1 Network Sampling 187
		8.3.2 Constraining Adaptive Samples 191

Contents

Chapter 9   Analytical Methods 195
- 9.0   Abstract 195
- 9.1   Analytical Surveys: Contingency Tables 195
  - 9.1.1   Contingency 195
  - 9.1.2   Correlation, Regression Estimation 197
  - 9.1.3   Linearization 198
  - 9.1.4   Jack-knifing 202
  - 9.1.5   Bootstrap 204
  - 9.1.6   Permanent Random Numbers: Business Surveys 206
  - 9.1.7   Balanced Repeated Replication 209

A.1   Reviews and Further Openings 213

A.2   Case Studies 215

A.3   Exercises and Solutions Supplementaries 227

References 245

Author Index 255

Subject Index 261

List of Tables

2.1 Rational Choice of n in SRSWOR 18
2.2 Finding Sample Size in SRSWOR under Normal Approximation 19

9.1 Contingency Table of 1006 People Showing Their Frequencies in $4 \times 3 = 12$ Cells According to Their Financial Stature and Top Priority in Likings for Specific Kinds of Sport 196

Preface

At last this is my venture to address a textbook on sample surveys to an international readership.

This text is meant for students taking undergraduate-level university courses in statistics and also for those taking courses at the graduate and master levels in statistics.

The present author also wrote an elementary-level textbook called *Essentials of Survey Sampling* published by Prentice Hall of India in New Delhi in January 2010, which is supposed to be printed as a revised second edition in early 2014.

Survey Sampling: Theory & Methods is by the present author in collaboration with Professor H. Stenger of Mannheim University, Germany, with the first edition published in 1992 by Marcel Dekker, New York and a thoroughly revised and enhanced second edition published by Chapman & Hall/CRC, Taylor & Francis Group, Boca Raton, FL in 2005.

All three texts present a sizable amount of material on various aspects of survey sampling. Yet, there is a need for what may be considered an international-level textbook on survey sampling despite there being quite a few publications available in the international market on the present subject.

As I have been teaching this subject for numerous years at the Indian Statistical Institute and have considerable experience with this topic, having published profusely in numerous journals, I wish to share with those around the world my work and knowledge gained as a veteran teacher.

Applied Statistics Unit
Indian Statistical Institute,
203, B.T. Road,
Kolkata - 700 108, India

Arijit Chaudhuri
email: arijitchaudhuri1@rediffmail.com
November 2013

Acknowledgments

I am grateful to the Director of the Indian Statistical Institute for giving me an opportunity to continue to work so many years since retirement to teach and publish. Also, I appreciate the congenial atmosphere created by my colleagues in the Applied Statistics Division of the Institute for productive pursuits.

About the Author

Arijit Chaudhuri is Honorary Visiting Professor at the Indian Statistical Institute (ISI) after serving as a CSIR Emeritus Scientist and earlier as a Professor in ISI. He has a Ph.D. in Statistics from Calcutta University. He is an author of *Randomized Response and Indirect Questioning Techniques in Surveys* (Chapman & Hall, CRC Press, 2011), *Essentials of Survey Sampling* (Prentice Hall of India, 2010), and co-author of *Indirect Questioning in Sample Surveys* jointly with TC Christofides (Springer-Verlag, 2013), *Survey Sampling Theory and Methods* (1992 1st. edition, Marcel Dekker and 2nd. edition jointly with H. Stenger, Chapman & Hall, CRC Press, 2005), *Randomized Response: Theory & Techniques* (Marcel Dekker, 1988) jointly with Rahul Mukherjee, and *Unified Theory and Strategies of Survey Sampling* (North-Holland, 1988) jointly with late JWE Vos. The author has widely travelled with academic assignments in universities and statistical offices in the United States, Canada, England, Australia, Sweden, Germany, the Netherlands, South Africa, Japan, Turkey, Cuba, Cyprus, and Israel. He is the founder-chairman of a registered "Advanced Survey Research Centre" (website: www.asrc.net.in).

Arijit Chaudhuri
email: arijitchaudhuri1@rediffmail.com

1 Exposure to Sampling

Abstract. Introduction. Concepts of population, sample, and sampling.

1.0 ABSTRACT

A survey means organized observation with the purpose of reaching conclusions in a scientific manner. Observation relates to a totality, called a Population or a Universe, and a part thereof is a Sample. It is possible and useful to survey a sample to make inferences concerning parameters that mean characteristics of a population. In survey sampling a parameter typically is an unknown real number. Using observed real values for a selected sample, a derived real number is proposed as a possible value of the unknown parametric value or as a point estimator. As an alternative, an interval with this point estimator within it along with two numbers on either side of it is claimed to contain within itself the unknowable parametric value with a reasonable claim for such an assertion to hold true. This is Interval Estimation. An organizer of a Sample Survey is assigned a task to explain how to choose a sample appropriately to justify the specification from the observed sample values a point estimator and an interval estimator in a way acceptable to a scientific community.

1.1 INTRODUCTION

The dominating topic in Survey Sampling is estimation. A real-life problem demanding practical application is here the motivating factor. But to formulate, develop, and study a related theory, certain abstractions are naturally needed. We stumble first upon the conceptualization of what is called a Population or a Universe. This denotes the totality of all objects of interest in a given context. For example, all the building structures on a street in Kolkata constitute a Population. So, all the 50 states plus the National Capital Territory, Washington, D.C., form a universe in the context of all the components of the United States. A serious study of certain features of the constituent components of such a Population would be a tremendous task. So, handling only a few of these parts of a Population should appear reasonable while being cognizant of the general characteristics of all the elements of such Populations. To grasp an essential idea in such a situation, it is judged imperative to hit upon the concept of a Sample which, nontechnically speaking, is but a Part of a Population. For the individuals in the sample, respective values of one or more variables of interest are ascertained to the extent possible. Some suitable functions of the values on the sampled individuals are taken as suitable Statistics, combined judiciously, if necessary, with other available variate-values for the sampled units, or all the population members are employed to estimate the values of parameters representing characteristics of interest defined for the population of all the individuals concerned. If a statistic is worked out as a real number to be the value of a parameter of interest defined as a real number, then the statistic is treated as a point estimator, for the real-valued parameter. Sometimes an interval around the value of this point estimator taking some sampled variate-values suitably combined with certain other suitable constants, is constructed, claiming to contain within itself the parameter value with a high probability. This interval is called a Confidence Interval. The probability associated as above with such an interval is called the Confidence Coefficient.

The problem to be addressed by the Investigator is how to choose an appropriate sample and procedures for Point and Interval estimation for real-valued parameters defined on Populations of interest.

1.2 CONCEPTS OF POPULATION, SAMPLE, AND SAMPLING

By $U = (1,\ldots,i,\ldots,N)$ we denote a Population of a known number N of individuals also called members, units, or elements, in the context of Sample Surveys or Survey Sampling. Each element of U, say, i, is supposed to be identifiable and assigned labels as i for identification and referencing. For a population of all the villages in a given district in a province in a country, Names will be the identifier but will be substituted as label i, which bears the values $1, 2, 3, \ldots$ etc.

The crop-fields in a given village in their turn must also be tagged with such labels of positive integers for identification and referrals, though thus tagged every element is a tangible, concrete object in a Survey Population which is finite with a known number N of objects. If we consider a pond in a city or town or a village the number of fishes in it is of course finite at every instant of time, but this number for all practical purposes must be considered as unknown. So, the theory we are going to establish cannot cover such a concept as a Population of fishes in a given pond. This is because no individual fish in a pond can be identified and tagged without causing damage to its life and liberty. However, taking special care for such contingencies, a modified theory will be described in brief to cover such populations which are finite but composed of unknown and unidentifiable entities.

Treating $i_1, i_2, \ldots, i_n$ each as one of the labels i in U, we shall denote by the sequence $(i_1,\ldots,i_j,\ldots,i_n) = s$, a Sample from U. Here, the order in which the labels occur in s is important and such an s is called an "ordered" sample from U. These labels in s need not all be "distinct." Yet number of labels in s is recognized as n, and this n is called the "size" of the Sample s. But by $v(s)$ we denote the number of Distinct units in s, and this $v(s)$ is called the "Effective Size" of the sample s. Of course, $1 <= v(s) <= n$. By Sampling we mean the act of selecting a sample from the population. In order that a scientific theory may be developed for sample selection and estimation of

Population Parameters using the samples drawn and surveyed, it is useful if one chooses a sample s with a pre-assigned probability, denoted, say, as $p(s)$. Since it is a probability, of necessity, we must have

(i) $0 <= p(s) <= 1$

(ii) $\sum_s p(s) = 1$.

Here $\sum_s$ denotes summing over all possible choices of a sample s from U.

A sample is more usefully defined as a set $s^* = \{i_1, \ldots, i_n\}$ of distinct labels $i_1, i_2, \ldots, i_n$ which are the n distinct entities of U. Here it is immaterial in which order the respective labels of U are written in s^*. Here n is taken as the Size of the sample s^* and it is obviously its effective size as well. Here $\binom{N}{n}$ is the total number of possible such unordered samples of n distinct units drawable from the Population U of N units. By $p(s^*)$ we mean the selection-probability of s^* from U, and we need

(i) $0 <= p(s^*) <= 1$

(ii) $\sum_{s^*} p(s^*) = 1$.

Here $\sum_{s^*}$ denotes sum over all the $\binom{N}{n}$ possible number of ways defining the samples like s^* that may be chosen from U.

Any such function p defined on the totality of all possible samples like s or s^* from U described above with the two specified properties is called a Sampling Design. In practice, for simplicity, we shall write s to denote either a Sequence-type or a Set-type sample avoiding the cumbrous symbol s^* unless it is crucial to stress that we mean to imply an "unordered" sample of only "distinct units."

We shall throughout, unless emphasized otherwise, mean to use a sample s to be chosen with a certain probability $p(s)$ employing a design (more elaborately a sampling design) p. Thus, s is a random variable.

In surveying a sample our concern will be to observe the values y_i for a main real variable y of interest for the respective units i in the sample s actually chosen to be surveyed. Of course the real values y_i are defined for every $i(=1, 2, \ldots, N)$ in the population U defining the vector $\underline{Y} = (y_1, \ldots, y_i, \ldots, y_N)$. Likewise, other real variables x, z, w, etc., are with respective values x_i, z_i, w_i, etc., for i in U and the vectors

$$\underline{X} = (x_1, \ldots, x_i, \ldots, x_N),$$
$$\underline{Z} = (z_1, \ldots, z_i, \ldots, z_N),$$
$$\underline{W} = (w_1, \ldots, w_i, \ldots, w_N).$$

The prime interest in Sample Surveys is to suitably estimate the population total $Y = \sum_1^N y_i$ of y using the survey data denoted $d = (s, y_i | i \in s)$ along with knowledge of $\underline{Z}, \underline{W}$, etc., plus partial knowledge at least of $\underline{X}$ as, for example, the value of $X = \sum_1^N x_i$ being prima facie known. Let $t' = t(d) = t(s, \underline{Y})$ with the restriction that $t(s, \underline{Y})$ does not involve any y_i in $\underline{Y}$ unless i is in s (i.e., it is free of y_j for $j \notin s$).

Such a function t of d is called a statistic. Appreciating that $\underline{Y}$ is a vector of fixed but unknown values y_i for $i \in s$, this $t(s, \underline{Y})$ is a random variable because it is a function of the random variable s even though the other component in $t(s, \underline{Y})$ is a constant. So, we may define

$$E_p(t) = \sum_s p(s) t(s, \underline{Y})$$

as the expectation of t with respect to the design p which provides $t(s, \underline{Y})$ its multiplier in $E_p(t)$.

Such a function t of $t(s, \underline{Y})$ may be employed to estimate Y. The unknowable value of $t - Y = t(s, \underline{Y}) - Y$ is called the error in estimating Y by the value of $t(s, \underline{Y})$ for a sample s at hand. The expected value of this error viz.

$$E_p(t - Y) = B_p(t) = \sum_s p(s)(t(s, \underline{Y}) - Y)$$

is defined and called the Bias of t in estimating Y using the data

$$d = (s, y_i | i \in s)$$

on choosing the sample s on implementing the design p. The square error $(t - Y)^2$ has the expectation

$$E_p(t - Y)^2 = \sum_s p(s)(t(s, \underline{Y}) - Y)^2$$

called the Mean Square Error (MSE) of t in estimating Y.

Again $$V_p(t) = \sigma^2 = \sigma_p^2(t) = E_p(t - E_p(t))^2$$

is defined. This is called the variance of t in respect of the design p which has given the probability $p(s)$ to the sample s to be selected for being surveyed, yielding the data d and the estimator t for Y.

Clearly, $\qquad\qquad$ MSE $= V_p(t) + B_p^2(t)$

i.e., $\qquad\qquad$ MSE $=$ Variance $+$ Squared Bias.

The estimator t for Y is called a Point Estimator for Y, as it is just a value that is a real number proposed to represent the value Y which is just an unknown real number. The performance characteristics of t as a Point Estimator for

Y are the quantities $E_p(t)$, $B_p(t)$, $M_p(t)$, and $V_p(t) = \sigma_p^2(t)$. The quantity $\sigma_p(t) = +\sqrt{V_p(t)}$ is called the Standard Error of t.

Let us consider the expanded quantity

$$M_p(t) = \sum_s p(s)(t-Y)^2 = \sum_1 p(t)(t(s,\underline{Y}) - Y)^2 + \sum_2 p(s)(t(s,\underline{Y}) - Y)^2,$$

denoting by $\sum_1$ the sum over the samples s for which $|(t(s,\underline{Y}) - Y)|$ exceeds a positive number K, briefly

$$\sum_1 = \sum_{s:|t(s,\underline{Y})-Y| \geq K > 0} \quad \text{and} \quad \sum_2$$

denoting the sum over the samples s in the complementary set so that

$$\sum_2 = \sum_{s:|t(s,\underline{Y})-Y| < K}.$$

Then it follows that

$$MSE \geq K^2 \sum_{s:|t(s,\underline{Y})-Y| \geq K} = K^2 \text{Prob}[|t(s,\underline{Y}) - Y| \geq K]$$

writing Prob[.] for the probability of the event denoted by the symbol[.].

Hence, it follows that

$$\text{Prob}[|t(s,\underline{Y}) - Y| \geq K] \leq \frac{E_p(t-Y)^2}{K^2}$$

or
$$\text{Prob}[|t(s,\underline{Y}) - Y| \geq K] \leq \frac{\sigma_p^2(t) + B_p^2(t)}{K^2}$$

Choosing a positive number λ such that $K = \lambda \sigma_p(t)$ it follows that

$$\text{Prob}[|t - Y| \geq \lambda \sigma_p(t)] \leq \frac{1}{\lambda^2} + \frac{1}{\lambda^2}\left(\frac{B_p(t)}{\sigma_p(t)}\right)^2$$

or
$$\text{Prob}[t - \lambda \sigma_p(t) \leq Y \leq t + \lambda \sigma_p(t)] \geq (1 - \frac{1}{\lambda^2}) - \frac{1}{\lambda^2}\left(\frac{B_p(t)}{\sigma_p(t)}\right)^2$$

So, whatever may be the vector $\underline{Y}$ of real numbers $y_i, i \in U$, it follows that the random interval

$$CI = (t - \lambda \sigma_p(t), t + \lambda \sigma_p(t))$$

contains the unknown number $Y = \sum_1^N y_i$ within itself with a probability at least as high as

$$(1 - \frac{1}{\lambda^2}) - \frac{1}{\lambda^2}\left(\frac{B_p(t)}{\sigma_p(t)}\right)^2 = CC$$

This interval CI is called a Confidence Interval covering the parameter Y within itself with a Confidence Coefficient at least as high as this number CC. Reporting this about this CI in terms of CC is called Interval Estimation of Y. The situation simplifies greatly for an estimator t for Y ensuring $B_p(t) = 0$ for every $\underline{Y}$. Such a t is called an unbiased estimator for Y. In such a case $(t - \lambda\sigma_p(t), t + \lambda\sigma_p(t))$ provides a Confidence Interval for Y with a Confidence Coefficient at least as high as $(1 - \frac{1}{\lambda^2})$.

The quantity $2\lambda\sigma_p$ gives the width of the Confidence Interval. It is desirable to have a Confidence Interval with a small width. Thus it is desirable to employ for Y an unbiased point estimator t with $\sigma_p(t)$ small in magnitude for any prescribed choice of λ for which $(1 - \frac{1}{\lambda^2})$ should be as large as, say, equal to 0.95 or 0.99, giving us a CI with a CC at least as high as 95 or 99. The problem of sampling then for an investigator to solve is stipulating a sampling design p, throwing up the survey data d and a point estimator which is unbiased for Y, admitting a small $\sigma_p(t)$ so as to provide an accurate estimation rule, and as a by product yielding a CI with a desirably small width and a high Confidence Coefficient.

2 Initial Ramifications

Abstract. Introduction. Sampling design, sampling scheme. Random numbers and their uses in random sampling (SRS). Drawing simple random samples with and without replacement. Estimation of mean, total, ratio of totals/means: variance and variance estimation. Determination of sample sizes. Appendix to Chapter 2.

2.0 ABSTRACT

Only a few rudimentary concepts related to survey sampling are briefly set forth in this chapter. How to select a sample for which a selection-probability is specified to prescribe its performance characteristics is to be clearly narrated. Concepts of random samples and simple random samples are to be laid bare. How many samples are to be chosen and with what requirements also need to be clarified. How to measure accuracy in terms of unknowable features and how to assess its realization through measurements are briefly discussed.

2.1 INTRODUCTION

In Chapter 1 we noted that a finite Survey Population $U = (1, 2, \ldots, i, \ldots, N)$ containing N distinctly identifiable units labeled, respectively, as $1, 2, \ldots, i, \ldots, N$ is said to have N as its size which is a finite positive integer known to the investigator. A sample s from U may be either a sequence of a finite number of labels ordered successively as the first, second, etc., to the nth element $i_1, i_2, \ldots, i_n$ each of which is one of the labels of the units of U. Thus, $s = (i_1, \ldots, i_j, \ldots, i_n)$, and n is the size of this ordered sample and the elements in this sequence s need not all be distinct. Alternatively a sample s from U may denote a set of n distinct units of U with no regard for the order of succession in which these labels are inserted in the sample s which is a set of n distinct units of U. This number n is the size of the sample s. Here $1 \leq n \leq N$. In case a sample s is a sequence of ordered units n in number, not all of which need to be distinct from one another, this sample s, though it has the size as n, also has an Effective Size which is the number of distinct labels out of these total of n labels that are contained in s. How to draw samples with what rationales and how to use them with what purpose are issues to be settled.

2.2 SAMPLING DESIGN, SAMPLING SCHEME

By Sampling Design we mean a probability measure p that assigns to a sample s a selection-probability $p(s)$ with two properties as discussed in Chapter 1. Its principal role is to guide us in developing a unified theory of estimating from survey data Population parameters of theoretical and practical interest.

In contrast a Sampling Scheme specifies methods of actual selection of samples. When so laid down, a sampling scheme prescribes probabilities of selection of samples. Thus, a sampling scheme develops a sampling design. Correspondingly, given a sampling design, it is possible to work out a sampling scheme. Let us see some details.

If for every possible sample s from U there is given a procedure for its selection with a given probability $p(s)$, say, then a sampling design p is already formed. To grasp the idea of its converse also being true—that is, given a sampling design p assigning to a sample s its selection-probability $p(s)$—let us refer to Hanurav's (1962) classical device which gives us a corresponding procedure of choosing such a sample according to an actual scheme of selection by a one-by-one draw of units from the population.

Let $s = (i_1, \ldots, i_j, \ldots, i_n)$ denote a sample of which i_j is the unit chosen from U assigned the jth label for $j = 1, \ldots, n$. Let $p(s)$ values for each such s be given. Hanurav (1962) gives a draw-by-draw procedure for its selection.

Let $p(i_1)$ = probability of choosing the singleton sample (i_1),

$p(i_1, i_2)$ = probability of choosing the sample (i_1, i_2), etc., and finally,

$p(i_1, \ldots, i_j, \ldots, i_n)$ = given probability of choosing $s = (i_1, \ldots, i_j, \ldots, i_n)$

Let $\alpha_{i_1} = \sum_1 p(s)$, with $\sum_1$ as the sum over all samples with i_1 as its first element; $\alpha_{i_1 i_2} = \sum_2 p(s)$ with $\sum_2$ as the sum over all samples of which i_1, i_2 are the first two units and so on; and finally $\alpha_{i_1 i_2 \ldots i_n} = \sum_n p(s)$ with $\sum_n$ as the sum over all samples with $i_1, i_2, \ldots, i_n$ as the first n units in the samples. Then Hanurav's (1962) sampling scheme specifies the following:

1. Make the first draw from U with probability α_{i_1} to get i_1 as the first unit in the sample s.
2. Then, implement a Bernoullian trial with $\left(1 - \frac{\beta_{i_1}}{\alpha_{i_1}}\right)$ as the probability of "success" stipulating to make another draw only on realizing a "success," stopping further exercise in case of a "failure."
3. Then, if a 'success' results draw the unit i_2 with probability $\frac{\alpha_{i_1 i_2}}{\alpha_{i_1} - \beta_{i_1}}$.
4. Next perform another Bernoulli trial with probability of success $\left(1 - \frac{\beta_{i_1 i_2}}{\alpha_{i_1 i_2}}\right)$ with a similar stipulation as in step (2). If a failure results in step (4), one ends up getting the sample (i_1, i_2) with probability

$$\alpha_{i_1} \left(\frac{\alpha_{i_1} - \beta_{i_1}}{\alpha_{i_1}}\right) \frac{\alpha_{i_1 i_2}}{\alpha_{i_1} - \beta_{i_1}} \frac{\beta_{i_1 i_2}}{\alpha_{i_1 i_2}} = \beta_{i_1 i_2}.$$

With similar additional steps one chooses following this procedure a sample $s = (i_1, \ldots, i_j \ldots, i_n)$ with probability

$$\alpha_{i_1} \left(\frac{\alpha_{i_1} - \beta_{i_1}}{\alpha_{i_1}}\right) \times \frac{\alpha_{i_1 i_2}}{\alpha_{i_1} - \beta_{i_1}} \times \cdots \times \frac{\beta_{i_1 i_2 \cdots i_n}}{\alpha_{i_1 \cdots i_n}} = \beta_{i_1 i_2 \cdots i_n}.$$

Initial Ramifications

According to this selection scheme one ends up with a singleton sample $s = (i_1)$ with probability $p(i_1) = \alpha_{i_1}(\frac{\beta_{i_1}}{\alpha_{i_1}}) = \beta_{i_1}$. The Hanurav (1962) selection scheme as above obviously is not a commendable sampling procedure. But its merit consists in establishing Proposition 1. Given a sampling design, a sampling scheme is available corresponding to it.

Another such documentation is possible thanks to Chaudhuri (2010) as noted below on recognizing the exixtence of Random Numbers in the context of Survey Sampling.

2.3 RANDOM NUMBERS AND THEIR USES IN SIMPLE RANDOM SAMPLING (SRS)

A very large sequence of positive one-digit integers and the number zero when arranged side by side, even spread over different pages in a book in such a way that every such number anywhere in the sequence occurs with a common probability $\frac{1}{10}$ and moreover any group of consecutive $k\ (>1)$ such numbers occurs in the sequence Independently of each other and the probability of occurrence of any such k consecutive single positive integers or zeroes is equal to $1/10^k$. Such Random Numbers are easy to construct using computers, PCs, laptops, or even cell phones. But such constructions are always implemented following arithmetical rules and procedures. So, the sequences of numbers so constructed are called Pseudo Random numbers. They attain the status of Random numbers provided they satisfy statistical tests for randomness. For example, on taking a big enough sequence of such consecutive pseudo numbers it is usual to count the frequencies of number $0, 1, 2, \ldots, 8, 9$, say, f_i for $i = 0, 1, \ldots, 9$.

Suppose that M is the total number of such single digits or zero considered, i.e., $M = \sum_{i=0}^{9} f_i$. Then $\frac{f_i}{M} = r_i$, the proportion of the ith digit found, $i = 0, 1, \ldots, 9$ and may be called the relative frequency of i. If the pseudo numbers are really random numbers, since every digit i occurs with the common probability $\frac{1}{10}$, it follows that the expected value of this r_i is $\frac{1}{10}$ for each of these 10 relative frequencies.

Consequently, $$\chi_9^2 = \sum_{i=0}^{9} \left(r_i - \frac{1}{10}\right)^2 \Big/ \left(\frac{1}{10}\right)$$

follows the chi-square distribution with 9 degrees of freedom because $\sum_0^9 r_i$ equals unity causing a reduction by 1 in the degrees of freedom in the variability of $r_i's, i = 0, 1, \ldots, 9$. So, a χ^2-test is available to check if the Pseudo Random numbers may really be treated as random numbers. More such tests are available in the Statistics literature. Conventionally every Table of Random numbers is accompanied by the facts of several tests for randomness applied on them with positive results.

Chaudhuri (2010) illustrates as follows how random numbers may be used in drawing samples from a finite survey population $U = (1, \ldots, N)$ of a given number of units when a Sampling Design for it is prescribed. Suppose T is the total number of possible samples identified by labels $1, 2, \ldots, T$ such that positive selection probabilities $p_1, p_2, \ldots, p_T$ are respectively associated with them such that $p_1 + p_2 + \ldots + p_T = 1$. Let K be a sufficiently large positive number yielding Kp_j as a positive integer equal to I_j so that $I = I_1 + \ldots + I_T$ is also a positive integer. Let $C_j = I_1 + \ldots + I_j$ and obviously $C_T = I$. Now, let a random number R be read such that $1 \leq R \leq I$.

If $C_{j-1} < R \leq C_j$, $j = 1, \ldots, T$, taking $C_0 = 0$, then the sample j is to be selected. Its selection-probability is

$$\frac{C_j - C_{j-1}}{I} = p_j, \ j = 1, \ldots, T \qquad \text{as desired.}$$

Let us proceed to illustrate further uses of Tables of Random Numbers in working out further sample selection schemes.

2.4 DRAWING SIMPLE RANDOM SAMPLES WITH AND WITHOUT REPLACEMENT

In our context of finite survey population $U = (1, \ldots, i, \ldots, N)$ we have sufficiently discussed probability samples. When out of the totality of all possible samples every sample is given an equal probability of selection, it is called a Simple Random Sample (SRS).

There are two types of such simple random samples as follows. In one, every unit i of U is given the same probability $\frac{1}{N}$ of selection on the first draw; the draws are independently repeated a given number of n times. Consequently, such a sample is of the sequence type, the labels chosen are ordered, and they need not all be distinct. Writing such a Sequence-type sample as $s = (i_1, i_2, \ldots, i_n)$, each i_j being one of the labels of U, $j = 1, \ldots, n$, its selection-probability is $p(s) = \frac{1}{N^n}$. Such a sample is called a Simple Random Sample (SRS) with Replacement (WR). For such an SRSWR of n units, ordered but permitted to be repeated more than once, the sampling design p and the scheme of sampling are thus clearly stated.

Another form of Simple Random Sampling (SRS), without Replacement (WOR), considers all possible sets of distinct labels, say n in numbers, out of those in $U = (I, \ldots, N)$, $\binom{N}{n}$ being the numbers of possible such sets, assigns the probability of selection as $\frac{1}{\binom{N}{n}}$ to each set of unordered distinct units of U. Such a sampling is known as Simple Random Sampling without Replacement (SRSWOR). SRSWOR sampling Design is thus clearly stated. To specify a corresponding Sampling Scheme there are two alternative ways. The respective unordered sample of n distinct units of U may be denoted as $1, 2, \ldots, T = \binom{N}{n}$, and choosing a Random Number R from 1 to $\binom{N}{n}$, one may

Initial Ramifications

choose the Rth such a sample. Its selection-probability will be

$$\frac{1}{T} = \frac{1}{\binom{N}{n}} \quad \text{for every} \quad R = 1, 2, \ldots, T = \binom{N}{n}.$$

Alternatively, we may on a first draw, using a table of random numbers, choosing R between 1 and N, take one label from $U = (I, \ldots, N)$, then omitting this label chosen, draw a second random number from the remaining $(N-1)$ units of U and repeat this procedure to the nth draw. Ignoring the order in which the respectively chosen labels appear, one gets the final sample with the realized probability:

$$\left\{ \frac{1}{N} \cdot \frac{1}{N-1} \cdot \frac{1}{N-2} \cdots \cdot \frac{1}{(N-n+1)} \right\} n! = \frac{(N-n)!n!}{N!} = \frac{1}{\binom{N}{n}}.$$

So, such a procedure of sample selection is also SRSWOR. Such a sample is denoted as $s^* = \{i_1, i_2, \ldots, i_n\}$ which is a set of "unordered" n distinct units of U.

In the next section we show application of SRSWR and SRSWOR in the estimation of parameters defined on U with respect to one or more real variables y, x, z, w, etc.

2.5 ESTIMATION OF MEAN, TOTAL, RATIO OF TOTALS/MEANS: VARIANCE AND VARIANCE ESTIMATION

The simplest parameters one needs to estimate in the context of Survey Sampling are the Population Total $Y = \sum_1^N y_i$ and also the mean $\bar{Y} = \frac{Y}{N}$, since the population size N is known.

Suppose s is an SRSWR chosen in $n (\geq 1)$ draws from $U = (1, \ldots, i, \ldots, N)$. Let y_r be the value of y for the unit chosen on the rth draw with probability $\frac{1}{N}$ independently for each draw, $r = 1, \ldots, n$.

Let
$$\bar{y} = \frac{1}{n} \sum_{r=1}^{n} y_r.$$

Then,
$$E_p(y_r) = \sum_{i=1}^{N} y_i \operatorname{Prob}(y_r = y_i) = \frac{1}{N} \sum_1^N y_i = \bar{Y}.$$

Thus, $\bar{y}$ is an unbiased estimator for $\bar{Y}$.

Also the variance of y_r for SRSWR is

$$V_p(y_r) = E_p(y_r^2) - (\bar{Y})^2 = \frac{1}{N} \sum_1^N y_i^2 - (\bar{Y})^2 = \frac{1}{N} \sum_1^N (y_i - \bar{Y})^2 = \sigma^2, \quad \text{say}.$$

Then, $$V_p(\bar{y}) = \frac{1}{n^2} \sum_{r=1}^{n} V_p(y_r) = \frac{\sigma^2}{n}.$$

Let $$s^2 = \frac{1}{(n-1)} \sum_{r=1}^{n} (y_r - \bar{y})^2.$$

$$E_p(s^2) = E_p \frac{1}{(n-1)} \left[\left(\sum_{1}^{n} y_r^2 - n\bar{y}^2 \right) \right] = \frac{1}{n-1} \left[\sum_{1}^{n} E_p\left(y_r^2\right) - nE_p\left(\bar{y}\right)^2 \right]$$

$$= \frac{1}{n-1} \left[\sum_{1}^{n} \left(V_p(y_r) + (\bar{Y})^2 \right) - n \left(V_p(\bar{y}) + (\bar{Y})^2 \right) \right]$$

$$= \frac{1}{n-1} \left[n\sigma^2 + n(\bar{Y})^2 - \sigma^2 - n(\bar{Y})^2 \right] = \sigma^2.$$

So, $v = \dfrac{s^2}{n}$ has $E_p(v) = \dfrac{\sigma^2}{n}.$

Hence, v is an unbiased estimator for $V(\bar{y})$. Needless to say,

$$\hat{Y} = N\bar{y} \quad \text{is an unbiased estimator for} \quad Y, V_p(\hat{Y}) = N^2 \frac{\sigma^2}{n},$$

and $N^2 \dfrac{s^2}{n}$ is an unbiased estimator for $V_p(\hat{Y})$.

Suppose s is an SRSWOR in n draws from a finite population:

$$U = (1, \ldots, N) \quad \text{and} \quad \bar{y} = \frac{1}{n} \sum_{i \in s} y_i$$

is the sample mean of the values y_i for i in s.

Then,

$$E_P(\bar{y}) = \frac{1}{\binom{N}{n}} \sum_{s} \left(\frac{1}{n} \sum_{i \in s} y_i \right) = \frac{1}{\binom{N}{n}} \frac{1}{n} \left[\sum_{s} \sum_{i \in s} y_i \right] = \frac{1}{\binom{N}{n}} \frac{1}{n} \left[\sum_{i=1}^{N} y_i \left(\sum_{s \ni i} 1 \right) \right]$$

$$= \frac{1}{\binom{N}{n}} \frac{1}{n} \sum_{i=1}^{N} y_i \binom{N-1}{n-1} = \frac{1}{N} \sum_{1}^{N} y_i = \bar{Y}.$$

So, $\bar{y}$ is a unbiased estimator for $\bar{Y}$. Obviously $N\bar{y}$ is an unbiased estimator of Y, and this is called the Expansion estimator for a Population total. Now,

Initial Ramifications

the variance of $\bar{y}$ is

$$V_p(\bar{y}) = E_p(\bar{y} - \bar{Y})^2 = E_p\left[\frac{1}{n}\sum_{i \in s}(y_i - \bar{Y})\right]^2$$

$$= E_p\left[\frac{1}{n^2}\left\{\sum_{i \in s}(y_i - \bar{Y})^2 + \sum\sum_{i \neq j \in s}(y_i - \bar{Y})(y_j - \bar{Y})\right\}\right]$$

$$= \frac{1}{n^2}\frac{1}{\binom{N}{n}}\left[\sum_s\sum_{i \in s}(y_i - \bar{Y})^2 + \sum_s\sum\sum_{i \neq j \in s}(y_i - \bar{Y})(y_j - \bar{Y})\right]$$

$$= \frac{1}{n^2}\frac{1}{\binom{N}{n}}\left[\sum_{i=1}^{N}(y_i - \bar{Y})^2\left(\sum_{s \ni i}\right)\right.$$

$$\left. + \sum_{i \neq j=1}^{N}\sum^{N}(y_i - \bar{Y})(y_j - \bar{Y})\left(\sum_{s \ni i,j}1\right)\right]$$

$$= \frac{1}{n^2}\frac{1}{\binom{N}{n}}\left[\binom{N-1}{n-1}\sum(y_i - \bar{Y})^2\right.$$

$$\left. + \binom{N-2}{n-2}\sum\sum_{i \neq j}(y_i - \bar{Y})(y_j - \bar{Y})\right]$$

$$= \frac{1}{n^2}\left[\frac{n}{N}\sum_1^N(y_i - \bar{Y})^2 + \frac{n(n-1)}{N(N-1)}\sum\sum_{i \neq j}(y_i - \bar{Y})(y_j - \bar{Y})\right]$$

$$= \frac{1}{nN}\sum_1^N(y_i - \bar{Y})^2 + \frac{n-1}{nN(N-1)}\left[\left\{\sum_1^N(y_i - \bar{Y})\right\}^2 - \sum(y_i - \bar{Y})^2\right]$$

$$= \frac{1}{nN}\left[1 - \frac{n-1}{N-1}\right]\sum_1^N(y_i - \bar{Y})^2 = \frac{N-n}{Nn}\frac{1}{N-1}\sum(y_i - \bar{Y})^2$$

$$= \frac{N-n}{Nn}S^2.$$

Writing conventionally, $\quad S^2 = \dfrac{1}{N-1}\sum_1^N(y_i - \bar{Y})^2,$

it follows that $\quad \dfrac{N-1}{N}S^2 = \sigma^2.$

So, $\quad V_p(\bar{y}) = \dfrac{N-n}{(N-1)}\dfrac{\sigma^2}{n}.$

For every $n \geq 2$, $\bar{y}$ based on SRSWOR has a smaller variance than $\bar{y}$ based on SRSWR with the same number n of draws in both.

Now, for
$$s^2 = \frac{1}{(n-1)} \sum_{i \in s} (y_i - \bar{Y})^2,$$

based on an SRSWOR in n draws from U we have

$$E_p(s^2) = \frac{1}{(n-1)} E_p \left[\sum_{i \in s} \left\{ (y_i - \bar{Y}) - (\bar{y} - \bar{Y})^2 \right\} \right]$$

$$= \frac{1}{(n-1)} \left[E_p \left\{ \sum_{i \in s} (y_i - \bar{Y})^2 - n (\bar{y} - \bar{Y})^2 \right\} \right]$$

$$= \frac{1}{n-1} \left[\frac{1}{\binom{N}{n}} \sum_{s} \sum_{i \in s} (y_i - \bar{Y})^2 - n V_p (\bar{y}) \right]$$

$$= \frac{1}{(n-1)} \left[\frac{\binom{N-1}{n-1}}{\binom{N}{n}} \sum_{i=1}^{N} (y_i - \bar{Y})^2 - n \frac{N-1}{Nn} S^2 \right]$$

So, $\quad v = \dfrac{N-n}{nN} s^2$ is an unbiased estimator for $V_p(\bar{y}) = \dfrac{N-n}{Nn} S^2$,

and $N \left(\dfrac{N-n}{n} \right) s^2$ is an unbiased estimator for $V_p(N\bar{y}) = N \left(\dfrac{N-n}{n} \right) S^2$.

Let us next consider estimating the ratio $R = \frac{Y}{X} = \frac{\bar{Y}}{\bar{X}}$ of the totals or means of two variables y and x on taking an SRSWOR of size n from a finite population $U = (1, \ldots, N)$.

It is interesting to consider the situation when X or $\bar{X}$ is known for a reason soon to be explained.

Let $\bar{x} = \frac{1}{n} \sum_{i \in s} x_i$ be the mean of the same SRSWOR of size n from U on which $\bar{y}$ is also based. Obviously,

$$E_p(\bar{x}) = \bar{X}, V_p(\bar{x}) = \frac{N-n}{Nn} \frac{1}{N-1} \sum_{i=1}^{N} (x_i - \bar{X})^2.$$

Also let $\text{cov}_p(y, x) = \dfrac{1}{N-1} \sum (y_i - \bar{Y})(x_i - \bar{X}) \equiv$ Covariance between y and x.

Let $z_i = y_i - x_i$, $\bar{z} = \bar{y} - \bar{x}$, $\bar{Z} = \bar{Y} - \bar{X}$

$$\text{cov}_p(y, x) = \frac{1}{N-1} \left[\sum_{1}^{N} (y_i - \bar{Y})(x_i - \bar{X}) \right] = E_p (y - \bar{Y})(x - \bar{X})$$

Initial Ramifications

Now,
$$V_p(\bar{z}) = \frac{1}{N-1}\sum_1^N (z_i - \bar{Z})^2 \frac{N-n}{Nn}$$
$$= \frac{1}{N-1}\left[\sum_1^N \{(y_i - \bar{Y}) - (x_i - \bar{X})\}^2\right] \frac{N-n}{Nn}$$
$$= \frac{N-n}{Nn}\left[\frac{1}{N-1}\sum_1^N (y_i - \bar{Y})^2 + \frac{1}{N-1}\sum (x_i - \bar{X})^2 \right.$$
$$\left. - \frac{2}{N-1}\sum_1^N (y_i - \bar{Y})(x_i - \bar{X})\right]$$

Also, $V_p(\bar{z}) = V_p(\bar{y}) + V_p(\bar{x}) - 2\mathrm{cov}_p(\bar{y}, \bar{x})$.
So, $\mathrm{cov}_p(\bar{y}, \bar{x}) = $ the covariance between $\bar{y}$ and $\bar{x}$ which equals
$$\frac{N-n}{Nn}\frac{1}{N-1}\sum_1^N (y_i - \bar{Y})(x_i - \bar{X}).$$

An unbiased estimator for this $\mathrm{cov}_p(\bar{y}, \bar{x})$ is then
$$\frac{N-n}{Nn}\frac{1}{n-1}\sum_{i \in s} (y_i - \bar{y})(x_i - \bar{x}) \text{ with the following proof.}$$

It is clear that
$$E_p\left[\frac{1}{n-1}\sum_{i \in s}(z_i - \bar{z})^2\right] = \frac{1}{N-1}\sum_{i=1}^N (z_i - \bar{Z})^2, \text{ and we also know that}$$
$$E_p \frac{1}{n-1}\sum_{i \in s}(y_i - \bar{y})^2 = \frac{1}{N-1}\sum_{i=1}^N (y_i - \bar{Y})^2 \text{ and}$$
$$E_p \frac{1}{n-1}\sum_{i \in s}(x_i - \bar{x})^2 = \frac{1}{N-1}\sum_1^N (x_i - \bar{X})^2.$$

So, it follows at once that
$$E_p \frac{1}{n-1}\sum_{i \in s}(y_i - \bar{y})(x_i - \bar{x}) = \frac{1}{N-1}\sum_1^N (y_i - \bar{Y})(x_i - \bar{X}).$$

So,
$$E_p\left[\frac{N-n}{Nn}\frac{1}{n-1}\sum_{i \in s}(y_i - \bar{y})(x_i - \bar{x})\right] = \frac{N-n}{Nn}\frac{1}{N-1}\sum_1^N (y_i - \bar{Y})(x_i - \bar{X}).$$

Let us employ for $R = \frac{Y}{X} = \frac{\bar{Y}}{\bar{X}}$ the estimator $\hat{R} = \frac{\bar{y}}{\bar{x}}$ based on SRSWOR of size n from a population for which values y_i, x_i for $i \in s$ are gathered but $\bar{Y}$ is unknown though $\bar{X}$ is known.

Let
$$\frac{\bar{y} - \bar{Y}}{\bar{Y}} = \epsilon \quad \text{and} \quad \frac{\bar{x} - \bar{X}}{\bar{X}} = \delta$$
so that
$$\bar{y} = \bar{Y}(1+\epsilon) \quad \text{and} \quad \bar{x} = \bar{X}(1+\delta).$$
Then
$$\hat{R} = \frac{\bar{Y}}{\bar{X}}(1+\epsilon)(1+\delta)^{-1} = R(1+\epsilon)(1+\delta)^{-1}.$$

Let n be so large that $|\delta| < 1$. Then, by expanding $(1+\delta)^{-1}$ by Taylor series we may write $(1+\delta)^{-1} = 1 - \delta + \delta^2$ approximately, neglecting the higher-order terms.

Then $\hat{R} \simeq R(1+\epsilon)(1-\delta+\delta^2)$ and approximately
$$B_p(\hat{R}) = E_p(\hat{R} - R) \simeq R\left[E_p\delta^2 - E_p\epsilon\delta\right]$$
or
$$B_p(\bar{R}) \simeq R\left[\frac{V_p(\bar{x})}{\bar{X}^2} - \frac{\text{cov}_p(\bar{x},\bar{y})}{\bar{X}\bar{Y}}\right] = \frac{1}{(\bar{X})^2}\left[RV_p(\bar{x}) - \text{cov}_p(\bar{x},\bar{y})\right].$$

If we ignore the term δ^2, then $B_p(\bar{R}) \simeq 0$. So, in practice, for a large sample, $\hat{R} = \frac{\bar{y}}{\bar{x}}$ is often supposed to approximately be an unbiased estimator for R. Consequently, $\bar{X}\hat{R} = \bar{X}\frac{\bar{y}}{\bar{x}}$ and $X\hat{R} = X\frac{\bar{y}}{\bar{x}}$ are supposed to approximately be unbiased estimators for $\bar{Y}$ and Y, respectively, and they are called Ratio Estimators.

Since $\hat{R}$ is not exactly unbiased for R to measure its accuracy in estimating R, it is proper to treat its mean square error about R, namely,
$$MSE(\hat{R}) = E_p(\hat{R} - R)^2.$$
This is approximately equal to
$$E_p\left(\hat{R}-R\right)^2 = R^2 E_p(\epsilon - \delta)^2 = R^2\left[V_p(\epsilon) + V_p(\delta) - 2\text{cov}_p(\epsilon,\delta)\right]$$
$$= R^2\left[\frac{V_p(\bar{y})}{\bar{Y}^2} + \frac{V_p(\bar{x})}{(\bar{X})^2} - \frac{2\text{cov}_p(\bar{y},\bar{x})}{\bar{Y}\bar{X}}\right]$$
$$= \frac{1}{(\bar{X})^2}[V_p(\bar{y}) + R^2 V_p(\bar{x}) - 2R\,\text{cov}_p(\bar{y},\bar{x})]$$
$$= \frac{1}{(\bar{X})^2}\frac{N-n}{Nn}\frac{1}{N-1}\left[\sum_1^N (y_i-\bar{Y})^2 + R^2\sum_1^N (x_i-\bar{X})^2 - 2R\sum_1^N (y_i-\bar{Y})(x_i-\bar{X})\right]$$
$$= \frac{1}{(\bar{X})^2}V_p(\bar{y}-R\bar{x})$$
$$= \frac{1}{(\bar{X})^2}V_p(\bar{y})\Big|_{y_i=y_i-Rx_i} = \frac{1}{\bar{X}^2}\frac{N-n}{Nn}\frac{1}{N-1}\sum_1^N(y_i-\bar{Y})^2\Big|_{y_i=y_i-Rx_i}$$
$$= \frac{1}{(\bar{X})^2}\frac{N-n}{Nn}\frac{1}{N-1}\sum_1^n\left[(y_i-\bar{Y}) - R(x_i-\bar{X})\right]^2.$$

Initial Ramifications

Though not exactly unbiased, a reasonable estimator for $MSE(\hat{R})$ is then

$$m\left(\hat{R}\right) = \frac{1}{(\bar{X})^2} \frac{N-n}{Nn} \frac{1}{(n-1)} \sum_{i \in s} \left[(y_i - \bar{y}) - \hat{R}(x_i - \bar{x})\right]^2$$

$$= \frac{1}{(\bar{X})^2} \frac{N-n}{Nn} \frac{1}{n-1}$$

$$\left[\sum_{i \in s}(y_i - \bar{y})^2 + \left(\hat{R}\right)^2 \sum_{i \in s}(x_i - \bar{x})^2 - 2\hat{R} \sum_{i \in s}(y_i - \bar{y})(x_i - \bar{x})\right].$$

It follows that $X\hat{R} = X\frac{\bar{y}}{\bar{x}}$ and $\bar{X}\bar{R} = \bar{X}\frac{\bar{y}}{\bar{x}}$ are, respectively, approximate unbiased ratio estimators for Y and of $\bar{Y}$. Their respective approximate Bias formulae are

$$\frac{N}{\bar{X}}[RV_p(\bar{x}) - \text{cov}_p(\bar{x}, \bar{y})] \text{ and } \frac{1}{\bar{X}}[RV_p(\bar{x}) - \text{cov}_p(\bar{x}, \bar{y})].$$

2.6 DETERMINATION OF SAMPLE SIZES

We have seen that based on an SRSWR in n draws from $U = (1, \ldots, i, \ldots, N)$, the sample mean $\bar{y} = \frac{1}{n} \sum_{r=1}^{n} y_r$ has $V_p(\bar{y}) = \frac{\sigma^2}{n} = \frac{1}{nN} \sum_{i=1}^{N}(y_i - \bar{Y})^2$. Thus, given $\underline{Y} = (y_1, \ldots, y_i, \ldots, y_N)$, $\bar{y}$ becomes more accurate as an estimator for $\bar{Y}$ if the number of draws or the sample size n is increased. But as the sampled units need not be distinct even if n is increased up to N, the sample mean $\bar{y}$ need not coincide with $\bar{Y}$ so that the $V_p(\bar{y}) = E_p(\bar{y} - \bar{Y})^2$ need not be zero. So, we may have to be satisfied with the conclusion that the more the sample size n increases, the more accurate is the sample mean as an estimator for $\bar{Y}$. But one cannot be sure if even with the sample size taken to be as large as the population size the accuracy will be high enough for the sample mean to match the population mean.

If n is increased further and beyond N then one may observe all the y_i-values for $i \epsilon U$, and in that case knowing $\bar{Y}$ one need not have to estimate this at all.

For a sample s taken by SRSWOR in n draws or of size n, the sample mean $\bar{y} = \frac{1}{n} \sum_{i \in s} y_i$ has the variance

$$V_P(\bar{y}) = \frac{N-n}{Nn} S^2 = \frac{N-n}{Nn} \frac{1}{N-1} \sum_{1}^{N}(y_i - \bar{Y})^2.$$

Here since the units in s are all distinct if n is taken as high as N, then $\bar{y}$ will equal $\bar{Y}$ and $V_p(\bar{y}) = E_p(\bar{y} - \bar{Y})^2$ will equal zero. In the above formula also, n equal to N will imply $V_p(\bar{y}) = 0$.

But when resources do not permit taking n so that an error in $(\bar{y} - \bar{Y})$ has to be put up with, there exist rational ways of choosing n in an SRSWOR.

TABLE 2.1
Rational Choice of n in SRSWOR

N (1)	100f (2)	α (3)	cv% (4)	n (upward to an integer) (5)
100	10	0.05	10	17
100	10	0.05	5	5
100	10	0.05	20	45

First, to choose n in an SRSWOR the problem to address is one of estimating $\bar{Y}$ by $\bar{y}$. Then one may specify that the error in $(\bar{y}-\bar{Y})$ may need to be controlled in a specified way. For example, one may stipulate that $|\bar{y}-\bar{Y}| \leq f\bar{Y}$, taking the fraction f as say 10. Since we are considering a probability sample it appears reasonable that we may ask for $\text{Prob}\big[|\bar{y}-\bar{Y}| \leq f\bar{Y}\big] \geq (1-\alpha)$ with α appropriately chosen as a fraction, say 0.05.

Now, Chebyshev's inequality tells us that $\text{Prob}\big[|\bar{y}-\bar{Y}| \leq \lambda\sigma_p(\bar{y})\big] \geq 1 - \frac{1}{\lambda^2}$. So, we may take $\alpha = \frac{1}{\lambda^2}$ and $f\bar{Y} = \lambda\sigma_p(\bar{y})$ giving us $f = \lambda\sqrt{\frac{N-n}{Nn}}\frac{S}{\bar{Y}}$.

But we know that $100\frac{S}{\bar{Y}} = CV$ is the coefficient of variation of y. So, we may choose

$$\left(\tfrac{1}{n} - \tfrac{1}{N}\right) = \alpha\left(\tfrac{100f}{CV}\right)^2 \text{ or } n = \frac{1}{\tfrac{1}{N} + \alpha(\tfrac{100f}{CV})^2} = \frac{N}{1 + N\alpha(\tfrac{100f}{CV})^2}.$$

So, we may prepare Table 2.1 to give us a rule to find n to suit a situation when we may anticipate the level of variability in the population vector $\underline{Y} = (y_1, \ldots, y_i, \ldots, y_N)$ and insist on certain reasonable magnitudes of f and α, of course using the knowledge of N.

An alternative solution as follows of the same problem is also possible, as was explained by Chaudhuri (2010).

The standardized pivotal $\frac{\bar{y}-\bar{Y}}{\sigma_p(\bar{y})}$ for large sample size may be supposed to have approximately the distribution $N(0,1)$ of the standardized normal deviate τ. Then, we may suppose that approximately

$$\text{Prob}\,[\,|\tau| \leq 1.96\,] \geq 1 - .05 = 0.95.$$

So,
$$\text{Prob}\left[\left|\frac{\bar{y}-\bar{Y}}{\sigma_p(\bar{y})}\right| \leq 1.96\right] \geq 0.95$$

or
$$\text{Prob}\left[\left|\frac{\bar{y}-\bar{Y}}{\sigma_p(\bar{y})}\right| \leq \frac{f\bar{Y}}{\sigma_p(\bar{y})}\right] \geq 0.95$$

TABLE 2.2
Finding Sample Size in SRSWOR under Normal Approximation

N (1)	100f (2)	α (3)	cv% (4)	n (upward to an integer) (5)
100	10	0.05	10	-
100	10	0.10	10	-
100	10	0.01	10	-

on taking
$$1.96 = \frac{f\bar{Y}}{\sigma_p(\bar{y})}$$

or approximately, for $\alpha = 0.05$,
$$\frac{100f}{100\frac{\sigma_{py}}{\bar{Y}}} = \frac{100f}{CV(\bar{y})} = 1.96$$

or
$$\frac{100f}{100\sqrt{(\frac{1}{n} - \frac{1}{N})}\frac{S}{\bar{Y}}} = \frac{100f}{\sqrt{(\frac{1}{n} - \frac{1}{N})}CV} \simeq 1.96$$

or
$$\frac{1}{n} = \frac{1}{N} + \frac{(100f)^2}{(CV)^2}\frac{1}{(1.96)^2}$$

or
$$n = \frac{N}{1 + \frac{N}{(1.96)^2}}$$
$$= \frac{N}{1 + \frac{N}{(1.96)^2}\left(\frac{100f}{CV}\right)^2}$$

We may recommend consulting Table 2.2 below in working out n, given, N, f, α and CV.

We may now deal with another incidental problem. Suppose an SRSWOR of size 7 has been drawn retaining the sampled labels as 17, 3, 28, 42, 7, 10, and 15. If, suppose, the population size N is not at hand, is it possible to reasonably give the value of N?

Freund (1994) offers the following solution. If an SRSWOR of size n is from a finite population of size N, then the Probability distribution of the largest label drawn, namely X, has the probability distribution

$$\text{Prob}[\,X = x\,] = \frac{\binom{x-1}{n-1}}{\binom{N}{n}}, x = n, n+1, \ldots, N$$

Then, one gets

$$\binom{N}{n} = \sum_{x=n}^{N}\binom{x-1}{n-1} \text{ and } E(X) = \frac{1}{\binom{N}{n}}\sum_{n}^{N} x\binom{x-1}{n-1} = \frac{n(N+1)}{(n+1)}$$

So, an unbiased estimator of N is

$$\hat{N} = \frac{n+1}{n}X - 1.$$

Thus, for the sample noted above $\hat{N} = \frac{8}{7} \times 42 - 1 = 47$.

A.2 APPENDIX TO CHAPTER 2

Distinct Units in Sampling with Replacement and Their Uses in Estimation. Horvitz-Thompson Estimator. Sufficiency, Likelihood. Non-existence Theorems.

A.2.1 MORE ON EQUAL PROBABILITY SAMPLING

In the mid-fifties Mr. S. Raja Rao of the Indian Statistical Institute (ISI), Kolkata, approached Professor C.R. Rao to find an explanation for his empirical finding that the numerical values of the variances of the sample means of the distinct units in SRSWR in several instances are coming out less than the variances of the means of all sample observations including the repeated ones. Prof. Rao showed little interest and asked him to approach Professor D. Basu. Basu immediately found the role of "sufficiency" in this context. It is thus that "sufficiency" came to be appreciated in the context of finite population sampling.

Basu (1958), as a matter of fact, observed that while $\bar{y} = \frac{1}{n}\sum_{K=1}^{n} y_K$ with y_K as the y - value for the unit chosen on the Kth draw out of n draws by SRSWR from a population $U = (1, \ldots, i, \ldots, N)$ has

$$E_p(\bar{y}) = \bar{Y} \text{ and } V_p(\bar{y}) = \frac{\sigma^2}{n}, \sigma^2 = \frac{1}{N}\sum_{1}^{N}(y_i - \bar{Y})^2, \bar{y}_v = \frac{1}{v}\sum_{i \in s_v} y_i,$$

the sample mean of the $v(1 \leq v \leq n)$ distinct units in the part s_v of the set of v distinct units in the sample s found in n draws has

$$E_p(\bar{y}_v) = \bar{Y} \text{ and } V_p(\bar{y}_v) = \left[E\left(\frac{1}{v}\right) - \frac{1}{N}\right]\frac{N\sigma^2}{N-1}.$$

Writing $$S^2 = \frac{1}{N-1}\sum(y_i - -Y)^2, Q = \frac{N+n-1}{Nn},$$

Basu (1958), Raj and Khamis (1958), and Asok (1980) observed the following:

$$V_p(\bar{y}) = \left(Q - \frac{1}{N}\right)S^2. \quad V_p(\bar{y}_v) = \left[E\left(\frac{1}{v}\right) - \frac{1}{N}\right]S^2.$$

Pathak (1961) and Korwar and Serfling (1970) showed that

$$E\left(\frac{1}{v}\right) = \frac{1}{N^n}\sum_{j=1}^{N} j^{n-1}.$$

Raj and Khamis (1958) showed that $E(\frac{1}{v}) \leq Q$ which is easily proved by Asok (1980) following an inductive argument. It then follows that $V_p(\bar{y}) \geq V_p(\bar{y}_v)$.

For an SRSWOR in n draws the sample mean $\bar{y}^* = \frac{1}{n}\sum_{i \in s} y_i$ has

$$E_p(\bar{y}^*) = \bar{Y} \quad \text{and} \quad V_p(\bar{y}^*) = \frac{N-n}{Nn}S^2 = \frac{N-n}{N-1}\frac{\sigma^2}{n},$$

for every $n \geq 2$, $V_p(\bar{y}^*) < V_p(\bar{y}) = \frac{\sigma^2}{n}$.

So, SRSWOR yields a smaller variance for $\bar{y}^*$ than the variance for $\bar{y}$ yielded by SRSWR.

This comparison is unfair according to Basu (1958) because $\bar{y}^*$ involves effective sample size n while $\bar{y}$ involves $v(\leq n)$ as the effective sample size. Basu (1958) prefers for comparison versus $\bar{y}$ based on n draws a $\bar{y}^*$ based on $E(v)$ as the number of draws ignoring the fractional part in it.

We already know that

$$E(v) = N\left[1 - \left(\frac{N-1}{N}\right)^n\right].$$

Also $\bar{y}^*$ based on number of draws as $E(v)$ has the variance

$$V(\bar{y}*) = \frac{N - E(v)}{(N-1)E(v)}\sigma^2.$$

Basu (1958) has shown

$$\frac{\sigma^2}{n} > \frac{N - E(v)}{(N-1)E(v)}\sigma^2.$$

Thus SRSWOR yields a better sample mean based on number of draws $E(v)$ which is less than n than the sample mean based on SRSWR in n draws though $n > E(v)$.

Pathak (1961) considered

$$C_m(n) = m^n - \binom{m}{1}(m-1)^n + \ldots + (-1)^{m-1}\binom{m}{m-1}1^n$$

and showed the recursive relation

$$C_m(n) = m[C_m(n-1) + C_{m-1}(n-1)]$$

and showed
$$N^n = \sum_{m=1}^{N} C_m(n) \binom{N}{m}$$

and proved
$$E_p(\frac{1}{v}) = \frac{1}{N^n}(1^{n-1} + 2^{n-1} + \ldots + N^{n-1})$$

for all positive integers N, n.

Then, he derived, for SRSWR,

$$V_p(\bar{y}_v) = \frac{1}{N^n}\left[1^{n-1} + 2^{n-1} + \ldots + (N-1)^{n-1}\right] S^2 = E\left(\frac{1}{v} - \frac{1}{N}\right) S^2,$$

and he approximated $V_p(\bar{y}_v)$ by $V_p(\bar{y}_v) \simeq \left[\frac{1}{n} - \frac{1}{2N} + \frac{n-1}{12N^2}\right] S^2$

observing that
$$E_p(v) = N\left[1 - (1 - \frac{1}{N})^n\right] = \sum_{1}^{N} \pi_i$$

for SRSWR in n draws from $U = (1, \ldots, N)$

$$E_p(v^2)$$
$$= N\left[1 - \left(\frac{N-1}{N}\right)^n\right] + N(N-1)\left[1 - 2\left(\frac{N-1}{N}\right)^n + \left(\frac{N-2}{N}\right)^n\right],$$

$$V_p(v) = N\left(\frac{N-1}{N}\right)^n - N^2\left(\frac{N-1}{N}\right)^{2n} + N(N-1)\left(\frac{N-2}{N}\right)^n.$$

Pathak (1962) noted that $\bar{y}_v^* = \frac{v}{E_p(v)}\bar{y}_v$ is the Horvitz and Thompson's estimator based on SRSWR and worked out

$V_p(\bar{y}_v^*) =$

$$\frac{S^2\left[N\left\{1 - \left(\frac{N-1}{N}\right)^n\right\} - \left\{1 - \left(\frac{N-1}{N}\right)^n\right\} - (N-1)\left\{1 - 2\left(\frac{N-1}{N}\right)^n + \left(\frac{N-2}{N}\right)^n\right\}\right]}{\left\{N^2\left[1 - \left(\frac{N-1}{N}\right)^n\right]^2\right\}}$$
$$+ \frac{(\bar{Y})^2\left[N\left(\frac{N-1}{N}\right)^n - N^2\left(\frac{N-1}{N}\right)^{2n} + N(N-1)\left(\frac{N-2}{N}\right)^n\right]}{N^2\left[1 - \left(\frac{N-1}{N}\right)^n\right]^2}.$$

We noted already that

$s^2 = \frac{1}{(n-1)}\sum_{K=1}^{n}(y_k - \bar{y})^2$ is an unbiased estimator for $\sigma^2 = \frac{1}{N}\sum_{1}^{N}(y_i - \bar{Y})^2$.

Initial Ramifications

Defining $\quad s_d^2 = \dfrac{1}{(v-1)} \sum_{i \in s}(y_i - \bar{y}_v)^2 \quad$ if $v > 1$,

$\qquad\quad = 0$ else, and

$$C_v(n) = v^n - \binom{v}{1}(v-1)^n + \ldots + (-1)^{v-1}\binom{v}{v-1}1^n \text{ and}$$

$$s_v^2 = \left[\dfrac{C_v(n) - C_v(n-1)}{C_v(n)}\right] s_d^2.$$

Pathak (1962) showed that s_v^2 is uniformly better than s^2 as an unbiased estimator for σ^2.

The following five unbiased estimators for $V_p(\bar{y}_v)$ are also given by Pathak (1962), namely,

i. $v_1(\bar{y}_v) = \left[\dfrac{1^{n-1} + \cdots + (N-1)^{n-1}}{N^n}\right] \dfrac{N}{N-1} s^2$

ii. $v_2(\bar{y}_v) = \left[\dfrac{1^{n-1} + \cdots + (N-1)^{n-1}}{N^n}\right] \dfrac{N}{N-1} \left[\dfrac{C_v(n) - C_v(n-1)}{C_v(n)}\right] s_d^2$

iii. $v_3(\bar{y}_v) = \dfrac{C_{v-1}(n-1)}{C_v(n)} s_d^2$

iv. $v_4(\bar{y}_v) = \left[\left(\dfrac{1}{v} - \dfrac{1}{N}\right) + \left(\dfrac{N-1}{N^n - N}\right)\right] s_d^2$

v. $v_5(\bar{y}_v) = \left[\left(\dfrac{1}{v} - \dfrac{1}{N}\right) + N^{1-n}\left(1 - \dfrac{1}{v}\right)\right] s_d^2 \quad$ if $v > 1$; v_4 and v_5

are given by Des Raj and Khamis (1958); it may be noted that

$$v_4 = v_5 \left(\dfrac{N^n}{N^n - N}\right).$$

Next we recall that to estimate $\bar{Y}$, Basu (1958) and Des Raj and Khamis (1958) considered also Inverse Sampling. In this a number v is first fixed, and it is deemed proper to go on having SRSWR sampling till the nth draw to produce the v distinct units from $U = (1, \cdots, i \cdots, N)$. Here n is a random variable, v is fixed, $\bar{y}_v$ is the mean $\frac{1}{v}\sum_{i \in s} y_i$ of the v distinct units drawn, and $\bar{y}_n = \frac{1}{n}\sum_{K=1}^{n} y_K$ is the mean of the sample of all the units chosen in the n draws.

In the sample s chosen in n draws, the ith distinct unit appeared λ_i times so that $\sum_{i=1}^{v-1} \lambda_i = (n-1)$ because the vth distinct unit must appear only once on the nth draw itself. Chikkagoudar (1966) has given us the following results for the Inverse Sampling by SRSWR: $\sum' \dfrac{(n-1)!}{\lambda_1! \cdots \lambda_{(v-1)}!}$ vide Pathak (1961)

equals
$$(v-1)^{n-1} - (v-1)(v-2)^{n-1} + \cdots + (-1)^{v-2}\binom{v-1}{v-2} = \Delta^{v-1}x^{n-1}\big|_{x=0}$$

writing $\sum'$ the sum over $\lambda_1, \cdots, \lambda_{v-1}$ subject to $\sum_{i=1}^{v-1}\lambda_i = (n-1)$ and Δ is the difference operator such that $\Delta f(x) = f(x+1) - f(x)$.

Noting $E_p(\lambda_i | n, u_1, \ldots, u_v) \to$ the v distinct units) $= \frac{n-1}{v-1}, i = 1, \ldots, v-1$, Chikkagoudar (1966) finds

$$V_p(\bar{y}_n) = S^2 \binom{N-1}{v-1} \Delta^{v-1} \left[\frac{1}{x} + \frac{N(x-3)}{x^2} \log \frac{N}{N-x} + \frac{2}{x}\sum_{n=1}^{\infty}\frac{1}{n^2}\left(\frac{x}{N}\right)^{n-1} \right]_{x=0}.$$

Writing $s_n^2 = \frac{1}{(n-1)} \sum_{K=1}^{n}(y_k - \bar{y}_n)^2$, he derives for $V_p(\bar{y}_n)$ his unbiased estimator

$$\hat{V}(\bar{y}_n | n) = \left[\left\{ \frac{N-n}{Nn}\Delta^{v-1}x^{n-1} + \frac{(n-1)(n-2)}{n}\Delta^{v-1}x^{n-2} \right\}_{x=0} \right]$$
$$X \left[\left\{ n\Delta^{v-1}x^{n-1} - (n-2)\Delta^{v-1}x^{n-2} \right\}_{x=0} \right]^{-1} s_n^2.$$

Lanke (1975), however, has given the formula

$$V_P(\bar{y}_n) = NS^2 \left[NE_n(\frac{1}{n}) - E_n\frac{3n+1}{(n+1)^2} \right]$$

with E_n as the expectation operator for the distribution of n with respect to the Inverse SRSWR considered here.

For $\bar{y}_v = \frac{1}{v}\sum_{i \in d} y_i$, d is the set of distinct units chosen, $E_p(\bar{y}_v) = \bar{Y}$ and $V_p(\bar{y}_v) = \left(\frac{1}{v} - \frac{1}{N}\right) S^2$. An unbiased estimator for this is

$$v(\bar{y}_v) = \left(\frac{1}{v} - \frac{1}{N}\right) \frac{1}{(v-1)} \sum_{i \in d}(y_i - \bar{y}_v)^2.$$

A.2.2 HORVITZ-THOMPSON ESTIMATOR

For any sampling design p the quantities of frequent uses are $\pi_i = \sum_{s \ni i} p(s)$, called the corresponding inclusion-probability of the unit i in a sample and $\pi_{ij} = \sum_{s \ni i,j} p(s)$, called the inclusion probability of distinct units i and $j (i \neq j)$ in a sample; π_{ii} is clearly the same as π_i. The following theorem is crucial in Survey Sampling.

Theorem 2.1

A "necessary and sufficient condition" for the existence of an unbiased estimator for a finite population total from a sample chosen according to any design p is $\pi_i > 0$ for every i in U. ∎

PROOF
Necessity. Let $t = (s, \underline{Y})$ be an unbiased estimator for Y. Then

$$Y = \sum_{1}^{N} y_i = E_p(t) = \sum_s p(s) t(s, \underline{Y}) = \sum_{s \ni i} p(s) t(s, \underline{Y}) + \sum_{s \not\ni i} p(s) t(s, \underline{Y}).$$

Let, if possible, for a particular i in U, $\pi_i = 0$. Then, $p(s) = 0$ for $s \ni i$.

Then,
$$y_i + \sum_{j \neq i} y_j = \sum_{s \not\ni i} p(s) t(s, \underline{Y}).$$

This cannot hold for every possible vector $\underline{Y} = (y_1, \cdots, y_i, \cdots, y_N)$ since the Right-Hand Side (RHS) quantity does not have the value y_i at all. So, for the existence of an unbiased estimator for the population total π_i cannot be zero for any i in U. ∎

Sufficiency. Let $\pi_i > 0 \; \forall \; i$. Let $I_{si} = 1/0$ according as $i \in s$ or $i \notin s$, respectively, equivalently as $s \ni i$ or $s \not\ni i$, respectively.

Then,
$$E_p(I_{si}) = \sum_s p(s) I_{si} = \sum_{s \ni i} p(s) = \pi_i.$$

Taking $\pi_i > 0 \; \forall \; i \in U$, we may consider the estimator

$$t = \sum_{i \in s} \frac{y_i}{\pi_i} = \sum_{i=1}^{N} y_i \frac{I_{si}}{\pi_i}$$

Then,
$$E_p(t) = \sum_{i=1}^{N} y_i E_p(\frac{I_{si}}{\pi_i}) = \sum_{1}^{N} y_i = Y.$$

So, $\pi_i > 0$ implies the existence for Y,

the unbiased estimator
$$t = \sum_{i \in s} \frac{y_i}{\pi_i}.$$

This completes the proof.

The estimator
$$t = \sum_{i \in s} \frac{y_i}{\pi_i} = \sum_{1}^{N} \frac{y_i}{\pi_i} I_{si}$$

is due to Horvitz and Thompson (1952), and we should denote this Horvitz-Thompson (HT) estimator by the symbol

$$t_{HT} = \sum_{i \in s} \frac{y_i}{\pi_i} = \sum_{i=1} y_i \frac{I_{si}}{\pi_i} \text{ for } \pi_i > 0 \; \forall \; i \in U$$

which is unbiased for Y.

Let us now note the following details.

Writing $I_{sij} = I_{si} I_{sj}$ which equals 1 if i and j both are in s i.e. $I_{si} = 1$ and $I_{sj} = 1$, and equals 0 when i, or j, or both are not in s.

Thus $I_{sij} = 1$ if $s \ni i,j$ and $I_{sij} = 0$ if $s \not\ni i$, or $s \not\ni j$, or $s \not\ni i,j$.

Now,
$$\pi_{ij} = \sum_{s \ni i,j} p(s) = \sum_s p(s) I_{sij}.$$

Let us note further

$$\sum_1^N \pi_i = \sum_1^N \left(\sum_s p(s) I_{si} \right) = \sum_s p(s) \left(\sum_1^N I_{si} \right)$$
$$= \sum_s p(s) v(s) = E_p(v(s)) = v \text{ (say)}$$

$$\sum_{\substack{j=1 \\ j \neq i}}^N \pi_{ij} = \sum_{j \neq i} \left[\sum_s p(s) I_{sij} \right] = \sum_s p(s) \left[I_{si} \sum_{j \neq i} I_{sj} \right]$$
$$= \sum_s p(s) I_{si}(v(s) - 1) = \sum_{s \ni i} v(s) p(s) - \sum_{s \ni i} p(s)$$
$$= \sum_{s \ni i} v(s) p(s) - \pi_i; \quad \sum_{i=1}^N \sum_{j \neq i} \pi_{ij}$$
$$= \sum_s p(s) v(s) \left(\sum_i I_{si} \right) - \sum_i \pi_i$$
$$= \sum_s p(s) v^2(s) - \sum_i \pi_i$$
$$= E_p v^2(s) - E_p v(s)$$
$$= V_p(v(s)) + v^2 - v$$
$$= V_p(v(s)) + v(v-1).$$

If, in particular, for a design p, the value of $v(s)$ is a constant $= n$, say, for every s with $p(s) > 0$, then the above results reduce, respectively, to

$$\sum \pi_i = n, \sum_{j \neq i} \pi_{ij} = (n-1)\pi_i \quad \text{and} \quad \sum\sum_{i \neq j} \pi_{ij} = n(n-1).$$

Initial Ramifications

Next let us observe the following:

$$E_p(I_{si}) = \pi_i, V_p(I_{si}) = E_p(I_{si}^2) - E_p^2(I_{si}) = E_p(I_{si})(1 - E_p(I_{si})) = \pi_i(1 - \pi_i)$$

because $I_{si}^2 = I_{si}$. Also,

$$\text{cov}_p(I_{si}, I_{sj}) = E_p(I_{si} \, I_{sj}) - E_p(I_{si})(I_{sj}) = E_p(I_{sij}) - \pi_i \, \pi_j = \pi_{ij} - \pi_i \, \pi_j.$$

So, it follows that

$$V_p(t_{HT}) = V_p \left(\sum_i^N \frac{y_i}{\pi_i} I_{si} \right)$$

$$= \sum_i \left(\frac{y_i}{\pi_i} \right)^2 V_p(I_{si}) + \sum\sum_{i \neq j} \left(\frac{y_i \, y_j}{\pi_i \, \pi_j} \right) \text{cov}_p(I_{si} I_{sj})$$

$$= \sum_1^N \left(\frac{y_i}{\pi_i} \right)^2 \pi_i(1 - \pi_i) + \sum\sum_{i \neq j} \frac{y_i \, y_j}{\pi_i \, \pi_j}(\pi_{ij} - \pi_i \pi_j)$$

$$= \sum_i y_i^2 \frac{1 - \pi_i}{\pi_i} + \sum\sum_{i \neq j} y_i y_j \left(\frac{\pi_{ij} - \pi_i \pi_j}{\pi_i, \pi_j} \right)$$

$$= \sum_i \frac{y_i^2}{\pi_i} + \sum\sum_{i \neq j} y_i y_j \frac{\pi_{ij}}{\pi_i \pi_j} - \sum y_i^2 - \sum\sum_{i \neq j} y_i y_j$$

$$= \sum_i \frac{y_i^2}{\pi_i} + \sum\sum_{i \neq j} y_i y_j \frac{\pi_{ij}}{\pi_i \pi_j} - Y^2$$

$$= E_p(t_{HT}^2) - Y^2 = E_p(t_{HT} - Y)^2.$$

To work out alternative variance formulae let us check the following:

$$\sum\sum_{i<j} \left(\frac{y_i}{\pi_i} - \frac{y_j}{\pi_j} \right)^2 (\pi_i \pi_j - \pi_{ij})$$

$$= \frac{1}{2} \sum\sum_{i \neq j} \left(\frac{y_i^2}{\pi_i^2} + \frac{y_j^2}{\pi_j^2} - \frac{2 y_i \, y_j}{\pi_i \, \pi_j} \right)(\pi_i \pi_j - \pi_{ij})$$

$$= \frac{1}{2} \left[\sum_i \frac{y_i^2}{\pi_i} \left(\sum_{j \neq i} \pi_j \right) + \sum_j \frac{y_j^2}{\pi_j} \left(\sum_{i \neq j} \pi_i \right) - 2 \sum\sum_{i \neq j} y_i y_j \right]$$

$$- \frac{1}{2} \left[\sum_i \frac{y_i^2}{\pi_i^2} \left(\sum_{j \neq i} \pi_{ij} \right) + \sum_j \frac{y_j^2}{\pi_j^2} \left(\sum_{i \neq j} \pi_{ij} \right) - 2 \sum\sum_{i \neq j} y_i y_j \frac{\pi_{ij}}{\pi_i \pi_j} \right]$$

$$= \sum_i \frac{y_i^2}{\pi_i}(v - \pi_i) - \sum\sum_{i \neq j} y_i y_j - \sum \frac{y_i^2}{\pi_i^2}\left(\sum_{s \ni i} v(s)p(s) - \pi_i\right)$$

$$+ \sum\sum_{i \neq j} y_i y_j \frac{\pi_{ij}}{\pi_i \pi_j}$$

$$= v \sum \frac{y_i^2}{\pi_i} + \sum\sum_{i \neq j} y_i y_j \frac{\pi_{ij}}{\pi_i \pi_j} - \sum\sum_{i \neq j} y_i y_j - \sum y_i^2$$

$$+ \sum \frac{y_i^2}{\pi_i} - \sum \frac{y_i^2}{\pi_i^2} \sum_{s \ni i} v(s)p(s)$$

$$= \left[\sum \frac{y_i^2}{\pi_i} + \sum\sum y_i y_j \frac{\pi_{ij}}{\pi_i \pi_j} - Y^2\right] + \sum \frac{y_i^2}{\pi_i}\left(v - \frac{1}{\pi_i}\sum_{s \ni i} v(s)p(s)\right)$$

$$= V_p(t_{HT}) - \sum \frac{y_i^2}{\pi_i}\left(\frac{1}{\pi_i}\sum_{s \ni i} v(s)p(s) - v\right).$$

So, $\sum\sum_{i<j} \left(\frac{y_i}{\pi_i} - \frac{y_j}{\pi_j}\right)^2 (\pi_i \pi_j - \pi_{ij}) + \sum_i \frac{y_i^2}{\pi_i} \beta_i,$

on writing $\beta_i = \frac{1}{\pi_i}\sum_{s \ni i} v(s)p(s) - v,$

is a different form of the variance of t_{HT}. One may note that in case every sample contains a fixed number of units, each distinct say n, then $\beta_i = 0$. So, in such a case

$$\sum\sum_{i<j} \left(\frac{y_i}{\pi_i} - \frac{y_i}{\pi_j}\right)^2 (\pi_i \pi_j - \pi_{ij}),$$

as given by Yates and Grundy (1953) is an alternative form of $V_p(t_{HT})$ for which the original form was given by Horvitz and Thompson (1952) themselves.

Before giving different forms of unbiased estimators of $V_p(t_{HT})$ let us consider the following general topic of interest.

Let t be an unbiased estimator for Y based on a given design p.

Then, $\qquad V_p(t) = E_p(t^2) - Y^2.$

Now, $\qquad Y^2 = \sum_i y_i^2 + \sum\sum_{i \neq j} Y_i Y_j.$

So, $\qquad \hat{Y_1^2} = \sum_1 y_i^2 \frac{I_{si}}{\pi_i} + \sum\sum_{i \neq j} Y_i Y_j \frac{I_{sij}}{\pi_{ij}}$

is an unbiased estimator for Y^2, provided $\pi_{ij} > 0 \; \forall i \neq j$ in U, and if so, $\pi_i > 0 \; \forall i \in U$ as well. Then, $v = t^2 - \hat{Y_1^2}$ is an unbiased estimator.

Alternatively,

$$\hat{Y_2^2} = \sum_i \frac{y_i^2 I_{si}}{\gamma_i p(s)} + \sum_{i \neq j}\sum \frac{y_i y_j I_{sij}}{\gamma_{ij} p(s)}$$

is another unbiased estimator for Y^2, writing $\gamma_i =$ number of samples containing i and $\gamma_{ij} =$ number of samples containing i and $j (j \neq i)$ of U.

Remembering these we may find the following unbiased estimation formulae for $V_p(t_{HT})$, assuming throughout $\pi_{ij} > 0 \ \forall i, j (i \neq j)$ in U:

$$v_1 = \sum_i y_i^2 \frac{1 - \pi_i}{\pi_i} \frac{I_{si}}{\pi_i} + \sum_{i \neq j}\sum y_i y_j \left(\frac{\pi_i \pi_j - \pi_{ij}}{\pi_i \pi_j}\right) \frac{I_{sij}}{\pi_{ij}}$$

$$v_2 = \sum_{i<j}\sum \left(\frac{y_i}{\pi_i} - \frac{y_j}{\pi_j}\right)^2 \frac{\pi_i \pi_j - \pi_{ij}}{\pi_{ij}} I_{sij} + \sum_i \frac{y_i^2}{\pi_i} \beta_i \frac{I_{si}}{\pi_i}$$

$$v_3 = \sum_i y_i^2 \frac{1 - \pi_i}{\pi_i} \frac{I_{si}}{\gamma_i p(s)} + \sum_{i \neq j}\sum y_i y_j \frac{(\pi_i \pi_j - \pi_{ij})}{\gamma_{ij} p(s)} I_{sij}$$

$$v_4 = t_{HT}^2 - \left[\sum \frac{y_i^2}{\pi_i} I_{si} + \sum_{i \neq j}\sum \frac{y_i y_j}{\pi_{ij}} I_{sij}\right]$$

$$v_5 = t_{HT}^2 - \left[\sum \frac{y_i^2 I_{si}}{p(s) \gamma_i} + \sum_{i \neq j}\sum \frac{y_i y_j I_{sij}}{p(s) \gamma_{ij}}\right].$$

In case a design p is so employed that every sample contains a fixed number n of units, each distinct, clearly,

$$\gamma_i = \binom{N-1}{n-1} \text{ and } \gamma_{ij} = \binom{N-2}{n-2}.$$

Now to answer the question as to how to discriminate among the variance-estimators of the three above types, namely, (1) those that are expressed as quadratic forms, (2) those that are expressed as squared estimator minus estimator of the squared population total, and (3) expressed otherwise.

It is desirable to have the variance-estimator as uniformly non-negative. It is very hard to ensure this in case (1) because the non-negativity condition for a quadratic form, that is, its Non-negative Definiteness uniformly in $\underline{Y}$, only in terms of the parameters of the sampling design is very hard to test; in case (2) it has been found empirically that the variance-estimator most often turns out negative in the sense that for samples with higher probabilities, they come up negative and positive for samples with relatively low probabilities. What happens in the remaining case will be addressed in the following discussion.

Clearly, as we have seen, for the Horvitz-Thompson Estimator (HTE) for Y,

$$V_p(t_{HT}) = E_p \left(\sum_1 y_i \frac{I_{si}}{\pi_i} - Y \right)^2$$

$$= \sum_i y_i^2 \left(\frac{1 - \pi_i}{\pi_i} \right) + \sum\sum_{i \neq j} y_i y_j \left(\frac{\pi_{ij} - \pi_i \pi_j}{\pi_i \pi_j} \right)$$

is identical with

$$\sum\sum_{i<j} \left(\frac{y_i}{\pi_i} - \frac{y_j}{\pi_j} \right)^2 (\pi_i \pi_j - \pi_{ij}) + \sum \frac{y_i^2}{\pi_i} \left[\frac{1}{\pi_i} \sum_{s \ni i} v(s) p(s) - v \right].$$

From this we deduce that

$$v_1 = \sum_i y_i^2 \left(\frac{1 - \pi_i}{\pi_i} \right) \frac{I_{si}}{\pi_i} + \sum\sum_{i \neq j} y_i y_j \left(\frac{\pi_{ij} - \pi_i \pi_j}{\pi_i \pi_j} \right) \frac{I_{sij}}{\pi_{ij}}$$

is one unbiased estimator for $V_p(t_{HT})$ and

$$v_2 = \sum\sum_{i<j} \left(\frac{y_i}{\pi_i} - \frac{y_j}{\pi_j} \right)^2 \frac{\pi_i \pi_j - \pi_{ij}}{\pi_{ij}} I_{sij} + \sum_i \frac{y_i^2}{\pi_i} \beta_i \frac{I_{si}}{\pi_i}$$

with $\beta_i = \frac{1}{\pi_i} \sum_{s \ni i} v(s) p(s_i) - v$ is another unbiased estimator

for $V_p(t_{HT})$ provided $\pi_{ij} \geq 0 \; \forall i, j (i \neq j)$ in U.

We stress that v_1 and v_2 are different. Even if $v(s) = n$ for every s with $p(s) > 0$ implying $\beta_i = 0$ rendering v_2 to be the Yates and Grundy's (1953) estimator, it is still not the same as v_1, the Horvitz and Thompson variance estimator.

For SRSWR in n draws let us check that

$$\pi_i = 1 - \left(\frac{N-1}{N} \right)^n;$$

also, to find π_{ij} in terms of n and N let us write A_i as the event that i is selected in the sample and A_i^C is its component. Then, by DeMorgan's law $(A_i \cap A_j) = \left(A_i^C \cup A_j^C \right)^C$ giving us

$$\pi_{ij} = 1 - [Pr(A_i^C) + Pr(A_j^C) - Pr(A_i^C \cap A_j^C)]$$

$$= 1 - \left[\left(\frac{N-1}{N} \right)^n + \left(\frac{N-1}{N} \right)^n - \left(\frac{N-2}{N} \right)^n \right]$$

$$= 1 - 2 \left(\frac{N-1}{N} \right)^n + \left(\frac{N-2}{N} \right)^n.$$

Initial Ramifications

Using these formulae it is well-nigh impossible to check v_1 to be a nonnegative definite quadratic form. Also, it is not easy to check if $v_2 \geq 0$ or not.

For SRSWOR in draws, one may easily check that for n draws

$$\pi_i = \frac{\binom{N-1}{n}}{\binom{N}{n}} = \frac{n}{N} \text{ and } \pi_{ij} = \frac{\binom{N-2}{n}}{\binom{N}{n}} = \frac{n(n-1)}{N(N-1)}.$$

Then, $\pi_i \pi_j - \pi_{ij} = \frac{n}{N}\frac{n}{N} - \frac{n}{N}\left(\frac{n-1}{N-1}\right) = \frac{n}{N}\frac{(N-n)}{N(N-1)} > 0 \ \forall N > n.$

Hence the Yates and Grundy estimator turns out

$$v_2 = \frac{N^2}{n^2} \frac{n(N-n)}{N^2(N-1)} \frac{N(N-1)}{n(n-1)} \sum\sum_{i<j \in n} (y_i - y_j)^2$$

$$= \frac{N}{n^2}(N-n)\frac{1}{n-1} \sum\sum_{i<j}(y_i - y_j)^2 \geq 0.$$

But the sign of v_1 is still hard to work out.

It will be our interest to extend these discussions when samples are chosen with unequal probabilities, and estimation may be by Horvitz and Thompson's method or by other methods.

A.2.3 SUFFICIENCY

For the elements of $\underline{Y} = (y_1, \ldots, y_i, \ldots, y_N)$ let us denote by

$$\Omega = \{\underline{Y} | -\infty < a_i \leq y_i \leq b_i < +\infty, i = 1, \ldots, N\}$$

the parametric space with a_i and b_i as possibly known.

Let for $\qquad d = (s, y_i | i \in s) \quad$ with $\quad s = (i_1, \ldots, i_j, \ldots, i_n),$

alternatively, $\qquad d = ((i_1, y_{i1}), \ldots, (i_j, y_{ij}), \ldots, (i_n, y_{in})).$

Let us write in particular, $\qquad \Omega_d = \{\underline{Y} \,|\, (y_{i1}, \cdots, y_{ij}, \cdots y_{in})$
as in "d" but $\qquad - \infty < a_i \leq y_i \leq b_i < +\infty \ \forall \ i \notin s \}$

to denote the part of the parametric space consistent with the observed data d. Let p denote a "non-informative" design for which $p(s)$ does not involve any co-ordinate of $\underline{Y}$. Contrarily, a design p is "informative" if it involves at least some y-values. For example, after the first unit i_1 is chosen one may observe the value y_{i1} and decide to choose the unit of U on the second draw with a probability $p(i_2|i_1, y_{i1})$ using the value of y_{i1}. Such a design is Informative. Let corresponding to an ordered sample $s = (i_1, \cdots i_j, \cdots i_n)$ of units of U not necessarily distinct, the unordered sample of distinct units be $s* = \{j_1, \cdots, j_K\}$ such that $j_1, \cdots, j_K$ are each one distinct element of s and $1 \leq K \leq n$.

Correspondingly to
$$d = ((i_1, y_{i1}), \cdot, (i_{n1}, y_{in})),$$
let us have
$$d^* = \{(j_1, y_{j1}), \cdots, (j_K, y_{jk})\}.$$

The set of units of s^* with the respective y-values not necessarily distinct for the distinct units of s. Then, $\Omega_d = \Omega_{d^*}$.

Let
$$I_{\underline{Y}}(d) = 1 \text{ if } \underline{Y} \in \Omega_d,$$
$$= 0 \text{ if } \underline{Y} \notin \Omega_d.$$

and
$$I_{\underline{Y}}(d*) = 1 \text{ if } \underline{Y} \in \Omega_{d*}$$
$$= 0 \text{ if } \underline{Y} \notin \Omega_d^*$$

obviously,
$$I_{\underline{Y}}(d) = I_{\underline{Y}}(d^*)$$
and
$$\Omega_d = \Omega_{d^*}.$$

Let $P_{\underline{Y}}(d)$ be the probability of observing the data point d when $\underline{Y}$ is the underlying parameter point. Similarly let $P_{\underline{Y}}(d^*)$ also be defined. Since d^* is derived on amalgamating different elements like d we may write

$$P_{\underline{Y}}(d) = P_{\underline{Y}}(d \cap d^*)$$
$$= P_{\underline{Y}}(d^*) P_{\underline{Y}}(d|d^*)$$

using the notations for intersection of data points or events and conditional probabilities.

Also,
$$P_{\underline{Y}}(d) = I_{\underline{Y}}(d) \, p(s)$$
and
$$P_{\underline{Y}}(d^*) = I_{\underline{Y}}(d^*) \, p(s^*).$$

Also, $p(s^*)$ is obtained on summing the values of $p(s)$ over the samples to each of which corresponds the same sample s^*.

So, we get
$$I_{\underline{Y}}(d) \, p(s) = P_{\underline{Y}}(d|d^*) \, I_{\underline{Y}}(d^*) \, p(s^*).$$

Hence,
$$P_{\underline{Y}}(d|d^*) = \frac{p(s)}{p(s^*)}.$$

Assuming only a non-informative design p is employed it follows that given the data point d, the amalgamated data point d^* is a sufficient statistic.

Initial Ramifications

Let us next consider the concept of a Minimal Sufficient Statistic. Every data point d is an element of the Data Space D, which is the totality of all possible and conceivable data points in a given context of sample selection from a population and surveying an observable sample and gathering the corresponding data points.

Every statistic $t = t(d)$ which is a function of d has the effect of Partitioning the Data Space into Partition Sets which are "mutually non-overlapping" with their union coinciding with the Data Space. If t_1 and t_2 are two statistics, such that every partition set induced by t_1 is contained in at least one partition set induced by t_2, then t_2 induces a thicker partitioning while t_1 induces a thinner partitioning than t_2. Between two such statistics the one that induces the thicker partitioning achieves 'greater summarization.' A desirable statistic is one that induces the thickest partitioning without losing any relevant information, A sufficient statistic sacrifices no information of relevance. So, a sufficient statistic inducing the thickest partitioning is the most desirable statistic and it is called the Minimal Sufficient Statistic. We shall show that in the present context d^* is the Minimal Sufficient statistic. This is proved as follows.

Let d_1^* be a sufficient statistic corresponding to d_1 just as is d^* to d and similarly d_2^* corresponding to d_2 while d_1 and d_2 are two different data points in the same context of choosing from U samples according to a common design p. Let $t(d_1)$ and $t(d_2)$ be statistics corresponding to d_1 and d_2 and t be a sufficient statistic.

We shall be through if we find $d_1^* = d_2^*$ if $t(d_1) = t(d_2)$.

Now we get
$$P_{\underline{Y}}(d_1) = P_{\underline{Y}}(t(d_1))\ P_{\underline{Y}}(d_1|t(d_1))$$
$$= P_{\underline{Y}}(t(d_2))\ P_{\underline{Y}}(d_1|t(d_1))$$
$$= P_{\underline{Y}}(d_2) \frac{P_{\underline{Y}}(d_1|t(d_1))}{P_{\underline{Y}}(d_2|t(d_2))}$$
$$= P_{\underline{Y}}(d_2)\ C,$$

where C is free of $\underline{Y}$.

Or $P_{\underline{Y}}(d_1^*) = P_{\underline{Y}}(d_2^*)\ C^1$ where C^1 is also a constant by hypotheses $\Rightarrow$ $I_{\underline{Y}}(d_1^*) = I_{\underline{Y}}(d_2^*) \Rightarrow d_1^* = d_2^*$ presuming P is Non-informative.

A.2.4 LIKELIHOOD

By $P_{\underline{Y}}(d) = p(s)\ I_{\underline{Y}}(d)$ we mean the probability of obtaining the data point when $\underline{Y}$ is the underlying parametric point. This same $P_{\underline{Y}}(d)$ may be written as $L_d(\underline{Y})$ to denote the "likelihood" of the parametric point $\underline{Y}$ given the observed data point d. Thus,

$$L_d(\underline{Y}) = p(s)\ I_{\underline{Y}}(d).$$

For a Non-informative design this likelihood function is Flat and we cannot discriminate among various competing parametric points possibly responsible for yielding the data point d at hand. So, Likelihood by itself is sterile in yielding a "valid Inference." With a Bayesian approach one may find a way out. We shall pursue this topic in a later chapter.

A.2.5 NON-EXISTENCE THEOREM

In Chapter 1 we remarked that a point estimator t for Y has a variance $V_p(t) = \sigma_p^2(t)$ as a measure of its accuracy in estimation. So, the less the value of $\sigma_p(t)$ for an estimator t which is unbiased for Y, the better is t. Also we noted that if t is an unbiased estimator for Y, then a desirably narrow confidence interval $(t - \lambda\, \sigma_p(t), t + \lambda\, \sigma_p(t))$ with a confidence coefficient as high as $\left(1 - \frac{1}{\lambda^2}\right)$ may be constructed with $\sigma_p(t)$ as small numerically as possible. So, for Y desirably an unbiased estimator t should be constructed with uniformly the least value of its variance $V_p(t) = \sigma_p^2(t)$.

But unfortunately we have the following two disappointing results.

Theorem 2.2

In the class of all unbiased estimators for Y there does not exist one with the uniformly least variance for any design unless it is a census design.

This Theorem is due to Basu (1971), and he has given us the following proof.

Let
$$\underline{Y} = (y_1, \cdots, y_i, \cdots, y_N) \quad \text{with} \quad Y = \sum_1^N y_i$$

and
$$\underline{A} = (a_1, \cdots, a_i, \cdots, a_N) \quad \text{with} \quad A = \sum_1^N a_i$$

be any two arbitrary points in the parametric space
$$\Omega = \{\underline{Y} | -\infty < c_i \le y_i \le d_i < +\infty, i = 1, \cdots, N\}.$$

Let a census design p_c be a design p such that $p(s) = 1$ when s coincides with U and $p(s) = 0$, otherwise.

Let $t = t(s, \underline{Y})$ be unbiased for Y, i.e., $E_p(t) = \sum_s p(s) t(s, \underline{Y}) = Y$ for every $\underline{Y}$ in Ω. ∎

Let t_0 have the uniformly least variance among all such unbiased estimators of Y. Thus, $V_p(t_0) \le V_p(t) \,\forall\, \underline{Y}$ for any t other than t_0 such that $E_p(t) =$

Initial Ramifications

$Y = E_p(t_0) \; \forall \; \underline{Y}$. Let $\underline{A} = (a_1, \cdots, a_i, \cdots, a_N)$ in Ω and t_A be an estimator for Y such that
$$t_A(s, \underline{Y}) = t_0(s, \underline{Y}) - t_0(s, \underline{A}) + A$$

Then,
$$E_p(t_A) = \sum_s p(s) t_A(s, \underline{Y})$$
$$= \sum_s p(s) t_0(s, \underline{Y}) - \sum_s p(s) t_0(s, \underline{A}) + A$$
$$= Y - A + A = Y \quad \forall \; \underline{Y}.$$

Thus, t_A is unbiased for Y.

Now, $\quad V_p(t_A) = E_p(t_A - Y)^2$

Also, $\quad V_p(t_A)|_{\underline{Y}=\underline{A}} = E_p [t_0(s, \underline{Y}) - t_0(s, \underline{A}) + A - Y]^2 \Big|_{\underline{Y}=\underline{A}}$

Then, $\quad V_p(t_0)|_{\underline{Y}=\underline{A}} \leq V_p(t_A)|_{\underline{Y}=\underline{A}} = 0$

So, $\quad V_p(t_0)|_{\underline{Y}=\underline{A}} = 0.$

Similarly, for any other arbitrary point $\underline{B} = (b_1, \cdots, b_i, \cdots, b_N)$ in Ω

one must have $\quad V_p(t_0)|_{\underline{Y}=\underline{B}} = 0$

on taking $\quad t_B(s, \underline{Y}) = t_0(s, \underline{Y}) - t_0(s, \underline{B}) + B$

with $\quad \underline{B} = (b_1, \cdots, b_i, \cdots, b_N)$ in Ω.

So, we must have $\quad V_p(t_0) = 0 \; \forall \; \underline{Y}$ in Ω.

This is not possible unless p is a census design, completing the proof due to Basu (1971).

Godambe (1955) defined a class of homogeneous linear unbiased estimators (HLUE) for Y of the form $t_b = \sum_{i \in s} y_i \, b_{si}$

Here b_{si} is supposed to be a constant involving no element of
$$\underline{Y} = (y_1, \cdots, y_i, \cdots, y_N) \quad \text{at all.}$$

Even, for example, $\quad b_{si} = \dfrac{X}{\sum_s x_i}$

rendering $\quad t_b = X \dfrac{\sum_s y_i}{\sum_s x_i} = X \dfrac{\bar{y}}{\bar{x}}$

still as a homogeneous linear estimator for Y even though $\sum_{i \in s} x_i$ is a random variable just as $\sum_{i \in s} y_i$ is so. Godambe restricts t_b to be an unbiased estimator for Y.

Consequently,

$$Y = E_p(t_b) = \sum_s p(s) \sum_{i \in s} y_i\, b_{si} = \sum_{i=1}^N y_i \left(\sum_{s \ni i} p(s) b_{si} \right) \quad \forall\, \underline{Y} \in \Omega$$

and demanding therefore, $\sum_{s \ni i} p(s) b_{si} = 1 \ \forall\, i \in U$.

In order to find one estimator in this class of homogeneous linear unbiased estimators (HLUEs) for Y with the uniformly minimum variance so as to render it a UMV (Uniformly Minimum Variance) HLUE, one needs to choose b_{si} such that $V_p(t_b)$ may be the minimum for every $\underline{Y}$. But Godambe (1955) gives us the Theorem 2.

In the class of HLUEs for Y there does not exist one with the UMV properties.

PROOF

$V_p(t_b) = E_p(t_b^2) - Y^2$. In order to choose b_{si} properly so as to minimize $V_p(t_b)$ one needs to solve $\frac{\delta}{\delta b_{si}} E_p(t_b^2) = 0$ subject to $\sum_{s \ni i} p(s) b_{si} = 1 \ \forall\, i$.

Suppose $p(s) > 0$ for a sample s such that $s \ni i$ and $y_i \neq 0$ for a particular i. So, we need to solve

$$0 = \frac{\delta}{\delta\, b_{si}} \left[\sum_s p(s) \left(\sum_{i \in s} y_i b_{si} \right)^2 - \sum_{i=1}^N \lambda_i \left(\sum_{s \ni i} p(s) b_{si} - 1 \right) \right]$$

with λ_i's introduced as Lagrange's undetermined multipliers. ■

This leads to
$$0 = 2 p(s)\, y_i \left(\sum_{i \in s} y_i\, b_{si} \right) - \lambda_i\, p(s)$$

or
$$\sum_{i \in s} y_i\, b_{si} = \frac{\lambda_i}{2 y_i}$$

Choosing $y_j = 0 \ \forall\, j \neq i$ in U, one gets $b_{si} = \frac{\lambda_i}{2\, y_i^2}$ demanding the form $b_{si} = b_i \ \forall\, s$ such that $s \ni i$.

So, one needs $1 = \sum_{s \ni i} p(s) b_i = b_i\, \pi_i \quad$ or $\quad b_i = \frac{1}{\pi_i}$,

recalling $\pi_i > 0 \ \forall i$ to ensure the existence of an unbiased estimator for Y. A UMV estimator must be of the form

$$\sum_{i \in s} \frac{y_i}{\pi_i} \quad \text{which must satisfy} \quad \sum_{i \in s} \frac{y_i}{\pi_i} = \frac{\lambda_i}{2 y_i^2}.$$

Initial Ramifications

Suppose a sample s_1 with $p(s_1) > 0$ contains $i, y_i \neq 0$ and another sample s_2 with $p(s_2) > 0$ also contains i with $y_i \neq 0$; then, we must have for the existence of a UMV HLUE,

$$\sum_{i \in s_1} \frac{y_i}{\pi_i} = \frac{\lambda_i}{2y_i^2} = \sum_{i \in s_2} \frac{y_i}{\pi_i}.$$

Godambe claims this is absurd for a general type of sampling design. Hence he closes his argument claiming the validity of the above Theorem 2.

But there may exist designs for which the above need not be construed as absurd. Vijaya Hege (1965), T.V. Hanurav (1966) and Jan Lanke (1975) identified such a design called a Uni-cluster Design (UCD) as an exceptional one satisfying the following condition:

Two samples s_1 and s_2 with $p(s_1) > 0$ and $p(s_2) > 0$ are either disjoint, i.e., $s_1 \cap s_2 = \phi$, the empty set, or they are equivalent, i.e., $s_1 \sim s_2$ in the sense that $i \in s_1 \Leftrightarrow i \in s_2$, i.e., every unit in either s_1 or s_2 must be in the other too.

For such a UCD of course the absurdity demonstrated by Godambe naturally does not hold. So, Godambe's Theorem 2 does not apply in the case of a UCD.

But the crucial question is whether a UCD admits a UMVE in the class of HLUEs for Y.

But before taking up this question let us discuss whether Godambe's non-existence theorem is implied by Basu's non-existence theorem as a special case. Godambe confines the theorem only to his Homogeneous Linear Unbiased Estimators but Basu covers the entire class of unbiased Estimators for a Population Total. Basu's class is wider than Godambe's. His result is more general than Godambe's. Yet Godambe's result cannot be deduced from Basu's as a special case. In fact Basu's result and his proof do not apply to Godambe's situation. As soon as one tries to appy Basu's proof starting with an HLUE one must fail because it cannot be applied without crossing the limit of a homogeneous class which rules out an added term to an HLUE involving $Y-$ free constant ingredients.

Now let us show that a Uni-cluster Design admits a UMV member in the HLUE class of estimators for Y.

For this we need to first enunciate and prove a Complete Class Theorem below invoking certain virtues of sufficient statistics in the context of survey sampling we have already discussed.

Let $s = (i_1, \cdots, i_j, \cdots, i_n)$ be an ordered sample with units not necessarily distinct taken from the population U and the corresponding observable survey data be $d = ((i_1, y_{i1}), \cdots (i_j, y_{ij}), \cdots (i_n, y_{in}))$.

Also, let the corresponding unordered sample of distinct units of s be

$$s^* = \{(j_1, y_{j1}), \cdots, (j_k, y_{jk})\}$$

with the resulting sufficient statistic
$$d^* = \{(j_1, y_{j1}), \cdots, (j_k, y_{jk})\}, \ 1 \leq k \leq n.$$

Let $t = t(s, \underline{Y})$ be a statistic

and $$t^* = t^*(s^*, \underline{Y}) = \frac{\sum_{s \to s^*} p(s) t(s, \underline{Y})}{\sum_{s \to s^*} p(s)}.$$

Here $p(s)$ is the probability of selecting s, $\sum_{s \to s^*}$ denotes summing over all samples s to each of which corresponds the same sample s^*. Also we may note that
$$t^* = t^*(s^*, \underline{Y}) = t^*(s, \underline{Y})$$
for every s to which s^* corresponds, and we may write
$$\sum_{s \to s^*} p(s) = p(s^*)$$
which is the selection probability of s^*.

It now follows that
$$E_p(t) = \sum_s p(s) \ t(s, \underline{Y})$$
$$= \sum_{s^*} \left[\sum_{s \to s^*} p(s) t(s, \underline{Y}) / p(s^*) \right] p(s^*)$$
$$= \sum_{s^*} t^*(s^*, \underline{Y}) p(s^*) = E_p(t^*).$$

Also, $$E_p(tt^*) = \sum_s p(s) t(s, \underline{Y}) t^*(s, \underline{Y})$$
$$= \sum_{s^*} t^*(s^*, \underline{Y}) \left[\sum_{s \to s^*} p(s) t(s, \underline{Y}) \right]$$
$$= \sum_{s^*} t^*(s^*, \underline{Y}) p(s^*) \left[\sum_{s \to s^*} p(s) t(s, \underline{Y}) / p(s^*) \right]$$
$$= E_p(t^*(s^*, \underline{Y}))^2 = E_p(t^*)^2.$$

So, $$E_p(t - t^*)^2 = E_p(t^2) + E_p(t^*)^2 - 2E_p(tt^*)$$
$$= E_p(t^2) - E_p(t^*)^2.$$

Hence, $V_p(t^*) = V_p(t) - E_p(t - t^*)^2.$
So, $V_p(t^*) \leq V_p(t)$
and $V_p(t^*) < V_p(t)$

Initial Ramifications

unless for every sample, t and t^* are the same, which may happen only if s coincides with s^*.

This is a Complete Class Theorem because given any estimator t there exists a better estimator t^* in the sense of having uniformly less variance with the latter estimator being independent of (1) the "order" and/or (2) the multiplicity with which the units occur in the original sample.

Now, suppose we employ a Uni-cluster Design p. In order for this p to admit a Uniformly Minimum Variance (UMV) HLUE of the form $t_b = \sum_{i \in s} y_i b_{si}$, with b_{si}'s free of $\underline{Y}$ and subject to $\sum_{s \ni i} p(s) b_{si} = 1 \ \forall \ i \in U$, a possible such UMV HLUE must be a member of the Complete Class of estimators satisfying the above two conditions (1) and (2).

Now, let s_{oi} be a sample s containing the unit i for which $\pi_i > 0$. Then, we must have
$$1 = \sum_{s \ni i} p(s) b_{si} = b_{soi} \sum_{s \ni i} p(s) = b_{soi} \pi_i$$
because in the UMV based on s_{oi} every b_{si} must be the same because otherwise this UMV HLUE cannot belong to the complete class. This gives $b_{soi} = \frac{1}{\pi_i}$ and t_{HT} as the unique UMV HLUE. Thus, the UCD admits the unique UMV member.

3 More Intricacies

Abstract. Introduction. Unequal probability sampling. PPS sampling strategies.

3.0 ABSTRACT

In practice samples are mostly chosen, surveyed, and put to analytical studies by researchers, health organizations, economists, social scientists, demographers, biologists, physiologists, criminologists, epidemiologists, geologists, and others who are involved in social, animal, and human problems but are themselves not experts in sophisticated Sampling technologies. Market researchers and advertising agencies also fall in this category. Many such users of sample survey data draw only simple random samples even without caring for the distinction between sampling with or without replacement. But the statistician per se and especially the survey sampling experts often choose samples with varying probabilities on most occasions. They do this to derive profits. We shall narrate briefly how they achieve their gains.

3.1 INTRODUCTION

The classical survey sampling theory is design based. A sample drawn from a population is used to provide a sample-based statistic as an estimator for a population parameter. A sampling design assigns to every sample a selection-probability and thus renders the sample the status of a random variable giving it a probability distribution. Thereby a statistic also becomes a random variable with its design-based expectation, Bias vis-à-vis a parameter it seeks to estimate, Mean Square Error, and a variance. In order to control the variance of an estimator which provides a proper measure of accuracy of an estimator, it follows of course that the selection-probability of a sample is a crucial quantity to hit upon in an appropriate manner. So, equal-probability sampling need not be a very desirable scheme to employ in practice, because a varying probability scheme as an alternative may provide for a better and more accurate estimator. We describe various procedures of sample-selection cum estimator jointly called a sampling strategy, discussing the resulting consequences as far as practicable.

3.2 UNEQUAL PROBABILITY SAMPLING STRATEGIES

The most celebrated classical varying probability sampling method was initiated rather fortuitously coincidentally with the introduction of stratified sampling elegantly discussed by Neyman (1934), who in fact laid Survey Sampling on an indisputably firm theoretical foundation. Let us elaborate.

Suppose a finite population $U = (1, \cdots, N)$ is large enough with considerable variation among the elements of $\underline{Y} = (y_1, \cdots, y_i \cdots, y_N)$. In such a case often the population is split up into a number of mutually exclusive parts called strata which together exhaust the population and which are separately and independently sampled. If from each such part called a stratum an SRSWOR sample is chosen, then let us observe the following.

Suppose $H (\geq 2)$ strata are formed with the hth stratum $(h = 1, \cdots, H)$ composed of N_h $(2 \leq N_h, h = 1, \cdots, H)$ units such that

$$\sum_{h=1}^{H} N_h = N$$

and from the h stratum of N_h units an SRSWOR of n_h units

$$\left(2 \leq n_h \leq N_h, \ h = 1, \cdots, H, \sum_{h=1}^{H} n_h = n\right) \quad \text{can be taken.}$$

Then inclusion-probability of a unit of hth stratum will be

$$\frac{n_h}{N_h}; \quad h = 1, \cdots, H.$$

Unless $\dfrac{n_n}{N_n}$ is a constant for every $h (= 1, \cdots, H)$

the units of the population will have varying probabilities of inclusion. Hence, one virtually adopts an Unequal Probability Sampling, as a matter of fact.

A forerunner to the concept of a sophisticated varying probability sampling scheme, in fact, is Mahalanobis' (1938, 1946) technique of random selection of crop fields in villages in his rural survey of paddy in undivided Bengal in the late thirties and early forties of the last century. In such a survey for a randomly selected village, maps of paddy fields are sketched. Then a co-ordinatograph is used taking randomly selected x and y co-ordinates along the horizontal and the vertical axes, respectively, to choose a point (x, y) on the map of the paddy field with a plan to draw a circle with a small pre-determined radius around the chosen point. A desired number of independent (x, y) points are thus chosen on the various paddy fields of the villages. Every time a point falls in a given paddy field, a small circle around that point is harvested. Obviously a crop-field with a relatively large area has a larger probability of being harvested than the ones with smaller geographical areas. The area of a paddy field is actually a measure of its size. Thus though the co-ordinatograph chooses the points (x, y) by simple random sampling with replacement, the paddy fields are selected for being harvested clearly with various probabilities, in fact proportionately to their size-measures with replacement. This is a historical background for the well-known sampling with Probabilities Proportional to Sizes (PPS) with Replacement (WR) introduced by Hansen and Hurwitz (1943), who understood by "Size" of a unit as any non-negative-valued variable x well-correlated with the variable y of real interest and not necessarily as the physical measurement of the area or volume of any population entity. Let us now deal with the statistical details of PPSWR sampling and the related estimation theory.

3.3 PPS SAMPLING

Let $U = (1, \cdots, i, \cdots, N)$ denote a finite population,

$$\underline{Y} = (y_1, \cdots, y_i, \cdots, y_N),$$

the vector of values y_i for i in U of a real variable

$$y, \underline{X} = (x_1, \cdots, x_i, \cdots, x_N),$$

the vector of positive-valued $x_i, i \in U$, to be called the size-measures of i in U which are the values of a real variable x well-correlated with y. The x_i's are, if necessary, converted into integers, easily on multiplication by a suitable positive-integer power of 10. In order to select a unit from U, let a random integer R between 1 and $X = \sum_1^N x_i$ be drawn. Then, the cumulative totals $C_1 = x_1, C_2 = x_1 + x_2, \cdots, c_j = x_1 + \cdots + x_j, \cdots C_N = x_1 + x_2 + \cdots + x_N$ are calculated and C_0 is taken as zero. Then, if

$$C_{i-1} < R \leq C_i, i = 1, \cdots, N$$

the unit i is taken into the sample. The probability that the unit i may be selected into a sample according to this scheme is

$$p_i = \frac{C_i - C_{i-1}}{X} = \frac{x_i}{X}.$$

In order to draw more samples, more units are taken into the sample on following this rule independently as many times as desired.

This is called Probability Proportional to Size sampling With Replacement (WR). As an alternative to this Cumulative Total method, Lahiri (1951) gave us the following easier method of PPS sampling of a single unit which avoids calculation of these cumulative totals $C_i, i = 1, \cdots, N$. Instead, a number M is taken as an integer exceeding the largest of the positive integers $x_i, i = 1, \cdots, N$. Then to select one unit, first a random integer I between 1 and N is chosen and independently another random integer R between 1 and M is chosen.

If $R \leq x_i$, then i is taken into the sample. Else, the choice (I, R) is dropped and the similar exercise is continued until eventually, by this rule one unit is actually selected. Let us show that this method selects the unit with the probability
$$p_i = \frac{x_i}{X}, i \in U.$$

For this method
$$\text{Prob } (i \text{ is selected}) = \frac{1}{N}\frac{x_i}{M} + Q\left(\frac{1}{N}\frac{x_i}{M}\right) + Q^2\left(\frac{1}{N}\frac{x_i}{M}\right) + \cdots$$

on writing $\quad Q = 1 - \frac{1}{N}\frac{\sum_1^N x_i}{M} = 1 - \frac{\bar{X}}{M} \quad$ which is less than 1.

So, $\quad\text{Prob } (i \text{ is selected}) = \frac{1}{N}\frac{x_i}{M}\left(1 + Q + Q^2 + \cdots\right)$
$$= \frac{1}{N}\frac{x_i}{M}(1-Q)^{-1}$$
$$= \frac{1}{N}\frac{x_i}{M}\frac{M}{\bar{X}} = \frac{x_i}{X} = p_i.$$

So, Lahiri's selection method is a PPS method of selection equivalent to the Cumulative Total Method of PPS sampling.

Hansen and Hurwitz (1943) have given the following estimator for Y based on PPSWR:
$$t_{HH} = \frac{1}{n}\sum_{r=1}^n \frac{y_r}{p_r}.$$

Here n is the number of draws with replacement with probabilities proportional to size (PPS) of units i bearing size-measures x_i such that
$$p_i = \frac{x_i}{X}, i \in U, \left(0 < p_i < 1, \sum_1^N p_i = 1\right),$$

y_r is the y-value and p_r is the p_i-value, respectively, of a unit selected on the rth draw, $r = 1, \cdots, n$.

Then, $\quad E_p\left(\frac{y_r}{p_r}\right) = \sum_{i=1}^N \left(\frac{y_i}{p_i}\right) p_i = Y$

implying $\quad E_p = (t_{HH}) = Y,$

i.e., t_{HH} is unbiased for Y.

Again,
$$V_p\left(\frac{y_r}{p_r}\right) = E_p\left(\frac{y_r}{p_r}\right)^2 - Y^2 = \sum_1^N \frac{y_i^2}{p_i^2} p_i - Y^2$$

$$= \sum_1^N \frac{y_i^2}{p_i} - Y^2 = \sum_1^N p_i \left(\frac{y_i}{p_i} - Y\right)^2$$

$$= \sum_{i<i'}^N \sum^N p_i p_{i'} \left(\frac{y_i}{p_i} - \frac{y_{i'}}{p_{i'}}\right)^2 = V, \quad \text{say}.$$

Hence,
$$V_p(t_{HH}) = \frac{1}{n^2} \sum_{r=1}^n V_p\left(\frac{y_r}{p_r}\right) = \frac{V}{n}.$$

Next, we may note
$$E_p\left(\frac{y_r}{p_r} - \frac{y_{r'}}{p_{r'}}\right)^2 = E_p\left\{\left(\frac{y_r}{p_r} - Y\right) - \left(\frac{y_{r'}}{p_{r'}} - Y\right)\right\}^2.$$

So,
$$v = \frac{1}{2n^2(n-1)} \sum_{r \neq r'}^n \sum^n \left(\frac{y_r}{p_r} - \frac{y_{r'}}{p_{r'}}\right)^2$$

has
$$E_p(v) = \frac{1}{n^2(n-1)} n(n-1) V = \frac{V}{n}.$$

Thus v is an unbiased estimator for $V(t_{HH})$.

An important criticism against PPSWR sampling is that the units of a PPSWR sample need not be distinct. Hence, t is not a function of the sufficient statistic which is the unordered set of distinct units that may be obtained from a PPSWR sample. Also, the complete class theorem of Chapter 2 may be applied to derive from t_{HH} as its conditional expectation given the unordered set of the distinct units in the PPSWR sample as an unbiased estimator for Y with a uniformly smaller variance than $V(t_{HH})$. Thus, t_{HH} as an unbiased estimator for Y is inadmissible. In a specified class "C" of estimators for Y, one is inadmissible if there exists a better estimator in "C". Between two unbiased estimators t_1 and t_2 for Y, one has t_1 as the better if

$$V_p(t_1) \leq V_p(t_2) \; \forall \; \underline{Y} \quad \text{and} \quad V_p(t_1) < V_p(t_2)$$

for at least one $\underline{Y}$. In a class "C", say, of estimators, one is Admissible if there does not exist any other member which is better. Presently we shall discuss more about Admissible estimators for Y among (i) the HLUE class and separately among (ii) all unbiased estimators or the UE class.

Before that let us consider one more classical scheme of varying probability sampling. Five names are associated with this scheme which are Hajek (1949), Midzuno (1950), Lahiri (1951), and Sen (1953).

With the population $U = (1, \cdots, i, \cdots, N)$ are associated

$$\underline{Y} = (y_1, \cdots, y_i, \cdots, t_N)$$

with y_i the value of y for i in U and

$$\underline{X} = (x_1, \cdots, x_i, \cdots, x_N)$$

with x_i as a positive value of x well-correlated with y for the ith unit of U.

A first draw is made from U assigning a selection-probability $p_i = \frac{x_i}{X}$ to i following either the cumulative total method or Lahiri's (1951) method. Setting aside the unit i actually chosen, from the remaining $(N-1)$ units of U an SRSWOR of $(n-1)$ additional units is chosen either en masse or by the draw-by-draw method, as described in detail in Chapter 2. If s denotes such a sample of n distinct units suppressing the order of selection, then the selection-probability is

$$p(s) = \frac{\sum_{i \in s} x_i}{X} \frac{1}{\binom{N-1}{n-1}} = \frac{\sum_{i \in s} p_i}{\binom{N-1}{n-1}}.$$

The inclusion-probability of i is

$$\pi_i = p_i + (1 - p_i)\frac{n-1}{N-1} = \frac{N-n}{N-1}p_i + \frac{n-1}{N-1},$$

and the inclusion-probability of i, j $(i \neq j)$ is

$$\pi_{ij} = p_i \frac{n-1}{N-1} + p_j \frac{n-1}{N-2} + (1 - p_i - p_j)\frac{(n-1)(n-2)}{(N-1)(N-2)}$$
$$= \frac{(n-1)(N-n)}{(N-1)(N-2)}(p_i + p_j) + \frac{(n-1)(n-2)}{(N-1)(N-2)}.$$

Lahiri (1951) was the first to point out to the Community of the survey samplers that the claim "ratio estimator" with its genesis described in detail by David (1978), namely,

$$t_R = \frac{(\sum_{i \in s} y_i)}{\sum_{i \in s} x_i} X,$$

if based on the above varying probability sampling design, say P_L with

$$P_L(s) = \left(\sum_{i \in s} p_i\right) / \binom{N-1}{n-1},$$

is unbiased for Y. This is obvious because

$$E_p(t_R) = \sum_s p_L(s)\left(\frac{\sum_{i\in s} y_i}{\sum_{i\in s} x_i}\right) X = \frac{1}{\binom{N-1}{n-1}}\sum_s\left(\sum_{i\in s} y_i\right) = Y.$$

To work out a formula for its variance and an unbiased estimator thereof, let us wait to first discuss Rao's (1979) and Ha'jek's (1959) general results concerning Mean Square Errors and their estimation.

A context of ratio estimation is again enlightened by the following interesting contribution by Hartley and Ross (1954). The ratio estimator treated so far involves the ratio of summed values of y to the summed values of x, both in the sample, based either on SRSWOR or on Lahiri's (1951) sampling scheme. The ratio estimator t_R based on Lahiri's scheme p_L is often called Lahiri's ratio estimator.

Hartley and Ross (1954) start with the sample sum of the individual ratios of y to x based on SRSWOR in, say, n draws. They observe the following:

$$\operatorname{cov}_p\left(\frac{y}{x}, x\right) = E_p(y) - E_p\left(\frac{y}{x}\right)E(x)$$

$$= \frac{1}{N}\sum_1^N y_i - \left(\frac{1}{N}\sum_{i=1}^N \frac{y_i}{x_i}\right)\left(\frac{1}{N}\sum_1^N x_i\right)$$

$$= \bar{Y} - \bar{R}\,\bar{X}, \quad \text{writing} \quad r_i = \frac{y_i}{x_i},\ \bar{R} = \frac{1}{N}\sum_{i=1}^N r_i.$$

On the other hand,

$$\operatorname{cov}_p\left(\frac{y}{x}, x\right) = \frac{1}{N}\sum_{i=1}^N \left(\frac{y_i}{x_i} - \bar{R}\right)(x_i - \bar{X}).$$

Writing
$$\bar{r} = \frac{1}{n}\left(\sum_{i\in s} r_i\right)$$

this $\operatorname{cov}_p\left(\frac{y}{x}, x\right)$

is unbiasedly estimated by

$$\left(\frac{N-1}{N}\right)\frac{1}{(n-1)}\sum_{i\in s}(r_i - \bar{r})(x_i - \bar{x})$$

$$= \left(\frac{N-1}{N}\right)\frac{1}{(n-1)}\left(\sum_{i\in s} y_i - \bar{r}\sum_{i\in s} x_i\right)$$

$$= \frac{n(N-1)}{N(n-1)}(\bar{y} - \bar{r}\bar{x}).$$

This is also an unbiased estimator for $\bar{Y} - \bar{R}\bar{X}$. Hence, as usual supposing x_i's are known,
$$\bar{X}\bar{r} + \frac{n(N-1)}{N(n-1)}(\bar{y} - \bar{r}\bar{x})$$
unbiasedly estimates $\bar{Y}$. Hence,
$$X\bar{r} + \frac{n(N-1)}{(n-1)}(\bar{y} - \bar{r}\bar{x})$$
unbiasedly estimates Y and
$$\bar{r} + \frac{n(N-1)}{X(n-1)}(\bar{y} - \bar{r}\bar{x})$$
unbiasedly estimates $R = \frac{Y}{X}$.

Each of these three respective unbiased estimators is known as Hartley-Ross's (1954) unbiased Ratio-Type Estimators. This is because though $\bar{r}$ is the sample-mean of ratio's
$$\frac{y_i}{x_i}, i \in s,$$
the final estimator is not in the form of ratios.

Next let us consider PPS Systematic Sampling. Here also the relevant entities are
$$U = (1, \cdots, i, \cdots, N),$$
$$\underline{Y} = (y_1, \cdots, y_i, \cdots, y_N),$$
and
$$\underline{X} = (x_1, \cdots, x_i, \cdots, x_N)$$
with every x_i as a positive number known and suitably converted into an integer. In order to draw a sample of an intended size n,

take
$$K = \left[\frac{X}{n}\right] \text{ or } K = \left[\frac{X}{n}\right] + 1.$$

Now take a Random integer R between 1 and X and calculate
$$a_j = (R + jk) \bmod (X), j = 0, 1, \cdots, (n-1)$$
and also
$$C_i = \sum_{j=1}^{i} x_j$$
and take
$$C_0 = 0.$$
If
$$C_{i-1} < a_j \leq C_i,$$
take i into the sample s, also, if $a_j = 0$ take N into s. This is called Circular PPS Systematic sampling. If $\frac{X}{n}$ is an integer equal to K, then, take the random

R between 1 and K,

calculate $\quad b_j = R + jk, j = 0, 1, \cdots, (n-1),$

check if $\quad C_{i-1} < b_j \leq C_i,$

and take i into s. This is called Linear PPS Systematic Sampling.

If no size-measure x_i's are available, then, replacing X by N one gets Circular Systematic Sampling if $\frac{N}{n}$ is not an integer and a Linear Systematic Sampling scheme if $\frac{N}{n}$ is an integer deducing the schemes respectively from the above schemes,

For the above Circular PPS Systematic Sampling, clearly each "equal-probability" draw of R from 1 to X produces a sample for which besides this random element R every other aspect is deterministically fixed. So, even without caring for the actual draw of R one may straightaway write down all the X possible samples together. Then, one may count n_i = number of samples containing i and n_{ij} = number of samples containing $i, j (i \neq j)$.

Then, $\quad \pi_i = \dfrac{n_i}{X} \quad$ and $\quad \pi_{ij} = \dfrac{n_{ij}}{X}$

are the first- and second- order inclusion-probabilities of

$$i \quad \text{and} \quad (i, j), (i \neq j) \quad \text{in samples.}$$

Obviously, $\pi_i > 0 \, \forall \, i$ and Horvitz and Thompson's estimator are available. But π_{ij} turns out zero for numerous pairs $(i, j), i \neq j$. So, unbiased variance-estimation is not possible. Chaudhuri and Pal (2003), following Das (1982) have given the following modification of PPS Circular Systematic sampling.

Take a random integer R between 1 and $X(X-1)$ and calculate

$$K = \left[\dfrac{X(X-1)}{n}\right]$$

and $\quad a_j = (R + jk) \bmod (X(X-1)), j = 0, 1, \cdots, (n-1)$

and then proceed as in the original PPSCSS. Chaudhuri and Pal (2003) have shown that for this modified PPSCSS, $\pi_{ij} > 0 \, \forall \, i, j \, (i \neq j)$. Another problem with each kind of systematic sampling is that intended sample-size n may not be realized. Depending upon N, n, and $x_i, i \in U$ for systematic sampling a_j- values may not produce a distinct unit i to go into the sample for every $j = 0, 1, \cdots, n-1$. The resulting numbers may go into loop hampering choice of distinct units. So, the realized effective size of a sample may fall short of n. Kunte Sudhakar (1978) was the first to point this out. He showed that a necessary condition that from a finite population of size N an equal probability circular systematic sample of an intended size n may be realized with each unit distinct is that N and n should be mutually co-prime. Some further works on this also have appeared in the literature.

Again, an easy way to get an exact unbiased variance estimator for an unbiased estimator for a finite population total is to take two independent systematic samples, estimate the population total as the average of the two unbiased population-total estimators from the two independent samples, and get one-fourth of the squared difference of these two estimators as the estimator for the variance of this estimator for the population total.

Next we consider another unequal probability sampling scheme together with estimator for the population total and the variance estimator as given by Rao, Hartley, and Cochran (1962).

As usual with varying probability sampling schemes, this one also starts with the three entities with the earlier connotations, namely,

$$U = (1, \cdots, i \cdots, N), \underline{Y} = (y_1, \cdots, y_i \cdots, y_N), \underline{X} = (x_1, \cdots, x_i \cdots, x_N).$$

In order to select a sample s of n units each distinct, first the population U is divided at random into n disjoint groups of a certain number of units totaling N. For this, positive integers $N_1, N_2, \cdots, N_i, \cdots, N_n$ are chosen such that

$$\sum_n N_i = N, \sum_n$$

denoting summation over the n groups formed in applying the Rao-Hartley-Cochran sampling scheme. Then take an SRSWOR of N_1 from U to form the first group, from the remaining $(N - N_1)$ units of U take an SRSWOR of N_2 units to get the second group, and so on from the finally remaining $(N - N_1 - \cdots N_{n-1})$ units of U take the final or nth SRSWOR of N_n units to construct the nth group. Letting $p_i = \frac{x_i}{X}$, the normed size-measures of the N units of U, let

$$Q_i = p_{i1} + \cdots + p_{ij} + \cdots + p_{iN_i}$$

on writing p_{ij} for the normed size-measure of the jth unit falling in the ith group, with $j = 1, \cdots, N_i$ and $i = 1, \cdots, n$. Now from each of these n groups formed take exactly one unit independently across the groups so that the ijth unit of the $i$thgroup is assigned the selection-probability

$$\frac{p_{ij}}{Q_i}, j = 1, \cdots, N_i, i = 1, \cdots, n.$$

This is the Rao-Hartley-Cochran (RHC) sampling scheme. For Y an unbiased estimator is

$$t_{RHC} = \sum_n y_{ij} \frac{Q_i}{p_{ij}}.$$

By E_1 we shall denote the operator for taking the expectation over the random formation of the n groups, by E_2 we shall denote the operator for calculation of expectation with respect to the independent selection of one unit from each of the n groups with probabilities

$$\frac{p_{ij}}{Q_i}, j = 1, \cdots, N_i; i = 1, \cdots, n.$$

More Intricacies

Thus, $E_p = E_1 E_2$ and similarly, the overall variance operator is
$$V_p = E_1 V_2 + V_1 E_2$$
writing V_1, V_2 the variance operators corresponding to E_1, E_2.

So, $E(t_{RHC}) = E_1 \sum_n E_2 \left(y_{ij} \dfrac{Q_i}{p_{ij}} \right)$

$$= E_1 \sum_n \left(\sum_{j=1}^{N_i} y_{ij} \right) = E_1 \left(\sum_1^N y_i \right) = E_1(Y) = Y$$

$$V_p(t_{RHC}) = E_1 V_2 \left(\sum_n y_{ij} \dfrac{Q_i}{p_{ij}} \right) + V_1 \left(E_2 \sum_n y_{ij} \dfrac{Q_i}{p_{ij}} \right)$$

$$= E_1 \sum_n \left[\sum_{j<j'=1}^{N_i} \sum^{N_i} \dfrac{p_{ij} p_{ij'}}{Q_i Q_i} \left(\dfrac{y_{ij}}{\frac{p_{ij}}{Q_i}} - \dfrac{y_{ij'}}{\frac{p_{ij'}}{Q_i}} \right)^2 \right] + V_1(Y)$$

$$= E_1 \sum_n \left[\sum_{j<j'=1}^{N_i} \sum^{N_i} p_{ij} p_{ij'} \left(\dfrac{y_{ij}}{p_{ij}} - \dfrac{y_{ij'}}{p_{ij'}} \right)^2 \right]$$

$$= \sum_n \left[\dfrac{N_i(N_i-1)}{N(N-1)} \sum_{i<i'}^N \sum^n p_i p_{i'} \left(\dfrac{y_i}{p_i} - \dfrac{y_{i'}}{p_{i'}} \right)^2 \right]$$

$$= \dfrac{\left(\sum_n N_i^2 - N \right)}{N(N-1)} V.$$

Now, Cauchy-Schwarz's inequality tells us that

$$\left(\sum_n 1 \right) \left(\sum_n N_i^2 \right) \geq \left(\sum_n N_i \right)^2 = N^2$$

implying
$$\sum_n N_i^2 \geq \dfrac{N^2}{n}$$

and hence the minimal choice of N_i should be

$$N_i = \dfrac{N}{n} \quad \text{for each} \quad i = 1, \cdots, n \quad \text{if} \quad \dfrac{N}{n} \quad \text{is an integer;}$$

if $\dfrac{N}{n}$ is not an integer, one should take the minimizing choice N_i as

$$N_i = \left[\dfrac{N}{n} \right], i = 1, \cdots, K$$

and
$$N_i = \left[\dfrac{N}{n} \right] + 1 \text{ for } i = K+1, \cdots, n$$

with K so chosen that uniquely,

$$N = \sum_n N_i = K\left[\frac{N}{n}\right] + (n - K)\left(\left[\frac{N}{n}\right] + 1\right).$$

For example,

$$N = 17, n = 5 \Rightarrow$$
$$17 = 3K + (5 - K)(4) = 20 - K \Rightarrow K = 3$$
$$N = 47, n = 8 \Rightarrow$$
$$47 = 5K + (8 - K)6 = 48 - K \, or \, K = 1.$$

In order to find an unbiased estimator for

$$V_p(t_{RHC}) = \frac{(\sum_n N_i^2 - N)}{N(N-1)} \sum\sum_{i<i\prime} p_i p_{i\prime} \left(\frac{y_i}{p_i} - \frac{y_{i\prime}}{p_{i\prime}}\right)^2$$

$$= \frac{(\sum_n N_i^2 - N)}{N(N-1)} \left(\sum_{i=1}^N \frac{y_i^2}{p_i} - Y^2\right)$$

$$= \frac{(\sum_n N_i^2 - N)}{N(N-1)} \sum p_i \left(\frac{y_i}{p_i} - Y\right)^2,$$

let us first observe that $\sum_n Q_i = \sum_{i=1}^n \sum_{j=1}^{N_i} p_{ij} = \sum_1^N p_i = 1.$

Further, for

$$a = \sum_n Q_i \left(\frac{y_{ij}}{p_{ij}} - t_{RHC}\right)^2,$$

we have

$$E_p(a) = E_1 \sum_n \left[E_2 \, Q_i \, \frac{y_{ij}^2}{p_{ij}^2}\right] + E_p(t_{RHC}^2) - 2E_p(t_{RHC}^2)$$

$$= E_1 \sum_n \left[\sum_{j=1}^{N_i} \frac{y_{ij}^2}{p_{ij}}\right] - E_p(t_{RHC}^2)$$

$$= \sum_n \left(\frac{N_i}{N}\right) \sum_1^N \frac{y_i^2}{p_i} - E_p(t_{RHC}^2)$$

$$= \left(\sum_1^N \frac{y_i^2}{p_i} - Y^2\right) - E_p(t_{RHC}^2 - Y^2)$$

$$= V - \frac{\sum_n N_i^2 - N}{N(N-1)} V = \frac{N^2 - \sum_n N_i^2}{N(N-1)} V.$$

So, $$\frac{(\sum_n N_i^2 - N)}{N^2 - \sum_n N_i^2} a \text{ has expectation equal to } V_p(t_{RHC}).$$

So, $$\frac{(\sum_n N_i^2 - N)}{N^2 - \sum_n N_i^2} \sum_n Q_i \left(\frac{y_{ij}}{p_{ij}} - t_{RHC}\right)^2$$

$$= \frac{\sum_n N_i^2 - N}{N^2 - \sum_n N_i^2} \sum_n \sum_n Q_i Q_{i\prime} \left(\frac{y_i}{p_i} - \frac{y_{i\prime}}{p_{i\prime}}\right)^2$$

is an unbiased estimator for $V_p(t_{RHC})$. By $\sum_n \sum_n$ we mean sum over pairs of $^nC_2$ groups formed for RHC scheme avoiding all repetitions; also $\frac{y_i}{p_i}$ is an abbreviation for the (y_i, p_i) values (y_{ij}, p_{ij}) for the single unit chosen from the ith group and similarly $(y_{i\prime}, p_{i\prime})$ the values of $(y_{i\prime k}, p_{i\prime k})$ for the single unit, say, $i\prime K$ from the $i\prime$th group.

Two advantages of the RHC scheme over the PPSWR scheme obviously are (1) the former produces the intended size for the sample of units each distinct and (2) $V_p(t_{RHC}) < V_p(t_{HH})$.

It is worth mentioning that like t_{HH} the t_{RHC} also admits a non-negative, unbiased variance estimator quite unconditionally.

Now let us consider another varying probability sampling scheme called Probability Proportional to Size Without Replacement, abbreviated PPSWOR.

As usual this also starts with the three ingredients

$$U = (1, \cdots, i, \cdots, N),$$
$$\underline{Y} = (y_1, \cdots, y_i, \cdots, y_N),$$
and
$$\underline{X} = (x_1, \cdots, x_i, \cdots, x_N)$$

with x_i's as positive integers, if necessary, converted so and

$$p_i = \frac{x_i}{X}, i \in U$$

are the normed positive size-measures of the units.

On the first draw, from U a unit i is chosen with probability p_i employing the cumulative total method or by Lahiri's method which is simpler though with a more sophisticated proof for its tenability given that a unit i_1 is chosen on the first draw with probability p_{i1}. Leaving this unit aside, a unit $i_2 (\neq i_1)$ of U is then selected with a probability

$$\frac{p_{i2}}{1 - p_{i1}},$$

noting that
$$\sum_{j \neq i_1}^{N} \frac{p_j}{(1 - p_{i1})} = 1,$$

it follows that on principle i_2 also may be thus chosen implementing the cumulative total method or by Lahiri's method. In this way successively distinct units $i_1, i_2, \cdots, i_n$ of U are drawn with typical probabilities

$$\frac{p_{ij}}{1 - p_{i1} - p_{i2} \cdots - p_i(j-1)} \quad \text{for } j = 2, \cdots, n.$$

Des Raj (1956) proposed the following estimators for Y based on the successively drawn distinct units of U by this PPSWOR sampling, namely,

$$t_1 = \frac{y_{i1}}{p_{i1}}, \; t_2 = y_{i1} + \frac{y_{i2}}{p_{i2}/(1 - p_{i1})}, \; \cdots$$

$$t_j = y_{i1} + y_{i2} + \cdots + \frac{y_{ij}}{p_{ij}/(1 - p_{i1} - \cdots - p_i(j-1))}, \; \cdots$$

$$t_n = y_{i1} + y_{i2} + \cdots + y_{i(n-1)} + \frac{y_{in}}{p_{in}/(1 - p_{i1} - \cdots - p_{i(n-1)})};$$

it follows that $E_p(t_1) = \sum_1^N \left(\frac{y_i}{p_i}\right) p_i = Y.$

Next we shall write $E_p = E_1 E_2$ to denote by E_2 the conditional expectation operator over the selection currently, given the outcomes of the draws prior to it, and by E_1 the expectation over the prior draws. Similarly, for the variance operators overall, current selection and prior selection V, V_2, and V_1 so that $V = E_1 V_2 + V_1 E_2$.

Thus,
$$E_2(t_2) = y_{i_1} + \sum_{\substack{j=1 \\ \neq i_1}}^{N} \frac{y_j}{p_j/(1 - p_{i_1})} \frac{p_j}{(1 - p_{i_1})}$$

$$= y_{i_1} + \left(\sum_{j \neq i_i}^{N} y_j\right) = y_{i_1} + (Y - y_{i_1}) = Y,$$

i.e., t_2 is conditionally unbiased for Y.

Hence, $\quad E_p(t_2) = E_1(E_2(t_2)) = E_1(Y) = Y,$

i.e., t_2 is unbiased for Y, unconditionally too.

Generally,

$$E_2(t_j) = y_{i_1} + \cdots + y_{i_{(j-1)}} + \sum_{\substack{j=1 \\ \neq i_1, \cdots, i_{(j-1)}}}^{N} y_j \frac{\frac{p_j}{\left(1 - p_{i_1} - \cdots - p_{i_{(j-1)}}\right)}}{p_j/\left(1 - p_{i_1} - \cdots - p_{i_{(j-1)}}\right)}$$

$$= \left(y_{i_1} + \cdots + y_{i_{(j-1)}}\right) + \left(Y - y_{i_1} - y_{i_2} - \cdots - y_{i_{(j-1)}}\right)$$

$$= Y,$$

More Intricacies

i.e., t_j is conditionally unbiased for Y.

But $E_p(t_j) = E_1(E_2(t_j)) = E_1(Y) = Y$, i.e., t_j is also unconditionally unbiased for Y. Hence, Des Raj (1956) proposes

$$t_D = \frac{1}{n} \sum_{j=1}^{n} t_j$$

as an unbiased estimator for Y. Next he observes as follows that $t_j, t_{j'}$ for every $j \neq j' (= 1, \cdots, n)$ are uncorrelated.

Suppose $j' < j$, then

$$E_p(t_j t_{j'}) = E_1 E_2(t_j t_{j'}) = E_1(t_{j'}) E_2(t_j)$$
$$= E_1(t_{j'}) Y = Y^2 = E_p(t_j) E_p(t_{j'}),$$

i.e., $$\text{cov}_p(t_j, t_{j'}) = 0 \ \forall \ j \neq j'.$$

So, $$V_p(t_D) = \frac{1}{n^2} \sum_{j=1}^{n} V_p(t_j).$$

Des Raj (1956) did not give any formula for $V_p(t_j), j = 1, \cdots, n$ and hence any for $V_p(t_D)$. But his observation that $t_j, t_{j'} (j \neq j')$ are uncorrelated led him to derive as follows an unbiased estimation formula for

$$V_p(t_D). \ E_p(t_j - t_{j'})^2 = E_p[(t_j - Y) - (t_{j'} - Y)]^2 = V_p(t_j) + V_p(t_{j'}).$$

$$E_p \left[\sum\sum_{j \neq j'} (t_j - t_{j'})^2 \right] = 2(n-1) \sum_{j=1}^{n} V_p(t_j).$$

Hence, $$v_D = \frac{1}{2n^2(n-1)} \sum\sum_{j \neq j'} (t_j - t_{j'})^2$$

is an unbiased estimator for $V_p(t_D)$.

Roychoudhury (1957) derived formulae for $V_p(t_D)$ in the following way.

Letting $$t_1 = \frac{y_i}{p_i}, \quad t_2 = y_i + \frac{y_j}{p_j}(1 - p_i)$$

and $$t_3 = y_i + y_j + \frac{y_k}{p_k}(1 - p_i - p_j)$$

based on an ordered sample $s = (i, j, k)$ chosen by PPSWOR sampling taken in the order i followed by j and j followed by k, Roychoudhury (1957) observed the following:

$$E_p(t_1) = \sum_1^N \frac{y_i}{p_i} p_i = Y, V_p(t_1) = \sum\sum_{i<i'} p_i p_{i'} \left(\frac{y_i}{p_i} - \frac{y_{i'}}{p_{i'}}\right)^2 = V$$

$$E_p(t_2) = E_1[E_2(t_2)] = E_1\left[y_i + \sum_{\substack{j=1 \\ \neq i}}^N \frac{y_j}{p_j}(1-p_i)\frac{p_j}{1-p_i}\right]$$

$$= E_1[y_i + Y - y_i] = E_1(Y)$$

$$V_p(t_2) = E_1[V_2(t_2)] + V_1[E_2(t_2)]$$

$$= E_1\left[\sum\sum_{j<j'(\neq i)} \left(\frac{p_j}{1-p_i}\right)\left(\frac{p_{j'}}{1-p_i}\right)\left\{\frac{y_j}{p_j}(1-p_i) - \frac{y_{j'}}{p_{j'}}(1-p_i)\right\}^2\right]$$

$$= E_1\left[\sum\sum_{j<j'(\neq i)} p_j p_{j'} \left(\frac{y_j}{p_j} - \frac{y_{j'}}{p_{j'}}\right)^2\right]$$

$$= \sum\sum_{i<i'} (1 - p_i - p_{i'}) p_i p_{i'} \left(\frac{y_i}{p_i} - \frac{y_{i'}}{p_{i'}}\right)^2 < V = V_p(t_1).$$

For simplicity, let $Q_j = \frac{p_j}{1-p_i}, j \neq i, R_{\substack{k \\ k \neq j \neq i}} = \frac{Q_k}{1-Q_j}.$

Now, $E_p(t_3) = E_1 E_2(t_3)$

$$= E_1\left[E_2\left(y_i + y_j + \frac{y_k}{p_k}(1 - p_i - p_j)\right)\right]$$

$$= E_1\left[(y_i + y_j) + \sum_{\substack{k=1 \\ (\neq i,j)}}^N \frac{y_k}{p_k}(1-p_i-p_j)\frac{p_k}{1-p_i-p_j}\right]$$

$$= E_1\left[(y_i + y_j) + \sum_{\substack{k=1 \\ \neq i,j}}^N \left\{\frac{y_k}{Q_k}(1-Q_j)\frac{Q_k}{1-Q_j}\right\}\right]$$

$$= E_1[(y_i + y_j) + (Y - y_i - y_j)] = E_1(Y) = Y$$

More Intricacies

$$V_p(t_3) = V_1[E_2(t_3)] + E_1[V_2(t_3)]$$

$$= E_1\left[E_2\sum\sum_{k<k'} R_k R_{k'}\left(\frac{y_k}{R_k} - \frac{y_{k'}}{R_{k'}}\right)^2\right]$$

$$= \left[\sum\sum_{\substack{k<k' \\ (i,j)}} (1 - Q_k - Q_{k'})Q_k Q_{k'}\left(\frac{y_k}{Q_k} - \frac{Y_{k'}}{Q_{k'}}\right)^2\right]$$

$$< V_2 = V_p(t_2).$$

Thus, in succession $\quad V_p(t_n) < V_p(t_n - 1) < \cdots < V_p(t_3) < V_p(t_2) < V.$

So, $\quad V_p(t_D) = \dfrac{1}{n^2}\sum_{j=1}^{n} V_p(t_j) < \dfrac{V}{n} = V_p(t_{HH}).$

The PPSWOR sampling scheme has an advantage over the PPSWR sampling scheme in the sense that the former produces one unbiased estimator better than the standard Hansen-Hurwitz estimator based on PPSWR.

This, however, does not enable us to claim the PPSWR scheme is worse than the PPSWOR scheme. This is because the PPSWR scheme admits two unbiased estimators, namely,

(1) Horvitz-Thompson estimator $\quad t_{HT} = \sum_{i \in s} \dfrac{y_i}{\pi_i}$

with variance $\quad V_p(t_{HT}) = \sum\sum_{i<j}(\pi_i \pi_j - \pi_{ij})\left(\dfrac{y_i}{\pi_i} - \dfrac{y_j}{\pi_j}\right)^2 + \sum \dfrac{y_i^2}{\pi_i}\beta_i,$

$$\beta_i = \dfrac{1}{\pi_i}\sum_{s \ni i} p(s)v(s) - v,$$

and

(2) the Rao-Blackwellized version of t_{HH},

namely, $\quad t^*_{HH} = \left(\sum_{s \to s^*} t_{HH}(s, \underline{Y}) \, p(s)\right) \Big/ p(s^*).$

Here $\quad \pi_i = 1 - (1 - p_i)^n$

and $\quad \pi_{ij} = 1 - (1 - p_i)^n - (1 - p_j)^n + (1 - p_i - p_j)^n$

for the PPSWR scheme.

We may remark here that t_{HT} based on PPSWR may not be desirable for the following reasons:

$$t_{HT} = \sum_{i \in s} \frac{y_i}{\pi_i} \quad \text{is good to estimate Y if}$$

$$V_p(t_{HT}) = E_p(t_{HT} - Y)^2 = \sum_s p(s) \left(\sum_{i \in s} \frac{y_i}{\pi_i} - Y \right)^2$$

may be under control for a scheme of sampling adopted. If $y_i \propto \pi_i$ and every s contains a fixed number of distinct units with $p(s) > 0$, then $V_p(t_{HT})$ becomes zero. From this one may be guided by the consideration that if π_i is closely proportional to y_i then $V_p(t_{HT})$ should be under control. On the contrary, if p_i is closely proportional to y_i then $V_p(t_{HH})$ should be under control. If the given normed size-measures may closely meet this demand, for this scheme $\pi_i = 1 - (1 - p_i)^n$ may not be anywhere close to y_i. So, the HT estimator may not be good enough for PPSWR sampling. A second point to note is that for the PPSWR scheme, not $p(s)$ or $\sum_{s \ni i} p(s) v(s)$ or $p(s^*) = \sum_{s \to s*} p(s)$ seem easy to work out for even a moderate value of n. So, t_{HH}^* does not seem practicably sound nor does t_{HT} for PPSWR sampling.

If for a PPSWOR sample in n draws s we obtain Des Raj estimator $t_D(s)$, then noting that s is an ordered sample, say,

$$s = (i_1, i_2, \cdots, i_n), \quad d = ((i_1, y_{i1}), \cdots, (i_n, y_{in}))$$

is not a sufficient statistic. But writing the unordered sample

$$s^* = \{j_1, \cdots, j_n\}$$

with every j_i as one of the n distinct units of s, the sufficient statistic is

$$d^* = \{(j_1, y_{j1}), \cdots, (j_n, y_{jn})\}.$$

Then, we write
$$t_D^*(s) = \left[\sum_{s \to s^*} t_D(s) p(s) \right] \bigg/ p(s^*),$$

and this is known as the Symmetrized Des Raj Statistic t_{SD}.

This was introduced by Murthy (1957). He showed that t_{SD} like t_D is unbiased for Y and $V_p(t_{SD}) < (t_D)$ based on PPSWOR samples. We have, however, shown more generally that $V_p(t^*(s^*)) \leq V_p(t(s))$ and strictly so unless $t^*(s^*)$ coincides with $t(s)$ with probability one, for a general class of sampling schemes. Next we consider Murthy's (1957) estimator based on a general class of sampling scheme given as follows by him, starting as usual with the three entities

$$U = (1, \cdots, i, \cdots, N),$$
$$\underline{Y} = (y_1, \cdots, y_i, \cdots, y_N),$$
and
$$\underline{X} = (x_1, \cdots, x_i, \cdots, x_N),$$

with x_i's as positive integers and $p_i = \frac{x_i}{X}$ as the positive normed size-measure for the ith unit. Let s be a sample of n distinct units chosen with probability $p(s)$, p_i be the selection-probability of the unit i selected on the first draw, $p(s|i)$ be the conditional probability of choosing s given that on the first draw i was chosen, and $p(s|ij)$ is the conditional probability of choosing s given that i and j ($i \neq j$) were selected on the first two draws no matter in which order, then, Murthy's (1957) estimator for Y is

$$t_M = \frac{1}{p(s)} \sum_{i \in s} y_i p(s|i).$$

To prove some of its properties let us observe that

$$p(s) = \sum_{i \in s} p_i p(s|i)$$

$$\sum_{s \ni i} p(s|i) = 1, \quad \sum_{s \ni ij} p(s|i,j) = 1$$

Then
$$E_p(t_M) = \sum_s p(s) t_M(s) = \sum_{i=1}^N y_i \left(\sum_{s \ni i} p(s|i) \right) = Y,$$

i.e., t_M is unbiased for Y.

Murthy (1957) provided formulae for $V_p(t_M)$ and for an unbiased estimator thereof. Instead of presenting them here now, let us postpone the discussion to bypass into the following:

Ha'jek's (1959) approach was reformulated by Rao (1979) considering for Y the homogeneous linear estimator

$$t_b = \sum_{i \in s} y_i b_{si} = \sum_{i=1}^N y_i b_{si} I_{si}$$

with
$$I_{si} = 1/0 \quad \text{for } i \in s / i \notin s$$
and
$$b_{si} \quad \text{as quantities free of } \underline{Y} = (y_1, \cdots, y_i, \cdots, y_N).$$

Then, $E_p (t_b - Y)^2$ is the Mean Square Error (MSE) for t_b about Y to be written as

$$M = E_p \left(\sum_{i=1}^{N} y_i (b_{si} I_{si} - 1) \right)^2$$

$$= \sum_{i=1}^{N} \sum_{j=1}^{N} y_i y_j d_{ij}$$

writing
$$d_{ij} = E_p (b_{si} I_{si} - 1)(b_{sj} I_{sj} - 1).$$

Ha'jek's (1959) and Rao's (1979) approach is to express this M into a form from which an "unbiased estimator" may be derived for which easily one may derive Necessary Conditions for such an estimator's uniform non-negativity irrespective of the y-variate values but only on observing conditions the features of t_b and of the design p need to satisfy.

Let there exist constants w_i independent of $\underline{Y}$ such that $w_i \neq 0 \ \forall \ i \in U$ but $z_i = \frac{y_i}{w_i}$ may put M into the form

$$M = \sum_{i=1}^{N} \sum_{j=1}^{N} w_i w_j z_i z_j d_{ij}.$$

Ha'jek (1959) and Rao (1979) now demand the condition C that if $z_i = C \ \forall \ i \in U$, then $M = 0$.

Under this condition C they show that

$$M = -\sum_{\substack{i=1 \\ i<j}}^{N} \sum_{j=1}^{N} d_{ij} w_i w_j (z_i - z_i)^2$$

uniformly in $\underline{Y}$.

Before presenting their proof for this let us cite a few situations when the condition C "may hold" and also when it "may not hold."

Design	Estimator	Choice of w_i	Does C hold?
SRSWOR	$N\bar{y}$	unity	Yes
PPSWR	$\frac{1}{n}\sum_{r=1}^{n}\frac{y_r}{p_r}$	p_i	Yes
Murthy's	$\frac{1}{p(s)}\sum y_i p(s\|i)$	p_i	Yes
RHC's	$\sum_n y_i \frac{Q_i}{P_i}$	p_i	Yes
(i) Any Design	$\sum_{i\in s} \frac{y_i}{\pi_i}$	π_i	No
(ii) A Design giving every sample same effective size	$\sum_{i\in s} \frac{y_i}{\pi_i}$	π_i	Yes

More Intricacies

To check (i) and (ii) above one may note that for

$$z_i = \frac{y_i}{\pi_i} = C \quad \forall i,$$

$$t_{HT} = C \sum_{i \in s} i = Cv(s) \quad \text{and} \quad Y = C \sum \pi_i = Cv.$$

Thus, if (ii) holds then $Cv(s) = Cv$ and M equals zero, but if (i) holds $Cv(s)$ may not equal Cv for every s with $p(s) > 0$ rendering $M \neq 0$.

If the condition C holds, then

$$M = C^2 \sum_i^N \sum_j^N w_i w_j d_{ij} = 0.$$

To show that in such a case M may be written as

$$M = -\sum_{i<j} \sum d_{ij} w_i w_j (z_i - z_i)^2 = -\sum_{i<j} \sum d_{ij} w_i w_j \left(\frac{y_i}{w_i} - \frac{y_j}{w_i}\right)^2$$

let us consider the following algebraic result.

Let $Q = \sum_{i=1}^T \sum_{j=1}^T a_{ij} x_i x_j$ subject to $\sum_1^T \sum_1^T a_{ij} = 0,$

be a Non-negative Quadratic Form (NNDQF) in x_i's, $i = 1, \cdots, T$. Then, let $x_1 = A$, and $x_j = B$ for every $j = 2, \cdots, T$.

Then $\quad Q = A^2 a_{11} + 2AB \sum_2^T a_{ij} + B^2 \sum_2^T \sum_2^T a_{ij}$

and $\quad 0 = a_{11} + 2 \sum_2^T a_{ij} + \sum_2^T \sum_2^T a_{ij}.$

So, $\quad Q = a_{11} A^2 + 2 \left(\sum_2^T a_{ij}\right) AB + \left[-\left(a_{11} + 2 \sum_2^T a_{ij}\right)\right] B^2$

is now an NNDQF in A and B. So, we must have

$$\left| \begin{matrix} a_{11} & \sum_2^T a_{ij} \\ \sum_2^T a_{ij} & -(a_{11} + 2\sum_2^T a_{ij}) \end{matrix} \right| \geq 0$$

i.e., $\quad -\left[a_{11}^2 + 2a_{11} \sum_1^T a_{ij} + \left(\sum_2^T a_{ij}\right)^2\right] \geq 0$

i.e., $\quad -\left(\sum_1^T a_{ij}\right)^2 \geq 0 \Rightarrow \sum_1^T a_{ij} = 0$

Similarly, $$\sum_{j=1}^{T} a_{ij} = 0 \ \forall \ i.$$

Applying this result to $$M = \sum_{1}^{N}\sum_{1}^{N} w_i w_j d_{ij} z_i z_j$$

subject to $$\sum_{1}^{N}\sum_{1}^{N} w_i w_j d_{ij} = 0, \quad \text{it follows,}$$

because $$M = E_p \left(t_b - Y\right)^2 \geq 0 \ \forall \ \underline{Y},$$

that $$\sum_{j} d_{ij} w_i w_j = 0 \quad \forall \ i \in U.$$

Thus, $$d_{ii} w_i^2 = -\sum_{j \neq i} d_{ij} w_i w_j \quad \forall \ i \in U.$$

Hence, $$\sum_{1}^{N} d_{ii} w_i^2 = -\sum_{i \neq j} d_{ij} w_i w_j.$$

So, $$M = \frac{1}{2}\sum_{j} d_{jj} w_j^2 z_j^2 + \sum\sum_{i \neq j} d_{ij} w_i w_j z_i z_j$$
$$= -\frac{1}{2}\sum\sum_{i \neq j} d_{ij} w_i w_j \left(z_i - z_j\right)^2$$
$$= -\sum\sum_{i < j} d_{ij} w_i w_j \left(z_i - z_j\right)^2.$$

With the formula for M as a quadratic form
$$M = \sum_i\sum_i d_{ij} y_i y_j = \sum_i\sum_j d_{ij} w_i w_j z_i z_j$$
$$= \sum_i d_{ii'} w_i^2 z_i^2 + \sum\sum_{i \neq j} d_{ij} w_i w_j z_i z_j$$

More Intricacies

it is hard to ensure for its unbiased estimator

$$m = \sum d_{si} w_i^2 z_i^2 I_{si} + \sum\sum_{i \neq j} d_{sij} w_i w_j z_i z_j I_{sij}$$

subject to
$$\sum_s p(s) d_{si} I_{si} = d_{ii}$$

and
$$\sum p(s) d_{sij} I_{sij} = d_{ij} \ \forall \ i, j \ (i \neq j) \in U$$

so that m is uniformly non-negative for M. Because this m is also a quadratic form, in order that it may be a non-negative quadratic form we have to check for the non-negativity of each of its principal minors. That will be too tall an order. So, Ha'jek's (1959) and Rao's (1979) theorem, when condition C holds, gives us a convenient unbiased mean square error estimator with its condition for uniform non-negativity very easily checked.

Let
$$m = \sum_i \sum_j d_{sij} I_{sij} y_i y_j$$

be an unbiased estimator for M so that

$$M = E_p(m) = \sum_s p(s) \sum_i \sum_j d_{sij} I_{sij} y_i y_j$$

and let there exist constants

$$w_i (\neq 0 \ \forall \ i \in U) \quad \text{free of} \quad \underline{Y} = (Y_1, \cdots, Y_N)$$

so that $M = \sum_i \sum_j d_{ij} y_i y_j = \sum_i \sum_j d_{ij} w_i w_j z_i z_j \quad$ on writing $\quad z_i = \dfrac{y_i}{w_i}.$

Now if $z_i = C \forall i \Rightarrow M = 0$, then in order that the above m may be uniformly non-negative $\forall y_i \in \underline{Y}$, then m must be of the form

$$m = -\sum\sum_{i<j \in s} d_{sij} w_i w_j (z_i - z_j)^2.$$

The proof follows easily since

$$m = m(s) = \sum_i \sum_j d_{sij} I_{sij} w_i w_j z_i z_j$$

is required to be a non-negative quadratic form in z_i's for every s with $p(s) > 0$ and also equal to zero if $z_i = C \forall i \in U$ because $\sum_s p(s) m(s)$ equals zero under this condition C and for every s with $p(s) > 0$ we must have $m(s) \geq 0$.

Hence,
$$m(s) = \sum_i \sum_j d_{sij} I_{sij} w_i w_j z_i z_j$$

being a Non-negative Quadratic form subject to being zero if $z_i = C \,\forall\, i \in U$, it must be of the form

$$m = m(s) = -\sum\sum_{i<j} d_{sij} I_{sij} w_i w_j (z_i - z_j)^2$$

as is claimed.

A good application of these results is to consider Murthy's (1957) estimator

$$t_M = \frac{1}{p(s)} \sum_{i \in s} y_i\, p(s/i)$$

with

$$d_{ij} = E_p \left(\frac{p(s/i)}{p(s)} I_{si} - 1\right)\left(\frac{p(s/j)}{p(s)} I_{sj} - 1\right)$$

$$= \sum_{s \ni i,j} \frac{p(s/i)\, p(s/j)}{p(s)} - 1.$$

Now, writing

$$z_i = \frac{y_i}{p_i}$$

it follows that

$$z_i = C\,\forall\, i \in U \Rightarrow y_i = C\, p_i \,\forall\, i \in U \Rightarrow Y = C$$

so that

$$t_M = \frac{1}{p(s)} C \sum_{i \in s} p_i p(s/i) = C,$$

and hence,

$$t_M = Y \quad \text{or} \quad V_p(t_M) = 0,$$

it follows that identically,

$$V_p(t_M) = \sum\sum_{i<j} p_i p_j \left(\frac{y_i}{p_i} - \frac{y_j}{p_j}\right)^2 \left(1 - \sum_{s \ni i,j} \frac{p(s/i)\, p(s/j)}{p(s)}\right).$$

Writing

$$a_{ij} = p_i p_j \left(\frac{y_i}{p_i} - \frac{y_j}{p_j}\right)^2$$

it follows from Ha'jek (1958) and Rao's (1979) theorem that an unbiased estimator $v(t_M)$ in order to be uniformly non-negative among all quadratic estimators for $V_p(t_M)$ should be of the form

$$v(t_M) = \sum\sum_{i<j \in s} a_{ij} \left(\frac{p(s)\, p(s/i,j) - p(s/i)\, p(s/j)}{p^2(s)}\right).$$

More Intricacies

Murthy's estimator is unbiased for Y if based on Lahiri, Midzuno, and Sen's (LMS) scheme, because for this scheme

$$p(s|i) = \frac{1}{\binom{N-1}{n-1}} = p(s|j) \quad \text{and} \quad p(s|ij) = \frac{1}{\binom{N-2}{n-2}}$$

and

$$p(s) = \frac{\sum_{i \in s} P_i}{\binom{N-1}{n-1}} = \frac{\left(\sum_{i \in s} x_i\right)}{X \binom{N-1}{n-1}}.$$

Hence, t_M equals $t_R = X \dfrac{\sum_{i \in s} y_i}{\sum_{i \in s} x_i}$ which is the ratio estimator for Y.

Obviously, when based on the LMS scheme the ratio estimator t_R is exactly unbiased for Y with a variance equal to

$$V_p(t_R) = \sum\sum_{i<j} a_{ij} \left(1 - \frac{X}{\binom{N-1}{n-1}} \sum_{s \ni i,j} \left(\frac{1}{\sum_{i \in s} x_i}\right)\right)$$

and an exactly unbiased variance estimator

$$v(t_R) = \sum\sum_{i<j\in s} a_{ij} \left(\frac{X}{\sum_{i \in s} x_i}\right) \left(\frac{N-1}{n-1} - \frac{X}{\sum_{i \in s} x_i}\right).$$

Obviously, $v(t_M)$ and $v(t_R)$ are not easy to work out from large samples.

Let us now introduce and develop some consequences to the concept of Admissibility of an estimator for a populational total based on a given design. If given an estimator t in a class C of estimators for Y be such that within this class there exists no other estimator better than it, we shall call it Admissibile within the class.

We shall now consider two classes: namely, the class C_1 of all homogeneous linear unbiased estimators (HLUEs) for Y and the class C_2 of all unbiased estimators (UEs) for Y. We shall show that for any design admitting the Horvitz-Thompson Estimator (HTE) within C_1 the HTE is admissible within C_1 and similarly the result holds for C_2.

Let $$t_b = \sum_{i \in s} y_i b_{si} \quad \text{with } b'_{si}\text{s free of } \underline{Y}$$

but subject to $$\sum_{s \ni i} p(s) b_{si} = 1 \; \forall \; i.$$

Obviously, the HTE, namely,

$$t_H = \sum_{i \in s} \frac{y_i}{\pi_i} \quad \text{with} \quad \pi_i > 0 \; \forall \; i \in U$$

is one in this HLUE class C_1.

$$V_p(t_b) - V_p(t_H)\Big|_{\underline{Y}=(0,\cdots,0,y_i\neq 0,0,\cdots,0)} = y_i^2 \left(\sum_{s\ni i} p(s) b_{si}^2 - \frac{1}{\pi}\right).$$

Now, by Cauchy inequality

$$\left(\sum_{s\ni i} p(s)\right)\left(\sum_{s\ni i} p(s) b_{si}^2\right) \geq \left(\sum_{s\ni i} p(s) b_{si}\right)^2 = 1.$$

So, $$\sum_{s\ni i} p(s) b_{si}^2 \geq \frac{1}{\pi_i} \quad \forall i \in U.$$

Also, equality holds in this latest inequality only if $b_{si}\sqrt{p(s)} \propto \sqrt{p(s)} \; \forall \; s$ such that $p(s) > 0$ and $\pi_i > 0$. So, b_{si} must be of the form b_i subject to $b_i \sum_{s\ni i} p(s) = 1$, i.e., $b_i = \frac{1}{\pi_i}$, i.e., t_b equal to t_H. So, no estimator in the HLUE class C_1 may be better than t_H in terms of the variance criterion. So, the HTE is Admissibile for Y in the class C_1.

To show that $t_{HT} = \sum_{i\in s} \frac{y_i}{\pi_i}$ is admissible in the class C_2 of all unbiased estimators of Y let, if possible, there exist an unbiased estimator $t = t(s,\underline{Y})$ which is better than t_{HT}.

Let $h = h(s,\underline{Y})$ be an unbiased estimator of O so that we may write

$$t = t(s,\underline{Y}) = t_{HT} + h = t_{HT}(s,\underline{Y}) + h(s,\underline{Y}) = \sum_{i\in s} \frac{y_i}{\pi_i} + h(s,\underline{Y}).$$

Then, $$V_p(t) = V_p(t_{HT}) + V_p(h) + 2\text{cov}_p(t_{HT}, h)$$

Then, $V_p(t) \leq V_p(t_{HT}) \forall \underline{Y} \Rightarrow \sum_s h^2(s,\underline{Y}) p(s) \leq -2\sum_s p(s) t_{HT}(s,\underline{Y}) h(s,\underline{Y})$

since $$O = \sum_s h(s,\underline{Y}) p(s)$$

Since $t_{HT}(s,\underline{Y}) = 0$ if $\underline{Y} = (0,\cdots,0)$,

it follows that $\sum_s h^2(s,\underline{Y}) p(s)\Big|_{\underline{Y}=(0,\cdots,0)} = 0$.

Then, $h(s,\underline{Y}) p(s)\Big|_{\underline{Y}=(0,\cdots,0)} = 0$.

More Intricacies

Let $\underline{X}_i = \{\underline{Y} = (y_1, \cdots, y_N) |$ exactly i of the N co-ordinates of $\underline{Y}$ are non-zero$\}$.

Thus, $h(s, \underline{Y})p(s) = 0$ if $\underline{Y} \in \underline{X}_0$, $\forall\ s \in \varphi$, the sample space.

Now we shall prove the result by induction.

Let $h(s, \underline{Y})p(s) = 0\ \forall\ \underline{Y} \in \underline{X}_i$, for $i = 0, 1, \cdots, N-1$.

If $t = t(s, \underline{Y})$ is to be better than $t_{HT} = t_{HT}(s, \underline{Y})$, then we shall show that $h(s, \underline{Y})p(s) = 0$, $\forall\ s$ and $\forall\ \underline{Y} \in \underline{X}_{i+1}$. The result is shown to hold for $\underline{Y} \in \underline{X}_0$. Hence, by induction it will be possible to prove the result in general.

Let $\underline{Z} = (z_1, \cdots, z_i, \cdots, z_N) \in \underline{X}_{i+1}$ and let us assume that t is better than t_{HT} and $h(s, \underline{Y})p(s) = 0\ \forall\ \underline{Y} \in \underline{X}_i$.

Then, it follows that

$$0 = \sum_s p(s)\, h(s, \underline{Z}),\quad \underline{Z} = (z_1, \cdots, z_N)$$

and

$$\sum_s p(s)\, h^2(s, \underline{Z}) \leq -2 \sum_s p(s)\, t_{HT}(s, \underline{Z})\, h(s, \underline{Z}).$$

Let φ = Sample-space and $p(s) > 0\ \forall\ s \in \varphi$.

Let $\varphi_j = \{s|$ of the co-ordinates z_k in $\underline{Z}$ with $k \in s$, exactly j of them are non-zero$\}$.

Then $\varphi_j \bigcap \varphi_{j'}\ \forall\ j \neq j'$ is the empty set Φ

and $$\bigcup_{j=1}^{i+1} \varphi_j = \varphi.$$

So, $$0 = \sum_{j=0}^{i+1} \sum_{s \in \varphi_j} p(s)\, h(s, \underline{Z})$$

and $$\sum_{j=0}^{i+1} \sum_{s \in \varphi_j} p(s)\, h^2(s, \underline{Z}) \leq -2 \sum_{j=0}^{i+1} \sum_{s \in \varphi_j} p(s)\, t_{HT}(s, \underline{Z})\, h(s, \underline{Z}).$$

But by hypothesis $p(s)\, h(s, \underline{Z}) = 0\ \forall\ s \in \varphi_j, j = 0, 1, \cdots, i$.

So, $$0 = \sum_{s \in \varphi_{i+1}} p(s)\, h(s, \underline{Z})$$

$$\sum_{s \in \varphi_{i+1}} p(s)\, h^2(s, \underline{Z}) \leq -2 \sum_{s \in \varphi_{i+1}} p(s)\, t_{HT}(s, \underline{Z})\, h(s, \underline{Z}).$$

By definition of $\underline{Z}$, we have $\quad t_{HT}(s, \underline{Z}) = \sum_{i=1}^{N} \dfrac{z_i}{\pi_i} \ \forall \ s \in \varphi_{i+1}.$

This leads to $\quad p(s)\, h(s, \underline{Z}) = 0 \ \forall \ s \in \varphi_{i+1}.$

Hence, $\quad p(s)\, h(s, \underline{Z}) = 0 \ \forall \ s \in \varphi_j, j = 0, \cdots, i+1.$

This means $\quad p(s)\, h(s, \underline{Y}) = 0 \ \forall \ s \in \varphi \text{ and } \ \forall \ \underline{Y} \in \underline{X}_{i+1}.$

But already it is shown that $\quad p(s)\, h(s, \underline{Y}) = 0 \ \forall \ s \text{ and } \underline{Y} \in \underline{X}_0.$

So, $\quad p(s)\, h(s, \underline{Y}) = 0 \ \forall \ s \in \varphi \ \& \ \underline{Y} \in \Omega,$

$$\Omega = \{\underline{Y} \mid -\infty < a_i \le y_i \le b_i < +\infty, i = 1, \cdots, N\}.$$

So, if t is to be better than t_{HT} in the class C_2 then

$$t(s, \underline{Y})\, p(s) = t_{HT}(s, \underline{Y})\, p(s) + h(s, \underline{Y})\, p(s)$$
$$\Rightarrow t(s, \underline{Y}) = t_{HT}(s, \underline{Y})$$
$$\forall \ s \, . \, \ni \, . \, p(s) > 0, \forall \ \underline{Y} \in \Omega.$$

t and t_{HT} may differ only for s with $p(s) = 0$. We may ignore situations for which $p(s) = 0$.

Thus, no unbiased estimators other than t_{HT} for samples with positive selection-probabilities can be better than t_{HT}. This implies t_{HT} is admissible in the class C_2.

4 Exploring Improved Ways

Abstract. Introduction. Stratified sampling. Multi-stage sampling. Multi-phase sampling. Cluster sampling: Ratio and regression estimation. Controlled sampling.

4.0 ABSTRACT

It is evident by now that Survey Sampling as a subject for study has set before itself the singular objective of making rational inference about a finite collection of objects reflecting inherent variability among its numerous constituents through an appropriate selection of a manageable body of representative components. An in-depth study of these partial segments is so as to make a communicable appraisal of this totality, recognizing essential departures of the sampled segment's behavior from that of the totality, yet making a reasonable guess with a desire to assess the elements of discrepancies. This in its various aspects is meant to be examined in this chapter.

4.1 INTRODUCTION

If N in $U = (1, \cdots, i, \cdots, N)$ is quite large and in $\underline{Y} = (y_1, \cdots, y_i, \cdots, y_N)$ the variability among its co-ordinates is too large, then $\bar{Y}$ may not be accurately estimated by $\bar{y}$ from an SRS. Similarly, if variability of $\frac{y_r}{p_r}$'s for $r = 1, \cdots, n$ in PPSWR sampling or the variability among $\frac{y_i}{p_i}$'s for $i \in U$ or among $\frac{y_i}{\pi_i}$'s for $i \in U$ be appreciable, the RHC strategy, Murthy's strategy, or Horvitz-Thompson estimator together with schemes of selection of none-too big samples may not give us accurate estimates for Y or for $\bar{Y}$. A tentative way to tackle this issue is to split up the population into a manageable number of disjoint groups and to take samples from each of these groups in independent manners and employ estimation using group-wise samples so that the level of accuracy in estimation may be kept high within groups, resulting in commendably high overall accuracy. These mutually exclusive groups from each of which samples are independently drawn, the groups collectively being co-extensive with the entire population are called the "strata" and the sample selection thus resorted to is called Stratified Sampling. Its essential feature is that estimation uses samples from each stratum, variability of the overall estimator involving variability within respective strata in such a way that the overall variability in estimating is pooled across the variability occurring within each separate stratum, hence achieving control in measured overall level of variability keeping within strata variations in check. Stratification is of course technically and deliberately induced on the population by the survey sampler.

As a natural phenomenon the units in a population may occur showing tendencies of gregariousness or keeping close together displaying natural affinity. For examples, residential houses of people display natural neighborliness. People living in hostels or boarding houses constitute natural clusters. In estimating population characteristics when the population occurs in the form of such natural clusters, one may be inclined to select several such clusters and then survey the individuals in such selected clusters in assessing population features as it is operationally easier than taking individuals separately across the clusters disregarding their natural togetherness. Seemingly, variability of elements within clusters should be limited. So, unless adequate number and set of clusters is surveyed, the total population may be under-represented in samples camouflaging the total inherent variability.

If the strata are formed but only some of them are sampled, then these strata are regarded as first-stage units or clusters. If the sampled clusters are fully surveyed, then this is as good as cluster sampling. But if from these clusters or first-stage units (fsu) individuals are again sampled and surveyed we have what is called Two-stage Sampling. Similarly, stages of sampling may be extended giving rise to Multi-stage Sampling.

Some sophisticated sampling schemes and some estimation procedures need access to some preliminary materials. When they are found wanting, an exercise in sample selection prior to the intended sampling strategy to be adopted is often needed as a prelude to the final undertakings. This needs two-phase or multi-phase sampling in practice. Its application needs to be elaborated with specific illustrations as will be undertaken in this chapter.

Another topic demanding attention relates to situations when a planned selection scheme is difficult to be executed. This necessitates the advent of a new scheme called Controlled selection scheme to be dealt with in this chapter.

4.2 STRATIFIED SAMPLING

Let $U = (1, \cdots, i, \cdots, N)$ be a finite survey population of a known big number N of units. Let this be divided into $H(>1)$ parts consisting of

$$N_h \left(1 < N_h < N, \sum_{n=1}^{H} N_h = N \right) \text{ units,}$$

each disjoint with every other, called the strata U_h such that

$$U_h \cap U_{h'} = \Phi \ \forall \ h \neq h' (=1, \cdots, H) \quad \text{and} \quad \bigcup_{h=1}^{H} U_h = U.$$

Let from each such stratum samples of respective sizes

$$n_h \left(1 \leq n_h < N_h, \sum_{h=1}^{H} n_h = n \right)$$

Exploring Improved Ways

be independently drawn by SRSWOR method. This is called the stratified SRSWOR scheme of sampling. Let u_h be the SRSWOR in n_h draws from U_h, and let them be independently drawn across the strata, $h = 1, 2, \cdots, H$.

Let
$$\bar{y}_h = \frac{1}{n_h}\left(\sum_{i \in u_h} y_{hi}\right),$$

writing y_{hi} for the value of a real variable y for the unit $i(= 1, 2, \cdots, N_h)$ chosen from U_h $(h = 1, 2, \cdots, H)$. Then, this is an unbiased estimator for

$$\bar{Y}_h = \frac{1}{N_h}\sum_{i=1}^{N_h} y_{hi}$$

which is the hth stratum mean, $h = 1, \cdots, H$.

The population mean is

$$\bar{Y} = \frac{1}{N}\sum_{h=1}^{H}\sum_{i=1}^{N_h} y_{hi} = \frac{1}{N}\sum_{h=1}^{H} N_h \bar{Y}_h$$

$$= \sum_{h=1}^{H} W_h \bar{Y}_h, \text{ writing } W_h = \frac{N_h}{N}.$$

By construction of the strata we suppose N_h and W_h – values are known for $h = 1, \cdots, H$.

Writing
$$S_h^2 = \frac{1}{(N_h - 1)} \sum_{i=1}^{N_h} (y_{hi} - \bar{Y}_h)^2,$$

we find
$$V_p(\bar{y}_h) = \frac{N_h - n_h}{N_h n_h} S_h^2.$$

So,
$$\bar{y}_{st} = \sum W_h \bar{y}_h \text{ has}$$

$$E_p(\bar{y}_{st}) = \sum W_h \bar{Y}_h = \bar{Y},$$

i.e., $\bar{y}_{st}$ is unbiased for $\bar{Y}$

and
$$V_p(\bar{y}_{st}) = \frac{1}{N^2} \sum_{h=1}^{H} N_h^2 \frac{(N_h - n_h)}{N_h n_h} S_h^2$$

$$= \frac{1}{N^2} \sum_{h=1}^{H} N_h^2 \frac{S_h^2}{n_h} - \frac{1}{N^2} \sum_{h=1}^{H} N_h S_h^2.$$

Since
$$\left(\sum_{h=1}^{H} n_h\right) \left(\sum_{h=1}^{H} N_h^2 \frac{S_h^2}{n_h}\right) \geq \left(\sum_{h=1}^{H} N_h S_h\right)^2$$

and this minimum is attainable for the choice
$$\frac{N_h S_h}{\sqrt{n_h}} \propto \sqrt{n_h},$$

i.e., for $\quad n_h \propto N_h S_h$,

i.e., for $\quad n_h = \dfrac{n N_h S_h}{\sum_{h=1}^{H} N_h S_h}.$

This is called Neyman's (1934) optimal allocation rule.

So, with this optimal allocation of the total sample size $n = \sum_{h=1}^{H} n_h$ to the H strata the minimum value of $V_p(\bar{Y}_{st})$ becomes

$$V_{min}(\bar{y}_{st}) = \frac{1}{N^2} \left[\frac{1}{n} \left(\sum_{h=1}^{H} N_h S_h \right)^2 - \sum_{h=1}^{H} N_h S_h^2 \right]$$

$$= \frac{1}{n} \left(\sum W_h S_h \right)^2 - \frac{1}{N} \sum W_h S_h^2.$$

On the other hand on taking $n_h = n \frac{N_h}{N}$, the proportional allocation, $V_p(\bar{y}_{st})$ equals

$$\frac{1}{n} \left(\sum_{h=1}^{H} W_h S_h^2 \right) - \frac{1}{N} \sum_{h=1}^{H} W_h S_h^2 = \left(\frac{1}{n} - \frac{1}{N} \right) \sum_{h=1}^{H} W_h S_h^2.$$

From the consideration of cost involved in stratified SRSWOR sampling, there is another rule for allocation of the total sample-size n to the respective H strata.

Let C_0 denote an overhead cost in implementing a survey by stratified SRSWOR and C_h the cost per unit of survey-cum-selection so that the total cost is

$$C = C_0 + \sum_{h=1}^{H} C_h n_h.$$

Exploring Improved Ways

The resulting variance of $\bar{y}_{st}$ is

$$V = \frac{1}{N^2} \sum_{h=1}^{H} N_h^2 \frac{S_h^2}{n_h} - \frac{1}{N^2} \sum_{h=1}^{H} N_h S_h^2.$$

We may decide to choose the allocations n_h subject to $\sum_{h=1}^{H} n_h = n$ either to minimize the total cost for an intended level of accuracy to be determined by an attained value of the variance or minimize the variance and equivalently maximize the accuracy or efficiency for a pre-assigned total level of cost that may be permissible.

To solve either of these specifications one needs to minimize the value of

$$(C - C_0)\left(V + \frac{1}{N^2}\sum N_h S_h^2\right) = \left(\sum C_h n_h\right)\left(\sum W_h^2 \frac{S_h^2}{n_h}\right)$$

in deriving an optimal allocation rule.

By Cauchy inequality we get

$$\left(\sum C_h n_h\right)\left(\sum_h W_h^2 \frac{S_h^2}{n_h}\right) \geq \left(\sum W_h \sqrt{C_h} S_h\right)^2$$

giving us the allocation rule

$$n_h \propto \frac{W_h S_h}{\sqrt{C_h}}$$

leading to

$$n_h = \frac{n \frac{W_h S_h}{\sqrt{C_h}}}{\sum_{h=1}^{H} \frac{W_h S_h}{\sqrt{C_h}}} = \frac{n \frac{N_h S_h}{\sqrt{C_h}}}{\sum_{h=1}^{H} \frac{N_h S_h}{\sqrt{C_h}}}$$

Thus, this rule prescribes a higher scale of sampling per stratum of relatively bigger size, higher stratum-variability taking account of the relatively lesser cost reflected through the square root of the survey-sampling cost per stratum.

Unbiased estimation of the variance of $\bar{y}_{st}$, namely,

$$V_p(\bar{y}_{st}) = \sum W_h^2 \left(\frac{1}{n_h} - \frac{1}{N_h}\right) S_h^2$$

is very easy using

$$v(\bar{y}_{st}) = \sum W_h^2 \left(\frac{1}{n_h} - \frac{1}{N_h}\right) s_h^2$$

writing

$$s_h^2 = \frac{1}{(n_h - 1)} \sum_{i \in u_h} (y_{hi} - y_h)^2.$$

Next, let us examine how much accuracy we may gain through stratified SRSWOR rather than SRSWOR from the entire population by a sample of size n.

Suppose a stratified SRSWOR with strata-wise sample-sizes

$$n_h (h = 1, \cdots, H), \sum_{h=1}^{H} n_h = n$$

has been selected and surveyed giving the values y_{hi} for $i \in u_h, h = 1, \cdots, H$ producing the unbiased estimate for $\bar{Y}$ as

$$\bar{y}_{st} = \frac{1}{N} \sum_{h=1}^{N} N_h \bar{y}_h, \quad \bar{y}_h = \frac{1}{n_h} \sum_{i \in u_h} y_{hi}$$

and an unbiased estimate of its variance as

$$v(\bar{y}_{st}) = \frac{1}{N^2} \sum_{h=1}^{N} N_h^2 \left(\frac{1}{n_h} - \frac{1}{N_h} \right) s_h^2, \quad \text{with } s_h^2 = \frac{1}{(n_h - 1)} \sum_{i \in u_h} (y_{hi} - \bar{y}_h)^2.$$

Intending to judge if rather than this stratified SRSWOR, an overall SRSWOR of n units had been selected from U might have been comparably useful, let us note the following:

Let
$$S^2 = \frac{1}{N-1} \sum_{h=1}^{H} \sum_{i=1}^{N_h} (y_{hi} - \bar{Y})^2$$

$$= \frac{1}{N-1} \left\{ \sum_{h=1}^{H} \left[\sum_{i=1}^{N_h} y_{hi}^2 \right] - N (\bar{Y})^2 \right\}.$$

Now,
$$V_p(\bar{y}_{st}) = E(\bar{y}_{st})^2 - (\bar{Y})^2.$$

So, $(\bar{Y})^2$ has an unbiased estimator as

$$(\bar{y}_{st})^2 - v(\bar{y}_{st}) = \left(\hat{\bar{Y}} \right)^2.$$

For $\bar{y}$ = sample mean of n units in an SRSWOR the variance would be

$$V(\bar{y}) = \frac{N-n}{Nn} S^2.$$

So, an unbiased estimator of $V(\bar{y})$ derivable from the stratified SRSWOR data would be

$$\hat{V}(\bar{y}) = \frac{N-n}{Nn} \left[\frac{1}{N-1} \sum_{h=1}^{H} \left(\frac{N_h}{n_h} \sum_{i \in u_h} y_{hi}^2 - \left(\hat{\bar{Y}} \right)^2 \right) \right].$$

Now
$$G = \frac{\hat{V}(\bar{y}) - v(\bar{y}_{st})}{v(\bar{y}_{st})} \times 100$$

should provide us with a sample-based measure of the percentage gain in efficiency of stratified SRSWOR over SRSWOR. The higher the value of G, the greater is the efficiency.

Recalling that $$V_{SRSWR}(\bar{y}) = \frac{\sigma^2}{n} = \frac{1}{nN}\sum_1^N (y_i - \bar{Y})^2 = \frac{N-1}{Nn}S^2$$

and $$V_{SRSWOR}(\bar{y}) = \frac{N-n}{N}\frac{S^2}{n} = \frac{N-n}{N}\frac{1}{n}\frac{\sum_1^N(y_i-\bar{Y})^2}{(N-1)} = \frac{N-n}{N}\frac{S^2}{n}.$$

It is customary to call the factor

$$\frac{N-n}{N} = \left(1 - \frac{n}{N}\right) = 1 - f, \text{ writing } f = \frac{n}{N}$$

the Finite Population Correction factor (FPC or fpc) in order to emphasize that because of the Finiteness of the size of the Population $U = (1, \cdots, i, \cdots, N)$, we are getting an essential difference in these two variance formulae.

Thus, in stratified Simple Random Sampling

$$\frac{N_h - n_h}{N_h} = \left(1 - \frac{n_h}{N_h}\right) = (1 - f_h)$$

will be called the fpc in respect of the hth stratum ($h = 1, \cdots, H$).

Now, employing the Neyman allocation and ignoring the fpc the formula for the variance of $\bar{y}_{st}$ is

$$V_{opt}(\bar{y}_{st}) = \frac{1}{n}\left(\sum_{h=1}^H W_h S_h\right)^2.$$

Now let us consider a theoretical topic as to how to construct the strata.

Let the y-values be distributed over the ranges defining the H strata as $a_0 < a_1 < \cdots < a_{h-1} < a_h < a_{h+1} < \cdots < a_{H-1} < a_H$ such that the y-values over the range $a_{h-1} \leq y \leq a_h$ specify the hth stratum.

Supposing y-variable has a probability density function $f(.)$, we may define

$$W_h = \int_{a_{h-1}}^{a_h} f(t)dt \quad \text{giving} \quad \frac{\partial W_h}{\partial a_h} = f(a_h) ;$$

$$W_h S_h^2 = \int_{a_{h-1}}^{a_h} t^2 f(t)dt - \frac{\left[\int_{a_{h-1}}^{a_h} tf(t)dt\right]^2}{\int_{a_{h-1}}^{a_h} f(t)dt}$$

and $\mu_h = \int_{a_{h-1}}^{a_h} tf(t)dt \bigg/ \int_{a_{h-1}}^{a_h} f(t)dt$

$$\frac{\partial W_h}{\partial a_h} S_h^2 = S_h^2 \frac{\partial W_h}{\partial a_h} + 2W_h S_h \frac{\partial S_h}{\partial a_h} = a_h^2 f(a_h) - 2a_h \mu_h f(a_h) + \mu_h^2 f(a_h)$$

or $$2S_h^2 \frac{\partial W_h}{\partial a_h} + 2W_h S_h \frac{\partial S_h}{\partial a_h} = a_h^2 f(a_h) + S_h^2 f(a_h) - 2a_h \mu_h f(a_h) + \mu_h^2 f(a_h)$$

on noting
$$S_h^2 \frac{\partial W_h}{\partial a_h} = S_h^2 f(a_h)$$

and adding the LHS to the LHS and the RHS to the RHS and dividing both sides by $2S_h$ we get

$$\left(S_h \frac{\partial W_h}{\partial a_h} + W_h \frac{\partial S_h}{\partial a_h}\right) = \frac{1}{2} \frac{a_h^2 f(a_h)}{S_h} + f(a_h) - \frac{a_h \mu_h f(a_h)}{S_h} + \frac{\mu_h^2 f(a_h)}{2S_h}$$

$$= \frac{1}{2} \frac{f(a_h)}{S_h} \left[a_h^2 - 2a_h \mu_h + S_h^2 + \mu_h^2\right]$$

$$= \frac{1}{2} \frac{f(a_h)}{S_h} \left[(a_h - \mu_h)^2 + S_h^2\right]$$

or $$\frac{\partial (W_h S_h)}{\partial a_h} = \frac{1}{2} \frac{f(a_h)}{S_h} \left[(a_h - \mu_h)^2 + S_h^2\right].$$

Similarly, $$\frac{\partial (W_{h+1} S_{h+1})}{\partial a_h} = \left(\frac{1}{2}\right) \frac{f(a_h)}{S_{h+1}} \left[(a_h - \mu_{h+1})^2 + S_{h+1}^2\right].$$

So, $\frac{\partial}{\partial a_h} \left(\sum_1^H W_h S_h\right) = 0$ to find a_h for which $V_{\text{opt}}(\bar{Y}_{st})$

is to be the minimum we need.

This gives $$\frac{(a_h - \mu_h)^2 + S_h^2}{S_h} = \frac{(a_h - \mu_{h+1})^2 + S_{h+1}^2}{S_{h+1}}.$$

Since μ_h, S_h both depend upon a_h, it is impossible to solve this equation.

Dalenius and Hodges (1959) gave an approximate solution. Many follow-up actions have been reported in the literature. In this monograph we are not interested in pursuing that solution. Dalenius and Gurney (1951) are also a relevant reference.

Exploring Improved Ways

It is clear from our discussion on stratified SRSWOR that strata are to be so constructed that within strata variability should be as small as possible, and between strata variation should correspondingly be as high as possible. This is the principle, in fact, of stratification. For further clarification note the following:

$$T = TSS = \sum_{h=1}^{H} \sum_{i=1}^{N_h} \left(y_{hi} - \bar{Y}\right)^2 = \text{Total sum of squares}$$

$$\frac{T}{N-1} = \frac{1}{N-1} \sum_{i=1}^{N} \left(y_i - \bar{Y}\right)^2 = S^2$$

$V_p(\bar{y}) = \dfrac{N-n}{Nn} S^2 =$ the variance of the sample mean from an SRSWOR in n draws from U.

Now,
$$TSS = \sum_{h=1}^{H} \sum_{i=1}^{N_h} \left(y_{hi} - \bar{Y}\right)^2$$
$$= \sum_{h=1}^{H} \sum_{i=1}^{N_h} \left\{\left(y_{hi} - \bar{Y}_h\right) + \left(\bar{Y}_h - \bar{Y}\right)\right\}^2$$
$$= \sum_{h=1}^{H} \sum_{i=1}^{N_h} \left(y_{hi} - \bar{Y}_h\right)^2 + \sum_{h=1}^{H} N_h \left(\bar{Y}_h - \bar{Y}\right)^2$$
$$= WSS + BSS.$$

The quantity
$$WSS = \sum_{h=1}^{H} \sum_{i=1}^{N_h} \left(y_{hi} - \bar{Y}_h\right)^2,$$

called the Within Sum of Squares reflects the "variability within the strata." This quantity should be small if the strata are internally homogeneous.

The quantity
$$BSS = \sum_{h=1}^{H} N_h \left(\bar{Y}_h - \bar{Y}\right)^2,$$

called the Between sum of Squares, and its magnitude reflects the variability among the strata overall measures, namely, the "strata means among themselves."

Now
$$V_p(\bar{y}_{st}) = \frac{1}{N^2} \sum N_h^2 \left(\frac{1}{n_h} - \frac{1}{N_h}\right) \sum_{i=1}^{N_h} \left(y_{hi} - \bar{Y}_h\right)^2.$$

Thus, the Within Strata Variability measured by

$$\sum_{1}^{N_h}(y_{hi} - \bar{Y}_h)^2$$

determines the efficacy of $\bar{y}_{st}$ as an estimator for $\bar{Y}$. Unlike this $\bar{y}_{st}$ the overall sample mean from an SRSWOR from U has its variability as measured by a function of TSS. The portion BSS may be subtracted from TSS, and there will be greater elimination from the TSS to give us reduced variability to come into play in providing the measure of error in $\bar{y}st$ from a stratified SRSWOR.

To undertake stratified sampling, advantage is derived if enough BSS may be created for subtraction from TSS to yield a moderate WSS so as to end up with an effective level of accuracy in employing the $\bar{y}_{st}$ as an efficient estimator for $\bar{Y}$.

Just opposite is the situation when one resorts to Cluster Sampling.

4.3 CLUSTER SAMPLING

Cluster Sampling is guided by a principle just opposite to what applies to Stratified Sampling. A population may sometimes be divided into natural strata and into natural clusters. A stratum, like the district in a province or the street in a city block, which exists on its own, may suggest itself to be taken as one unit from each of which it may be judicious to independently select one or more constituent units from each of such several strata. The same type of entities may also be regarded as existing as natural clusters, and one may plan to select out of such clusters and then survey each of such chosen clusters completely in respect of their constituents. But surveyed subsequent data analysis of stratified survey sample observations is executed in a manner quite different from the analytical procedures to be applied to data realized from cluster samples. We shall describe the details in what follows, and it will be evident how clustering demands a separate principle to dictate its usage quite distinct from what applies to stratification. The analysis will shortly show that clusters should be so chosen or so constructed that variability among the elements within respective clusters should be as pronounced and as huge as possible, and on the contrary the clusters as a whole should be respectively as alike as possible. Thus, the principle of clustering is the reverse of that apposite to stratification. Let us now discuss single-stage Cluster Sampling in some details covering various aspects of it.

So far our narration covers the initial entity $U = (1, \cdots, i \cdots, N)$ of a known number N of units with identifiable labels i from 1 through N. Also we mentioned a few examples of such units. More technical is the term Sampling Units. By a Sampling Unit (SU) we mean certain entities out of which a suitable number of entities is to be chosen as a sample following a sophisticated selection procedure. Now the question is among a few possible alternative

Exploring Improved Ways

available choices out of a few possible ones which should be chosen as an appropriate SU and on following what criteria for the choice.

In a social survey in a province in a country each district may be taken as a stratum. Every village may be a selection unit in each district. Within the village one may like to take the households as the selection units. But instead, the houses, each accommodating several households, may be preferred as the units for selection, This is because (i) listing the houses rather than the individual households and (ii) preparing a Frame from which to choose the selection units will be easier. Also, importantly, the cost of surveying the sampled houses in all their constituent households may be much less than that for the individual households spread across the entire selected village. Thus, for example, bigger units may be more convenient than the smaller units in survey sampling from the twin consideration of (a) frame-preparation and (b) expenses involved.

But in sample surveys more crucial consideration is about the attainable precision level while going for the smaller units rather than for the larger units. It is understood that a cluster composed of the units in it will be surveyed in its entirety, ensuring Cluster Sampling. The question now is balancing these two opposing phenomena: the larger units with less cost and efficiency versus smaller units that are more expensive but more accurate in terms of variance in estimation. Let us go for an algebraic elaboration.

In case of agricultural surveys paralleling clusters of houses composed of elements as households, the plots or parcels of land composed of the constituent points of agricultural crop fields as their elements are regarded as Areal Units of specified and manageable geographical areas suitable for listing, frame-construction, and survey implementation.

Let N clusters of M elements each be defined specific to a finite survey population under study. Let y be a variable of interest with y_{ij} as the value for the jth element in the ith cluster, $j = 1, \cdots, M$ and $i = 1, \cdots, N$. Let us consider two rival strategies:

(I) Taking an SRSWOR s_1 of nM elements out of all the NM elements and employing the sample mean

$$t_1 = \frac{1}{nM} \sum_{i,j \in s_1} \sum y_{ij}$$

to unbiasedly estimate the population mean

$$\bar{Y} = \frac{1}{NM} \sum_{i=1}^{N} \sum_{j=1}^{M} y_{ij}.$$

(II) Taking an SRSWOR s_2 of n out of the N clusters, surveying all the M elements in each of them and employing the mean of these sampled clusters, namely,

$$t_2 = \frac{1}{n} \sum_{i \in s_2} \bar{y}_i, \quad \text{with} \quad \bar{y}_i = \frac{1}{M} \sum_{j=1}^{M} y_{ij},$$

to unbiasedly estimate the population mean

$$\bar{Y} = \frac{1}{N} \sum_{i=1}^{N} \bar{y}_i = \frac{1}{NM} \sum_{i=1}^{N} \sum_{j=1}^{M} y_{ij}.$$

Then, $\quad V_p(t_1) = \dfrac{NM - nM}{NM} \dfrac{S^2}{nM}, \quad S^2 = \dfrac{1}{NM-1} \sum_{i=1}^{N} \sum_{j=1}^{M} (y_{ij} - \bar{Y})^2.$

Let $\quad \bar{Y}_N = \dfrac{1}{N} \sum_{i=1}^{N} \bar{y}_i = \dfrac{1}{NM} \sum_{i=1}^{N} \sum_{j=1}^{M} y_{ij} = \bar{Y}.$

Then, $\quad V_p(t_2) = \dfrac{N-n}{Nn} \sum_{i=1}^{N} \dfrac{(\bar{y}_i - \bar{Y}_N)^2}{(N-1)} = \dfrac{N-n}{Nn} S_b^2, \quad \text{say},$

writing $\quad S_b^2 = \dfrac{1}{N-1} \sum_{i=1}^{N} (\bar{y}_i - \bar{Y}_N)^2.$

Now, $\quad \sum_{i=1}^{N} \sum_{j=1}^{M} (y_{ij} - \bar{Y})^2 = \sum_{i=1}^{N} \sum_{j=1}^{M} \{(y_{ij} - \bar{y}_i) + (\bar{y}_i - \bar{Y})\}^2$

$$= \sum_{i=1}^{N} \sum_{j=1}^{M} (y_{ij} - \bar{y}_i)^2 + M \sum_{i=1}^{N} (\bar{y}_i - \bar{Y})^2$$

$$= \sum_{i=1}^{N} (M-1) S_{Wi}^2 + M(N-1) S_b^2$$

writing $\quad S_{Wi}^2 = \dfrac{1}{M-1} \sum_{j=1}^{M} (y_{ij} - \bar{y})^2.$

So, $\quad V_p(t_1) = \dfrac{N-n}{nM} \left[\left(\dfrac{M-1}{N} \sum_{i=2}^{N} S_{Wi}^2 \right) + \dfrac{M(N-1)}{N} S_b^2 \right] / (NM - 1)$

$$= \dfrac{N-n}{nM} \left[(M-1) S_W^2 + \dfrac{M(N-1)}{N} S_b^2 \right] / (NM - 1)$$

writing
$$S_W^2 = \frac{1}{N}\sum_{i=1}^{N} S_{Wi}^2 = \frac{1}{N(M-1)}\sum_{j=1}^{M}(y_{ij}-\bar{Y}_1)^2$$
$$= \frac{1}{(NM-1)}\left[\frac{N-n}{n}\frac{M-1}{M}S_W^2 + (N-1)MV_p(t_2)\right].$$

Thus,
$$V_p(t_1) = \frac{N-n}{Nn}\frac{S^2}{M} \quad \text{and} \quad V_p(t_2) = \frac{N-n}{Nn}S_b^2$$
$$= \frac{N-n}{Nn}\frac{MS_b^2}{M}.$$

So, the efficiency of Cluster Sampling is
$$E = \frac{100 \times V_p(t_1)}{V_p(t_2)} = \frac{S^2}{MS_b^2} \times 100.$$

It is not obvious if this is greater or less than 100. But it is possible to set some clue on examining the following Table of Analysis of variance.

ANOVA Table

Source of Variation	Degrees of Freedom	Mean Square
Between Clusters	$N-1$	$\frac{M}{N-1}\sum_{i=1}^{N}(\bar{y}_i - \bar{Y})^2 = MS_b^2$
Within Clusters	$N(M-1)$	$\frac{1}{N(M-1)}\sum_i \sum_j (y_{ij} - \bar{y}_i)^2 = S_w^2$
Total Population	$NM-1$	$\frac{1}{NM-1}\sum_i \sum_j (y_{ij} - \bar{Y})^2 = S^2$

So,
$$(NM-1)S^2 = (N-1)MS_b^2 + N(M-1)S_W^2.$$

The smaller the Between Cluster sum of squares, and correspondingly the larger the Within Cluster variability, the higher the efficiency of Cluster sampling.

From our examples cited in cluster sampling and areal sampling, the nearby elements are clustered together to form the respective clusters and naturally in them the elements are likely to be like each other; hence, the Within Cluster variability is by construction expected to be relatively small vis-à-vis the Between Cluster variability resulting in reduced efficiency of cluster sampling. Moreover, the elements within the clusters are likely to be well and positively correlated. So, it seems important to look at the Intra-Cluster correlation among the element-values and examine how this affects the efficiency.

The Intra-Cluster Correlation Coefficient is defined as

$$\rho = \frac{\sum_{i=1}^{N} \sum_{j=1}^{M} \sum_{\substack{k=1 \\ j \neq k}}^{M} (y_{ij} - \bar{Y})(y_{ik} - \bar{Y}) \Big/ NM(M-1)}{\sum_{i=1}^{N} \sum_{j=1}^{M} (y_{ij} - \bar{Y})^2 \Big/ NM}.$$

The numerator of ρ may now be analyzed as

$$\frac{1}{NM(M-1)} \left[\sum_{i=1}^{N} \sum_{j \neq k} \sum \{(y_{ij} - \bar{y}_i) + (\bar{y}_i - \bar{Y})\} \{(y_{ik} - \bar{y}_i) + (\bar{y}_i - \bar{Y})\} \right]$$

$$= \frac{1}{NM(M-1)} \left[\sum_{1}^{N} \sum_{j \neq k} \sum (y_{ij} - \bar{y}_i)(y_{ik} - \bar{y}_i) + M(M-1) \sum_{1}^{N} (\bar{y}_i - \bar{Y})^2 \right]$$

$$= \frac{1}{NM(M-1)} \left[\sum_{1}^{N} \left\{ \left(\sum_{j} (y_{ij} - \bar{y}_i) \right)^2 - \sum_{j=1}^{M} (y_{ij} - \bar{y}_i)^2 \right\} + \frac{1}{N} \sum (\bar{y}_i - \bar{Y})^2 \right]$$

$$= -\frac{1}{NM} \sum_{i} S_{Wi}^2 + \frac{N-1}{N} S_b^2$$

and the denominator simplifies to $\dfrac{(NM-1)}{NM} S^2$.

So, $$\rho = \left(\frac{N-1}{N} S_b^2 - \frac{S_w^2}{M} \right) \Big/ \left(\frac{NM-1}{NM} \right) S^2.$$

Recalling, $(NM-1)S^2 = (N-1)MS_b^2 + N(M-1)S_w^2$

$$V_p(t_2) = \frac{N-n}{Nn} S_b^2$$

$$= \left(\frac{N-n}{Nn} \right) \left[\frac{(NM-1)S^2 - N(M-1)S_w^2}{(N-1)M} \right].$$

Also, $$(1-\rho) = \frac{S_w^2}{\frac{N-1}{N} S_b^2 - \frac{S_w^2}{M} + S_w^2} = \frac{S_w^2}{\frac{N-1}{N} S_b^2 + \frac{M-1}{M} S_w^2},$$

and $$\frac{NM-1}{NM} S^2 \rho = \left(\frac{N-1}{N} S_b^2 - \frac{S_w^2}{M} \right)$$

or $$(1-\rho) = \frac{S_w^2}{\frac{NM-1}{NM} \frac{NM-1}{NM} S^2 \rho + S_w^2} S^2 \rho + S_w^2$$

or
$$\frac{NM-1}{NM}S^2\rho + S_w^2 = \frac{S_w^2}{(1-\rho)}$$

or
$$S_w^2 = \frac{NM-1}{NM}S^2(1-\rho).$$

So,
$$S_b^2 = \frac{(NM-1)S^2 - N(M-1)S_w^2}{(N-1)M}$$
$$= \frac{(NM-1)S^2 - N(M-1)\frac{(NM-1)}{NM}S^2(1-\rho)}{(N-1)M}$$
$$= \frac{NM-1}{(N-1)M}\left[S^2 - \frac{M-1}{M}S^2(1-\rho)\right]$$
$$= \frac{NM-1}{M(N-1)}\frac{S^2}{M}[M - (M-1)(1-\rho)]$$
$$= \frac{NM-1}{M(N-1)}\frac{S^2}{M}[1 + (M-1)\rho].$$

Finally, $V_p(t_2) = \frac{N-n}{Nn}\frac{(NM-1)}{M(N-1)}\frac{S^2}{M}[1 + (M-1)\rho]$.

Comparing this with
$$V_p(t_1) = \frac{N-n}{Nn}\frac{S^2}{M},$$

we may observe that efficiency of cluster sampling is affected by the magnitudes of (i) N, the total number of clusters; (ii) M, the size of each cluster; and (iii) ρ, the intra-cluster correlation determined by the numerical values of the neighboring elements within the respective clusters.

Clearly,
$$MS_b^2 = S^2\frac{NM-1}{M(N-1)}[1 + (M-1)\rho]$$

so that the efficiency of cluster-sampling is measured by
$$E = \frac{100 \times S^2}{MS_b^2} = 100\frac{M(N-1)}{(NM-1)}\frac{1}{1+(M-1)\rho} \simeq \frac{100}{1+(M-1)\rho}$$

if N is large.

For a large number of clusters, the smaller the magnitude of $(M-1)\rho$, the higher is the efficiency of cluster sampling. So, it is desirable to have a small number of elements per cluster, and the neighboring elements in the clusters should be as dissimilar as possible.

So far we considered the algebra with the Analysis of Variance using the values supposed to be available for the entire population composed of N clusters arbitrarily defined, each consisting of a common number of M elements. These give us the values of S^2, S_b^2, S_w^2, and ρ. But in practice they will all need to be estimated from samples. To see how this may be accomplished one simple way is to start with the analogous ANOVA table based on the sample data as below.

ANOVA Table for a Sample

Source of Variation	Degrees of Freedom	Mean Square
Between Clusters	$n-1$	$\frac{1}{n-1}\sum_{i=1}^{n} M(\bar{y}_i - \bar{y}_n)^2 = Ms_b^2$
Within Clusters	$n(M-1)$	$\frac{1}{n(M-1)}\sum_{i=1}^{n}\sum_{j=1}^{M}(y_{ij} - \bar{y}_i)^2 = s_w^2$
Total Population	$nM-1$	$\frac{1}{nM-1}\sum_{i=1}^{n}\sum_{j=1}^{M}(y_{ij} - \bar{y}_n)^2 = s^2$

Here we suppose a sample of n clusters is chosen by SRSWOR, the sampled cluster means are denoted by $\bar{y}_i, i = 1, \cdots, n$, their mean by $\bar{y}_n$, and the rest is analogously denoted. So, this ANOVA table provides obvious estimates for S^2, S_b^2, S_w^2, and ρ.

In order to hit upon an appropriate cluster size M the following approach received attention from many early statisticians of eminence, namely, Fairfield Smith (1938), Mahalanobis (1940, 1942, 1944), Jessen (1942), Hendricks (1944), and others. The approach is model based. Based on extensive empirical studies one model postulates the relationship

$$S_w^2 = A\,M^g\,(g > 0).$$

Here S_w^2 of course is calculated using an initial choice of M; this model seeks to develop a more appropriate value for it. Here A and g are constants required to be empirically chosen. This model leads to

$$S_b^2 = \frac{(NM-1)S^2 - N(M-1)S_w^2}{(N-1)} = \frac{(NM-1)S^2 - N(M-1)AM^g}{(N-1)}$$
$$\simeq MS^2 - (M-1)AM^g.$$

Analogously, one may model

$$S^2 = A(NM)^g.$$

These lead to

$$\log S_w^2 = \log A + g\,\log M$$
$$\log S^2 = \log A + g\,log\,(NM),$$

and hence,
$$S_b^2 = AM^g\,[MN^g - (M-1)].$$

Now, using s^2, s_b^2, s_w^2 one may proceed to work out a decent choice for M.

The next point of importance is to introduce suitable cost functions to be combined with the variance functions so far discussed, so as to examine how variance function and cost function may appropriately be combined to hit upon a right plan to employ cluster sampling.

Our discussions so far tend to induce the impression that cluster sampling is likely to be less efficient with the increasing size of a cluster in the sense of the number of elements it contains. But the cost of cluster sampling turns out less compared to sampling of the elements with a parity in the total sample size, which is the total number of elements to be sampled and surveyed, remaining the same in both. Let us formally study the impact of a suitable cost function on the variance of an estimator of the population mean. For simplicity we restrict to SRSWOR of both the clusters and the elements, respectively, out of the total number of clusters formed and the total population of all the elements to be covered keeping in each cluster a common number of elements.

Following Jessen (1942) we may illustrate the empirically plausible cost function giving the total cost as

$$C = C_1 n M + C_2 \sqrt{n}.$$

Here C_1 represents the cost of covering an element within a cluster, C_2 the cost of covering a cluster, n the number of clusters to be sampled, and M the given number of elements per cluster. Further, for simplicity let us restrict to the simple variance function

$$S_w^2 = a M^b, \text{ with } a, b \text{ as unknown constants.}$$

Then,
$$S_b^2 = \frac{(NM - 1) S^2 - N (M - 1) a M^b}{M (N - 1)},$$

$$S^2 = a (NM)^b$$

$$V_p (t_2) = \frac{N - n}{Nn} S_b^2$$

and approximately $V_p(t_2)$ equals $\frac{1}{n}\left[S^2 - (M-1)aM^{b-1}\right]$, ignoring the

$$fpc = \frac{N - n}{N} = 1 - \frac{n}{N} = 1 - f,$$

treated as equal to unity.

Now writing this approximate variance as V and supposing the total cost to be incurred at $C = C_0$, one may optimally choose n and M so as to minimize the product VC with respect to n and M either fixing C at C_0 or fixing V at a value V_0. Sukhatme and Sukhatme (1954) gave a solution on taking

$$\phi = V + \lambda C \quad \text{and solving}$$

$$\frac{\partial \phi}{\partial n} = 0 \text{ and } \frac{\partial \phi}{\partial M} = 0,$$

treating λ as a Lagrangian undetermined multiplier.

As non-linear functions are involved, iterative numerical solutions only are available with not much difficulty in practice.

For more information on cluster sampling, instructive further references are Murthy's (1977) and Cochran's (1977) texts in adition to the ones by Sukhatme and Sukhatme (1954), Mukhopadhyay (1998), Singh and Chaudhary (1986), and an ICAR handbook by Singh, Singh, and Kumar (1994).

The study of cluster sampling extended by allowing unequal probability sampling of clusters and clusters of unequal number of elements deserves a treatment. But this is postponed to Appendix 3.

4.4 MULTI-STAGE SAMPLING

For convenience in sample-selection a flexible approach is to define sampling units of different successive stages. A country like India to be surveyed, because it is geographically and administratively composed of several states and union territories (UT) as in the United States, these in their turn comprising several districts which contain cities and villages composed of households, suggests advantages in selecting states and UTs as the first-stage units (fsu); next the districts as the second-stage units (ssu); next the third-stage units as the cities, towns, and villages; and finally the households as the fourth-stage units or the ultimate-stage units (usu).

Multi-stage sampling consists of selecting first a sample of fsu's from all the fsu's in the population, then in selecting samples of ssu's contained in the respective fsu's chosen independently across the selected fsu's and similarly repeating the procedure down the list of tsu's and so on until the usu's. The method of sample-selection at any of the stages may be any probability sampling procedure. A sample thus gathered is called a multi-stage sample, and the sampling procedure thus adopted is called Multi-stage Sampling.

Let us discuss the theory. The finite survey population $U = (1, \cdots, i, \cdots, N)$ is supposed to contain a known number N of fsu's. Let s be a sample of fsu's chosen with a probability $p(s)$. Let our problem be to estimate the population total $Y = \sum_1^N y_i$ of the values y_i for $i \in U$, of a real variable y.

Let
$$t = \sum_{i \in s} y_i b_{si} = \sum_{i=1}^{N} y_i b_{si} I_{si}$$

with b_{si} as constants independent of any y-value and $I_{si} = 1/0$ according as $i \in s$ or $i \notin s$, and moreover, let

$$\sum_{s \ni i} p(s) b_{si} = 1 \ \forall \ i \in U.$$

But we are facing a situation when y_i-values are not ascertainable for any i in U, but supposed to be unbiasedly estimable through sample selection in later stages from each i in s. We further postulate that the following conditions hold.

Let E_1, V_1 be the operators for taking expectation and variance in respect of sample selection in the first stage; E_L, V_L the operators for taking expectation and variance in respect of sampling at the later stages; and $E = E_1 E_L$ and $V = E_1 V_L + V_1 E_L$ the overall expectation and variance operators in sample selection over all the multi-stages. We further suppose that $\hat{y}_i$ is available subject to $E_L(\hat{y}_i) = y_i$, for every i, $V_L(\hat{y}_i) = V_i$ admits an estimator $\hat{V}_i$ such that $E_L(\hat{V}_i) = V_i$ or more generally $V_L(\hat{y}_i) = V_{si}$ for $i \in s$ and $\hat{V}_{si}$ exists satisfying $E_L(\hat{V}_{si}) = V_{si}$. It is of course implied that $\hat{y}_i$'s are independent of $\hat{y}_j$'s for $i \neq j \in s$ because of the condition of multi-stage sampling ensuring independence in selection from across the selected fsu's that are different from each other. Then, we may assert the following facts.

$$e = \sum_{i \in s} \hat{y}_i b_{si} = \sum_{i=1}^{N} \hat{y}_i b_{si} I_{si}$$

is an unbiased estimator for Y in the sense that

$$E(e) = E_1 \left[\sum_{i \in s} E_L(\hat{y}_i) b_{si} \right] = E_1 \left(\sum_{i \in s} y_i b_{si} \right) = E_1(t) = Y$$

$$V(e) = V_1 E_L(e) + E_1 V_L(e)$$
$$= V_1 \left[E_L \sum_{i \in s} \hat{y}_i b_{si} \right] + E_1 \left[\sum_{i \in s} V_{si} b_{si}^2 \right]$$
$$= V_1(t) + E_1 \left[\sum_{i=1}^{N} V_{si} b_{si}^2 I_{si} \right]$$
$$= \sum_{1}^{N} y_i^2 C_i + \sum \sum_{i \neq j} y_i y_j C_{ij} + E_1 \left[\sum_{i=1}^{N} V_{si} b_{si}^2 I_{si} \right]$$

writing $\quad C_i = \sum_{s \ni i} p(s) b_{si}^2 - 1 \quad$ and $\quad C_{ij} = \sum_{s \ni i,j} p(s) b_{si} b_{sj} - 1.$

Assuming that constants C_{si}, C_{sij} free of y's are available satisfying

$$\sum_{s \ni i} p(s) C_{si} = C_i \quad \text{and} \quad \sum_{s \ni i,j} p(s) C_{sij} = C_{ij},$$

we then have the following:

Theorem 4.1

$$v(e) = \sum_i^N (\hat{y}_i)^2 C_{si} I_{si} + \sum_{i \neq j}^N \sum^N \hat{y}_i \hat{y}_j C_{sij} I_{sij} + \sum_1^N v_{si} \left(b_{si}^2 - C_{si}\right) I_{si}$$

is an unbiased estimator for $V(e)$ in the sense that $Ev(e) = V(e)$. ∎

PROOF

$$E_L v(e) = \sum_1^N y_i^2 C_{si} I_{si} + \sum_{i \neq j} \sum y_i y_j C_{sij} I_{sij}$$

$$+ \sum_1^N V_{si} C_{si} I_{si} + \sum_1^N V_{si} \left(b_{si}^2 - C_{si}\right) I_{si}$$

$$Ev(e) = E_1 E_L v(e) = \sum y_i^2 C_i + \sum_{i \neq j} \sum y_i y_j C_{ij} + E_1 \sum V_{si} b_{si}^2 I_{si} = V(e).$$
∎

COROLLARY 4.1

In case $V_L(\hat{y}_i) = V_i$, then alternative unbiased estimators for $V(e)$ are

$$v_1(e) = \sum_i (\hat{y}_i)^2 C_{si} I_{si} + \sum_{i \neq j} \sum \hat{y}_i \hat{y}_j C_{sij} I_{sij} + \sum_i v_i b_{si} I_{si}$$

and

$$v_2(e) = \sum_i (\hat{y}_i)^2 C_{si} I_{si} + \sum_{i \neq j} \sum \hat{y}_i \hat{y}_j C_{sij} I_{sij} + \sum v_i \left(b_{si}^2 - C_{si}\right) I_{si}$$

$$V = E_1 V_L + V_1 E_L = E_L V_1 + V_L E_1.$$

PROOF

$$Ev_1(e) = E_1 \left[\sum_i E_L(\hat{y}_i)^2 C_{si} I_{si} + \sum_{i \neq j} \sum E_L(\hat{y}_i \hat{y}_j) C_{sij} I_{sij} \right.$$

$$\left. + \sum V_i C_{si} I_{si} + \sum_i E_L(v_i) b_{si} I_{si} \right]$$

$$= E_1 \left[\sum y_i^2 C_{si} I_{si} + \sum\sum_{i\neq j} y_i y_j C_{sij} I_{sij} + \sum_i V_i b_{si} I_{si} \right] + \sum V_i C_i$$

$$= \sum y_i^2 C_i + \sum\sum_{i\neq j} y_i y_j C_{ij} + E_1 \sum_i V_i b_{si}^2 I_{si}$$

$$= V(e)$$

because $E_1 \sum V_i b_{si} I_{si} = \sum V_i$ and $\sum V_i C_i = \sum V_i \left[E_1 \left(\sum b_{si}^2 I_{si} \right) - 1 \right]$

so that $\sum V_i C_i + E_1 \sum V_i b_{si} I_{si} = E_1 \left[\sum V_i b_{si}^2 I_{si} \right]$

$$Ev_2(e) = E_1 \left[\sum_i E_L(\hat{y}_i)^2 C_{si} I_{si} + \sum\sum_{i\neq j} E_2(\hat{y}_i \hat{y}_j) C_{si} I_{sij} \right]$$

$$+ E_1 \left[\sum_i E_L(v_i) \left(b_{si}^2 - C_{si} \right) I_{si} \right]$$

$$= E_1 \left[\sum y_i^2 C_{si} I_{si} + \sum\sum_{i\neq j} y_i y_j C_{sij} I_{sij} + \sum_i V_i C_{si} I_{si} \right]$$

$$+ E_1 \left[\sum_i V_i \left(b_{si}^2 - C_{si} \right) I_{si} \right]$$

$$= \sum_i y_i^2 C_i + \sum\sum_{i\neq j} y_i y_j C_{ij} + E_1 \sum V_i b_{si}^2 I_{si}. \quad \blacksquare$$

COROLLARY 4.2

In case $\quad t = t_{HT} = \sum_{i\in s} \dfrac{y_i}{\pi_i} = \sum_i \dfrac{y_i}{\pi_i} I_{si}$

and $\quad e = e_{HT} = \sum_{i\in s} \dfrac{\hat{y}_i}{\pi_i} = \sum_i \dfrac{\hat{y}_i}{\pi_i} I_{si},$

We have the theorem:

$$V(e_{HT}) = \sum_i y_i^2 \frac{1-\pi_i}{\pi_i} + \sum\sum_{i\neq j} \frac{y_i y_j}{\pi_i \pi_j} (\pi_{ij} - \pi_i \pi_j) + \sum_i \frac{V_i}{\pi_i}$$

and $v(e_{HT}) = \sum_i (\hat{y}_i)^2 \dfrac{1-\pi_i}{\pi_i} \dfrac{I_{si}}{\pi_i} + \sum\sum_{i\neq j} \hat{y}_i \hat{y}_j \left(\dfrac{\pi_{ij} - \pi_i \pi_j}{\pi_i \pi_j} \right) \dfrac{I_{sij}}{\pi_{ij}} + \sum \dfrac{v_i}{\pi_i} I_{si}.$

In this case V_{si} and v_{si} are not feasible. Only V_i and v_i are applicable.

PROOF

$$E_L(e_{HT}) = t_{HT}.$$

So,
$$E(e_{HT}) = E_1(t_{HT}) = Y.$$

So,
$$V(e_{HT}) = V_1(E_l(e_{HT})) + E_1(V_L(e_{HT}))$$
$$= V_1(t_{HT}) + E_1\left(\sum V_i \frac{I_{si}}{\pi_i^2}\right)$$
$$= V_1(t_{HT}) + \sum \frac{V_i}{\pi_i}$$
$$= \sum y_i^2 \frac{1-\pi_i}{\pi_i} + \sum\sum_{i\neq j} \frac{y_i y_j}{\pi_i \pi_j}(\pi_{ij} - \pi_i \pi_j) + \sum \frac{V_i}{\pi_i}.$$

Also,
$$Ev(e_{HT}) = E_1\left[\sum_i E_L(\hat{y}_i)^2 \frac{1-\pi_i}{\pi_i}\frac{I_{si}}{\pi_i}\right.$$
$$+ \sum\sum_{i\neq j} E_L(\hat{y}_i \hat{y}_j)\left(\frac{\pi_{ij} - \pi_i\pi_j}{\pi_i\pi_j}\right)\frac{I_{sij}}{\pi_{ij}}\right] + E_1 \sum_i \frac{E_L(v_i)}{\pi_i}I_{si}$$
$$= E_1\left[\sum_i y_i^2 \frac{1-\pi_i}{\pi_i}\frac{I_{si}}{\pi_i} + \sum\sum_{i\neq j} y_i y_j \left(\frac{\pi_{ij}-\pi_i\pi_j}{\pi_i\pi_j}\right)\frac{I_{sij}}{\pi_{ij}}\right.$$
$$\left.+\sum_i V_i \frac{1-\pi_i}{\pi_i}\frac{I_{si}}{\pi_i}\right] + E_1 \sum_i V_i \frac{I_{si}}{\pi_i}$$
$$= \sum_i y_i^2 \frac{1-\pi_i}{\pi_i} + \sum\sum_{i\neq j} y_i y_j \left(\frac{\pi_{ij}-\pi_i\pi_j}{\pi_i\pi_j}\right)$$
$$+\sum_i V_i \left(\frac{1-\pi_i}{\pi_i}\right) + \sum_i V_i$$
$$= \sum_i y_i^2 \frac{1-\pi_i}{\pi_i} + \sum\sum_{i\neq j} y_i y_j \frac{\pi_{ij}-\pi_i\pi_j}{\pi_i\pi_j} + \sum \frac{V_i}{\pi_i}. \quad \blacksquare$$

Suppose from $U = (1,\cdots,i,\cdots,N)$ the fsu's are selected by the PPSWR sampling scheme but the values y_r on a variable y of interest for the units chosen on the rth draw, $r = 1,\cdots,n$ are not ascertainable. But every time an fsu happens to be selected, it is supposed that independent samples of

Exploring Improved Ways

ssu's, tsu's, and usu's are chosen from this fsu in later stages. Taking the Hansen-Hurwitz (1943) estimator, say

$$t_{HH} = \frac{1}{n}\sum_{r=1}^{n}\left(\frac{y_r}{p_r}\right), \text{ in case } y_r \text{ is ascertainable,}$$

let

$$eHH = \frac{1}{n}\sum_{r=1}^{n}\left(\frac{\hat{y}}{p_r}\right) \text{ be an estimator chosen for } Y.$$

Let $V_L(\hat{y}_r) = V_r$ be the variance of $(\hat{y}_r), r = 1, \cdots, n$.
Then follows the theorem:

$$e_{HH} = \frac{1}{n}\sum_{r=1}^{n}\left(\frac{y_r}{p_r}\right)$$

is an unbiased estimator for Y with the variance

$$V(e_{HH}) = \frac{1}{n}\sum_{i=1}^{N}\frac{V_i}{P_i} + \frac{V}{n},$$

writing $\quad V = \sum_1 \frac{y_i^2}{p_i} - Y^2 = \sum\sum_{i<,j} p_i p_j \left(\frac{y_i}{p_i} - \frac{y_j}{p_j}\right)^2 = \sum p_i \left(\frac{y_i}{p} - Y\right)^2$

and

$$v = \frac{1}{2n^2(n-1)}\sum\sum_{r\neq r'}\left(\frac{\hat{y}_r}{p_r} - \frac{\hat{y}_{r'}}{p_{r'}}\right)^2$$

is an unbiased estimator for V.

PROOF

$$e_{HH} = \frac{1}{n}\sum_{r=1}^{n}\frac{\hat{y}_r}{p_r}$$

$$E(e_{HH}) = E_1[E_L(e_{HH})] = E_1\left[\frac{1}{n}\sum_{1}^{n}\frac{y_r}{p_r}\right]$$

$$= \frac{1}{n}\sum_{r=1}^{n}E_1\left(\frac{y_r}{p_r}\right) = \sum_{1}^{N}\frac{y_i}{p_i}p_i = Y$$

$$V(e_{HH}) = V_1\left[E_L(e_{HH})\right] + E_1\left[V_L(e_{HH})\right]$$

$$= V_1\left(\frac{1}{n}\sum_1^n \frac{y_r}{p_r}\right) + E_1\left[\frac{1}{n^2}\sum_1^n \frac{V_L(\hat{y}_r)}{p_r^2}\right]$$

$$= \frac{1}{n}V_1\left(\frac{y_r}{p_r}\right) + \frac{1}{n^2}\sum_1^n\left[\sum_1^N \frac{V_i}{p_i}\right]$$

$$= \frac{V}{n} + \frac{1}{n}\sum_1^N \frac{V_i}{p_i}$$

$$E(v) = E_1\, E_L\left[\frac{1}{2n^2(n-1)}\sum\sum_{r\neq r'}\left(\frac{\hat{y}_r}{p_r} - \frac{\hat{y}_{r'}}{p_{r'}}\right)^2\right]$$

$$= E_1\left[\frac{1}{2n^2(n-1)}\sum\sum_{r\neq r'}E_L\left\{\frac{\hat{y}_r - y_r}{p_r} - \frac{\hat{y}_{r'} - y_{r'}}{p_{r'}} + \left(\frac{y_r}{p_r} - \frac{y_{r'}}{p_{r'}}\right)\right\}^2\right]$$

$$= \frac{1}{2n^2(n-1)}\sum\sum_{r\neq r'}E_1\left[\left(\frac{V_r}{p_r^2} + \frac{V_{r'}}{p_{r'}^2}\right) + \left(\frac{y_r}{p_r} - \frac{y_{r'}}{p_{r'}}\right)^2\right]$$

$$= \frac{1}{2n^2(n-1)}\sum\sum_{r\neq r'}\left[2\sum_1^N \frac{V_i}{p_i}(n-1)\right]$$

$$+ \frac{1}{2n^2(n-1)}\sum\sum_{r\neq r'}\left[\sum\sum_{i\neq i'}\left(\frac{y_i}{p_i} - \frac{y_{i'}}{p_{i'}}\right)^2 p_i\, p_{i'}\right]$$

$$= \frac{1}{n}\sum_1^N \frac{V_i}{p_i} + \frac{V}{n} = V(e_{HH}). \quad\blacksquare$$

REMARK 4.1
The simplicity of this situation will not hold if only the distinct sample *fsu*'s are sampled again in the subsequent stages. ∎

Single-stage sampling conditions for non-negativity of unbiased estimators of Mean Square Errors or variances were discussed in detail in Chapter 3. There we referred to Ha'jek (1958), Rao (1979), and Chaudhuri and Pal (2002). How to extend those ideas to Multi-stage sampling needs is explained now. In this context we first recall Chaudhuri, Adhikary, and Dihidar's (2000) discussions in brief. The most crucial aspect of their work is that the operators E_1 and E_L are assumed to be commutative, i.e.,

$$E_1 E_L = E_L E_1 \text{ leading to } E_1 V_L + V_1 E_L = E_L V_1 + V_L E_1.$$

Exploring Improved Ways

For the initial single-stage-based estimator $t_b = \sum_i y_i\, b_{si}\, I_{si}$, it is important to express its variance revised from its Quadratic Form appearance into the form:

$$V_p(t_b) = -\sum\sum_{i<j} d_{ij}\, w_i w_j \left(\frac{y_i}{w_i} - \frac{y_j}{w_j}\right)^2$$

available on the existence of $w_i\,(\neq 0)$ ensuring the tenability of the condition C for which $z_i = \frac{y_i}{w_j} = $ a constant $\forall\, i \Rightarrow V_p(t_b) = 0\ \forall\ \underline{Y}$. We shall first note Chaudhuri et al.'s (2000) formulas of $V(e_b)$ under the above commutativity condition. Later, we shall take note of Chaudhuri and Pal's (2002) amendment that

$$V_p(t_b) = -\sum\sum_{i<j} d_{ij}\, w_i w_j \left(\frac{y_i}{w_i} - \frac{y_j}{w_j}\right)^2 + \sum \beta_i \frac{y_i^2}{w_i}$$

when C does not hold but $\exists\, w_i\,(\neq 0)\ \forall\, i$.

We should recall

$$d_{ij} = E_p(b_{si}\, I_{si} - 1)(b_{sj}\, I_{sj} - 1)$$

and

$$\beta_i = \sum_{j=1}^{N} d_{ij} w_j,\ i \in U = (1, \cdots, N).$$

Chaudhuri et al. (2000) cite two cases, namely: (1) Raj's (1956) unbiased estimator for Y based on PPSWOR sampling of n first-stage units given as

$$t_D = \frac{1}{n}\sum_{j=1}^{n} t_j,$$

where

$$t_1 = \frac{y_{i1}}{p_{i1}},$$

$$t_2 = y_{i1} + \frac{y_{i2}}{p_{i2}}(1 - p_{i1}),$$

$$t_j = y_{i1} + \cdot + y_{ij-1} \cdots + \frac{y_{ij}}{p_{ij}}(1 - p_{i1} \cdots - p_{ij-1})$$

$$j = 1 \cdots, n$$

based on the sample $s = (i_1, i_2 \cdots, i_{j-1}, i_j)$ drawn in order with probability

$$p_{i_1} \cdot \frac{p_{i_2}}{1 - p_{i_1}} \cdot \frac{p_{i_3}}{1 - p_{i_1} - p_{i_2}} \cdots \frac{p_{i_n}}{1 - p_{i_1} - \cdots - p_{i_{n-1}}}.$$

For this $V_p(t_D)$ is a tedious entity but its uniformly non-negative unbiased estimator is

$$v_D = \frac{1}{n(n-1)}\sum_{j=1}^{n}(t_j - t_D)^2.$$

The question is how to derive e_D and a simple formula for an unbiased estimator for $V(e_D)$ in multi-stage sampling. Deriving e_D seems simple but not

$\hat{V}(e_D)$. Second, they consider the Rao, Hartley, and Cochran (1962) strategy based on the RHC sampling scheme and the unbiased RHC estimator for Y described in Chapter 3. We know that in case of a one-stage RHC strategy,

$$V_p(t_{RHC}) = A \sum \left(\frac{y_i^2}{p_i} - Y^2\right) = A \sum\sum_{i<j} p_i p_j \left(\frac{y_i}{p_i} - \frac{y_j}{p_j}\right)^2$$

with
$$v_p(t_{RHC}) = \hat{V}_p(t_{RHC}) = B \sum_n Q_i \left(\frac{y_i}{p_i} - t_{RHC}\right)^2,$$

with
$$A = \frac{\sum_n N_i^2 - N}{N(N-1)}, \quad B = \frac{\sum_n N_i^2 - N}{N^2 - \sum_n N_i^2}$$

and the notations are as in Chapter 3.

If $t_D, t_{RHC}, V_p(t_D), V_p(t_{RHC}), \hat{V}_p(t_D), \hat{V}_p(t_{RHC})$ are to be extended to multi-stage sampling, then these variances are to be thrown into Quadratic Forms structures and then Rao's (1979) approach is to be tried. But Chaudhuri et al. (2000) verified that taking $\hat{y}_i$ as independent estimators of y_i from subsequent stages of sampling such that $E_L(\hat{y}_i) = y_i$, $V_L(\hat{y}_i) = V_i$ admitting v_i such that $E_L(v_i) = V_i$ then, for

$$e_{RHC} = \sum_n \hat{y}_i \frac{Q_i}{p_i},$$

$$E(e_{RHC}) = E_1 E_L(e_{RHC}) = E_1(t_{RHC}) = Y$$

and
$$E(e_{RHC}) = E_L E_1(e_{RHC}) = E_L(R) = Y,$$

writing
$$\underline{R} = (\hat{y}_1, \cdots, \hat{y}_i \cdots, \hat{y}_N), \quad R = \sum_1^N R_i.$$

Also, writing
$$\underline{V} = (V_1, \cdots, V_i, \cdots, V_N),$$

they observed
$$V(e_R) = E_1 V_L(e_R) + V_1 E_L(e_R)$$
$$= E_1 \left[\sum_n \left(\frac{Q_i}{p_i}\right)^2 V_i\right] + V_1(t_R)$$
$$= A \sum \frac{V_i}{p_i} + (1-A) \sum V_i + A \left(\sum \frac{y_i^2}{p_i} - Y^2\right)$$
$$= E_L V_1(e_{RHC}) + V_L E_1(e_{RHC}).$$

Following Chaudhuri, Adhikary, and Dihidar (2000) let us now discuss how to use unbiased estimators for Y along with unbiased variance estimators thereof in multi-stage sampling utilizing the above commutative property of expectation operators on noting applicability of other associated conditions.

Let us assume the following four conditions:

i. For y_i exists $\hat{y}_i$ satisfying $E_L(\hat{y}_i) = y_i$
ii. $V_L(\hat{y}_i) = V_i$
iii. $\hat{y}_i$'s are independently distributed
iv. There exist estimators v_i for V_i satisfying $E_L(v_i) = V_i$.

Note that if $V_L(\hat{y}_i)$ is of the form V_{si} then E_1 and E_L do not commute. The following results extend only Raj's (1956) estimation procedure to multi-stage sampling; Rao's (1975) approach does not extend. So Chaudhuri et al.'s (2000) following approach is needed.

Let
$$t_b = \sum y_i b_{si} I_{si}$$
such that
$$\sum_s p(s) b_{si} I_{si} = 1 \ \forall \ i \in U = (1, \cdots N),$$
$$d_{ij} = \sum p(s)(b_{si} I_{si} - 1)(b_{si} I_{si} - 1),$$
$$\exists \ w_i (\neq 0). \ni. \frac{y_i}{w_i} = a \text{ constant for every } i \in U,$$

E_1 commutes with E_L, and conditions (i) through (iv) hold.

Then,
$$e_b = \sum \hat{y}_i b_{si} I_{si} = t_b \Big|_{\underline{Y} = (\hat{y}_1, \cdots, \hat{y}_i, \cdots, \hat{y}_N)}$$

satisfies
$$E(e_b) = E_L E_1 (e_b) = E_L(\hat{Y}) = Y,$$

writing
$$\hat{Y} = \sum_1^N \hat{y}_i.$$

Again writing
$$\underline{\hat{Y}} = (\hat{y}_1, \cdots, \hat{y}_i, \cdots, \hat{y}_N),$$
$$\underline{v} = (v_1, \cdots, v_i \cdots, v_N),$$
$$t_b = t_b(s, \underline{Y}), e_b = e_b(s, \underline{\hat{Y}}),$$

one gets the following:
$$V(e_b) = E_L V_1(e_b) + V_L E_1(e_b)$$
$$= E_L \left[-\sum\sum_{i<j} d_{ij} w_i w_j \left(\frac{\hat{y}_i}{w_i} - \frac{\hat{y}_j}{w_j}\right)^2 \right] + V_L(\hat{Y})$$
$$= E_L E_1 \left[-\sum\sum_{i<j} d_{sij} I_{sij} \left(\frac{\hat{y}_i}{w_i} - \frac{\hat{y}_j}{w_j}\right)^2 \right] + \sum V_i$$

assuming availability of constants d_{sij} free of $\underline{Y}, \underline{\hat{Y}}, \underline{V}$

such that $\quad E_1 d_{sij} I_{sij} = d_{ij},$

like, for example, $\quad d_{sij} I_{sij} = \dfrac{d_{ij}}{\pi_{ij}} I_{sij}$

assuming $\quad \pi_{ij} = \sum\limits_{s \ni ij} p(s) > 0 \ \forall \ i, j (i \neq j \in U).$

Then, $\quad v(e_b) = -\sum\limits_{i<j}\sum d_{sij} I_{sij} \left(\dfrac{\hat{y}_i}{w_i} - \dfrac{\hat{y}_j}{w_j}\right)^2 + \sum b_{si} v_i I_{si}$

is an unbiased estimator of $V(e_b)$.

Since $\quad V_p(t_b) = -\sum\limits_{i<j}\sum d_{ij} w_i w_j \left(\dfrac{y_i}{w_i} - \dfrac{y_j}{w_j}\right)^2$

and $\quad \hat{V}_p(t_b) = -\sum\limits_{i<j}\sum d_{sij} I_{sij} \left(\dfrac{y_i}{w_i} - \dfrac{y_j}{w_j}\right)^2$

satisfies $\quad E_p \hat{V}_p(t_b) = V_p(t_b),$

we may write $\quad \hat{V}_1(e_b) = -\sum\limits_{i<j}\sum d_{sij} I_{sij} \left(\dfrac{\hat{y}_i}{w_i} - \dfrac{\hat{y}_j}{w_j}\right)^2$

so that $\quad E_1 \hat{V}_1(e_b) = -\sum\limits_{i<j}\sum d_{ij} w_i w_j \left(\dfrac{\hat{y}_i}{w_i} - \dfrac{\hat{y}_j}{w_j}\right)^2.$

So, we may write $\quad v(e_b) = \hat{V}_1(e_b) + \sum b_{si} v_i I_{si}$

satisfying $\quad Ev(e_b) = V(e_b).$

More generally, if (i) through (iv) hold, E_1 and E_2 commute,

$$t = t(s, \underline{Y}) \text{ has } E_1(t) = Y,$$

$$e = t\left(s, \underline{\hat{Y}}\right) = t(s, \underline{Y})\bigg|_{\underline{Y} = \underline{\hat{Y}}} \text{ has } E_L(e) = t,$$

then we have

$$E_1(e) = \hat{Y}, E(e) = E_L E_1(e) = E_L\left(\hat{Y}\right) = Y$$

$$V(e) = E_L V_1(e) + V_L E_1(e) = E_L V_1(e) + \sum V_i.$$

Exploring Improved Ways

Moreover, if $\exists$ a $v_1(t) = v_1(s, \underline{Y})$ such that $E_1 v_1(t) = V_1(t)$,

then, for
$$v_1(e) = v_1(t)\Big|_{\underline{Y}=\hat{\underline{Y}}}$$

and for
$$v(t) = v(t(s,y)) \ni E_1 v(t) = V_p(t),$$

if we write
$$v_2 = v_2(s, \underline{V}) = t(s, \underline{Y})\Big|_{\underline{y}=\underline{V}},$$

then
$$v(e) = v_1(e) + v_2(s, \underline{V}) \text{ satisfies}$$
$$E(e) = V(e),$$

i.e., $v(e)$ is an unbiased estimator for $V(e)$. This result covers Raj's (1956), Murthy's (1957), and RHC's (1962) estimators in particular in single-stage sampling and their counterparts in multi-stage sampling in respect of the estimators unbiased for Y and their unbiased variance estimators as well. Thus, for Raj (1956) we shall have

$$e_D = t_D\Big|_{\underline{Y}=\hat{\underline{Y}}}, \quad \hat{V}(e_D) = v_D\Big|_{\underline{Y}=\hat{\underline{Y}}} + t_D(s, \underline{V});$$

for Murthy (1957), we shall have

$$e_M = \frac{1}{p(s)} \sum_{i \in s} \hat{y}_i p(s|i),$$

$$\hat{V}(e_M) = \frac{1}{p^2(s)} \sum_{i<j} \sum p_i p_j \left(\frac{\hat{y}_i}{p_i} - \frac{\hat{y}_j}{p_j}\right)^2 [p(s)p(s|ij) - p(s|i)p(s|j)]$$
$$+ \frac{1}{p(s)} \sum_{i \in s} v_i p(s|i) \quad \text{with notations described in Chapter 3;}$$

for Rao, Hartley, and Cochran's (1962) case we have

$$e_{RHC} = \sum_n \hat{y}_i \frac{Q_i}{p_i}$$

has
$$E_1(e_{RHC}) = \sum_1^N \hat{y}_i, \quad E(e_{RHC}) = E_l \sum_1^N \hat{y}_i = Y.$$

$$\hat{V}(e_{RHC}) = B \sum_n Q_i \left(\frac{\hat{y}_i}{p_i} - e_{RHC}\right)^2 + \sum_n v_i \frac{Q_i}{p_1} \quad \text{has}$$

$$E_1 \hat{V}(e_{RHC}) = A \sum_n \sum_n Q_i Q_j \left(\frac{\hat{y}_i}{p_i} - \frac{\hat{y}_j}{p_j}\right)^2 + \sum_1^N v_i$$

$$E\hat{V}(e_{RHC}) = E_L E_1 \hat{V}(e_{RHC})$$
$$= E_L \left[A \sum \frac{(\hat{y}_i)^2}{p_i} - R^2 \right] + \sum V_i, \quad R = \sum_1^N \hat{y}_i$$
$$= A \left(\sum \frac{y_i^2}{p_i} - Y_2 \right) + A \left(\sum \frac{V_i}{p_i} - \sum V_i \right) + \sum V_i$$
$$= A \sum \frac{V_i}{p_i} + (1-A) \sum V_i + A \left(\sum \frac{y_i^2}{y_i} - Y^2 \right)$$
$$= V(e_{RHC}).$$

Let us now consider further extensions. Rao (1979) considered estimators

$$t_b = \sum y_i b_{si} I_{si} \quad \text{without} \quad \sum_s p(s) b_{si} I_{si} = 1 \forall i$$

and expressed the Mean Square Error (MSE) of t_b, namely $M = E_p(t_b - Y)^2$ in uni-stage sampling on supposing the condition C to hold, i.e., $\exists\, w_i\, (\neq 0)$ such that

$$\frac{y_i}{w_i} = \text{a constant } \forall i \in U \Rightarrow M(t) = \sum_i \sum_j d_{ij} w_i w_j y_i y_j = 0$$

with
$$d_{ij} = E_p(b_{si} I_{si} - 1)(b_{sj} I_{sj} - 1),$$
$$\Rightarrow M = -\sum_{i<j} \sum d_{ij} w_i w_j \left(\frac{y_i}{w_i} - \frac{y_j}{w_j} \right)^2.$$

To extend this to Multi-stage sampling let $e_b = \sum \hat{y}_i b_{si} I_{si}$, with $\hat{y}'_i s$ subject to (i)–(iv) leading to

$$M(e_b) = E(e_b - Y)^2 = E_1 E_L (e_b - Y)^2$$
$$= E_1 E_L \left[(e_b - E_L(e_b)) + E_L(e_b) - Y \right]^2$$
$$= E_1 V_L(e_b) + E_1 (t_b - Y)^2.$$

Heeding Rao's (1979) condition C and writing

$$m_1(t_b) = -\sum_{i<j} \sum d_{sij} I_{sij} w_i w_j \left(\frac{\hat{y}_i}{w_i} - \frac{\hat{y}_j}{w_j} \right)^2,$$

an unbiased estimator for $M(e_b)$ is

$$m(e_b) = m_1(e_b) - \sum_{i<j} \sum d_{sij} I_{sij} w_i w_j \left(\frac{v_i}{w_i} + \frac{v_j}{w_j} \right)^2 + \sum b_{si}^2 I_{si} v_i$$

as is derived following Chaudhuri et al. (2000).

In case everything else holds but

$$\frac{y_i}{w_i} = \text{a constant} \quad \forall\, i \not\Rightarrow \sum_i \sum_j d_{ij} w_i w_j y_i y_j = 0,$$

then following Chaudhuri and Pal (2002) we may observe the following:

Let $t_b = \sum y_i b_{si} I_{si}$ with b_{si}'s as constants free of $\underline{Y}$.

Let $w_i \neq 0\ \forall\, i \in U$ and $\alpha_i = \sum_j d_{ij} w_j$ for $i \in U$. Then Chaudhuri and Pal (2003) have shown in case of single-stage sampling that

$$M = M(t_b) = E_p(t_b - Y)^2 = -\sum\sum_{i<j} d_{ij} w_i w_j \left(\frac{y_i}{w_i} - \frac{y_j}{w_j}\right)^2 + \sum_i \alpha_i \frac{y_i^2}{w_i}.$$

In multi-stage sampling assuming all the conditions (i)–(iv) to hold one may note for

$$e_b = \sum \hat{y}_i b_{si} I_{si} \quad \text{that} \quad E_2(e_b) = t_b$$

and

$$\begin{aligned}
M(e_b) &= E(e_b - Y)^2 \\
&= E_1 E_L\left[(e_b - E_L(e_b)) + (E_L(e_b) - Y)\right]^2 \\
&= E_1 V_L(e_b) + E_1(t_b - Y)^2 \\
&= E_1 \sum V_i b_{si}^2 I_{si} - \sum\sum_{i<j} d_{ij} w_i w_j \left(\frac{y_i}{w_i} - \frac{y_j}{w_j}\right)^2 + \sum \alpha_i \frac{y_i^2}{w_i}.
\end{aligned}$$

For $M(t_b)$ an unbiased estimator is

$$m(t_b) = -\sum\sum_{i<j} d_{sij} I_{sij} w_i w_j \left(\frac{y_i}{w_i} - \frac{y_j}{w_j}\right)^2 + \sum \alpha_i \frac{y_i^2}{w_i} \frac{I_{si}}{\pi_i}, \quad \text{assuming } \pi_i > 0\ \forall\, i.$$

Let

$$m_1(e_b) = -\sum\sum_{i<j} d_{sij} I_{sij} w_i w_j \left(\frac{\hat{y}_i}{w_i} - \frac{\hat{y}_j}{w_j}\right)^2 + \sum_i \alpha_i \frac{(\hat{y}_i)^2}{w_i} \frac{I_{si}}{\pi_i}.$$

Then, $E_L[m_1(e_b)] = -\sum\sum_{i<j} d_{sij} I_{sij} w_i w_j \left[\left(\frac{y_i}{w_i} - \frac{y_j}{w_j}\right)^2 + \left(\frac{V_i}{w_i^2} + \frac{V_j}{w_j^2}\right)\right]$

$$+ \sum \alpha_i \frac{y_i^2}{w_i} \frac{I_{si}}{\pi_i} + \sum \alpha_i \frac{V_i^2}{w_i} \frac{I_{si}}{\pi_i}.$$

So, $Em_1(e_b) = E_1 E_L [m_1(e_b)]$

$$= -\sum\sum_{i<j} d_{ij} w_i w_j \left(\frac{y_i}{w_i} - \frac{y_j}{w_j}\right)^2 + \sum\sum_{i<j} d_{ij} w_i w_j \left(\frac{V_i}{w_i^2} + \frac{V_j}{w_j^2}\right)$$

$$+ \sum \alpha_i \frac{y_i^2}{w_i} + \sum \alpha_i \frac{V_i}{w_i}.$$

So, $m(e_b) = m_1(e_b) - \sum\sum_{i<j} d_{sij} I_{sij} w_i w_j \left(\frac{v_i}{w_i^2} + \frac{v_j}{w_j^2}\right)$

$$- \sum \alpha_i \frac{v_i}{w_i} \frac{I_{si}}{\pi_i} + \sum v_i b_{si}^2 \frac{I_{si}}{\pi_i}$$

$$= -\sum\sum_{i<j} d_{sij} I_{sij} w_i w_j \left[\left(\frac{\hat{y}_i}{w_i} - \frac{\hat{y}_j}{w_j}\right)^2 + \left(\frac{v_i}{w_i^2} + \frac{v_j}{w_j^2}\right)\right]$$

$$+ \sum \frac{I_{si}}{\pi_i} \left[\alpha_i \left\{\frac{(\hat{y}_i)^2}{w_i} - \frac{v_i}{w_i}\right\} + v_i b_{si}^2\right]$$

is an unbiased estimator of $M(e_b)$.

Next let us consider in Uni-stage sampling a non-homogeneous linear estimator $t'_b = a_s + \sum y_i b_{si} I_{si}$ for Y.

For its unbiasedness we need conditions

$$\sum p(s) a_s = 0 \text{ and } \sum p(s) b_{si} I_{si} = 1 \ \forall i.$$

But let us not insist on either of these conditions. Instead let us take

$$\Delta = E_p(t - Y) \neq 0$$

but the magnitude of Δ is negligible. In case of multi-stage sampling let $e'_b = a_s + \sum \hat{y}_i I_{si}$ and writing $\underline{R} = (\hat{y}_1, \cdots \hat{y}_i, \cdots \hat{y}_N)$ with $\hat{y}_i$'s subject to (i)–(iv) and $R = \sum \hat{y}_i$, though $d = E_1(e'_b - R)$ is not zero, let its magnitude be negligible. For these properties of Δ and the first-stage and the overall sample-sizes that are supposed to be large enough, let us then derive and discuss the following results. We have

$$M(e'_b) = E(e'_b - Y)^2 = E_L E_1 [(e'_b - E_1(e'_b)) + (E_1(e'_b) - Y)]^2$$

approximately equal to

$$E_L E_1 (e'_b - R)^2 + E_L (R - Y)^2 = E_L E_1 (e'_b - R)^2 + \sum V_i.$$

Let there exist $m_2(t'_b)$ such that $E_p m_2(t'_b)$ approximately equals $E_p(t'_b - Y)^2$.

Also, let

$$m_2(e'_b) = m_2(t'_b)\Big|_{\underline{Y}=\underline{R}=(\hat{y}_1,\cdots\hat{y}_N)}.$$

Then, an approximately unbiased estimator for $M(e'_b)$ may be taken as

$$v(e'_b) = m_2(e'_b) + a_s + \sum b_{si} v_i I_{si}$$

$$= m_2(e'_b) + t'_b\Big|_{\underline{Y}=\underline{V}=(v_1,\cdots,v_N)}.$$

Exploring Improved Ways

Let us now consider a special case of two-stage sampling with the first-stage units i bearing the values $y_i, i \in U = (1, \cdots i, \cdots N)$ containing the second-stage units $i_j, j = 1, \cdots M_i$. The values M_i are supposed to be unknown before the sample-selection. Let a sample s of fsu's be selected with probability $p(s)$ giving us the inclusion-probabilities π_i for i in U and $\pi_{ii'}$ for the pair i, i' in U. Let for an fsu i, if selected, out of the M_i ssu's now ascertained because i is sampled, a sample of m_i ssu's be selected by some procedure yielding an unbiased estimator $\hat{y}_i$ for y_i, and let this be independently repeated in respect of each of the fsu's thus selected.

Let $\bar{\bar{Y}} = \dfrac{\sum_1^N M_i \bar{Y}_i}{\sum_1^N M_i}$, writing $\bar{Y}_i = \dfrac{\sum_{j=1}^{M_i} y_{ij}}{M_i}, i \in U$,

and $\bar{\bar{Y}} = \dfrac{\sum_1^N Y_i}{\sum_1^N M_i}$, writing $Y_i = M_i \bar{Y}_i$, the ith fsu total of y-values.

Then, this $\bar{\bar{Y}}$ is the population mean. Since in this Ratio both the numerator and the denominator are unknown, a ratio estimator is needed to estimate it.

Let
$$Y = \sum_1^N Y_i$$

be unbiasedly estimated by
$$\hat{Y} = \sum_{i \in s} b_{si}\hat{y}_i = \sum_1^N \hat{y}_i b_{si} I_{si}$$

and
$$M = \sum_1^N M_i$$

be unbiasedly estimated by
$$\hat{M} = \sum_1^N M_i b_{si} I_{si}$$

So, let $\hat{\bar{\bar{Y}}} = \dfrac{\hat{Y}}{\hat{M}}$ and let

$$E\left(\hat{\bar{\bar{Y}}} - \bar{\bar{Y}}\right)^2 = E_1 E_L \left(\hat{\bar{\bar{Y}}} - \bar{\bar{Y}}\right)^2$$

$$= E_1 E_L \left[\dfrac{\hat{Y}}{\hat{M}} - \dfrac{Y}{M}\right]^2 = E_1 E_L \left[\dfrac{\sum \hat{y}_i b_{si} I_{si}}{\sum M_i b_{si} I_{si}} - \dfrac{Y}{M}\right]^2$$

$$= E_1 E_L \left[\dfrac{\sum (\hat{y}_i - y_i) b_{si} I_{si} + \sum y_i b_{si} I_{si}}{\sum M_i b_{si} I_{si}} - \dfrac{Y}{M}\right]^2$$

approximately equal

$$E_1\left[\frac{\sum V_L(\hat{y}_i)b_{si}^2 I_{si}}{(\hat{M})^2}\right] + E_1\left[\frac{\sum y_i b_{si} I_{si}}{\sum M_i b_{si} I_{si}} - \frac{Y}{M}\right]^2$$

approximately equal

$$E_1 E_L\left[\frac{\sum v_i b_{si}^2 I_{si}}{(\hat{M})^2}\right] + \frac{1}{M^2}E_1\left[\sum b_{si} I_{si}\left(y_i - \frac{Y}{M}M_i\right)^2\right].$$

So, a reasonable approximately unbiased estimator for this $E\left(\frac{\hat{Y}}{\hat{M}} - \frac{Y}{M}\right)^2$ is

$$\frac{\sum v_i b_{si}^2 I_{si}}{\left(\sum M_i b_{si} I_{si}\right)^2} + \frac{\hat{V}_1\left(\sum b_{si} I_{si}\hat{y}_i\right)}{\left(\sum M_i b_{si} Isi\right)^2}\bigg|_{y_i=\hat{y}_i-\frac{\hat{Y}}{\hat{M}}M_i, i\in s}.$$

Here $\hat{V}_1\left(\sum_{i\in s} b_{si} I_{si} y_i\right)$ for example equals

$$(1)\qquad \sum \hat{y}_i^2 C_i + \sum\sum_{i\neq j}\hat{y}_i\hat{y}_j C_{ij} \quad \text{with}$$

$$C_i = \sum p(s) b_{si}^2 I_{si} - 1, C_{ij} = \sum_s b_{si} b_{sj} I_{sij} - 1$$

or

$$(2)\qquad -\sum\sum_{i<j} d_{sij} w_i w_j \left(\frac{\hat{y}_i}{w_i} - \frac{\hat{y}_j}{w_j}\right)^2 + \sum \frac{(\hat{y}_i)^2 I_{si}}{w_i \pi_i}$$

following Horvitz and Thompson and Rao (1979) combined with Chaudhuri and Pal (2002), respectively.

This easily may be extended to multi-stage sampling vide Chaudhuri and Stenger (2005) and Chaudhuri (2010).

4.5 MULTI-PHASE SAMPLING: RATIO AND REGRESSION ESTIMATION

Neyman (1938) introduced the concept and technique of double or two-phase sampling to cater to the need to tackle the problems narrated below, and they extended to multiple or multi-phase sampling in due course.

First let us mention the role of double sampling in the context of stratified sampling. Let a population $U = (1,\cdots,i,\cdots,N)$ be composed of a known

Exploring Improved Ways

number N of units with widely varying values of a variable y so that we find it important to distinguish among the individuals with y-values across the respective ranges $(a_{h-1}, a_h), h = 1, \cdots, H$ defining H separate strata. But suppose at the beginning we do not know which and how many of them bear y-values in the respective distinct ranges. Let these unknown numbers be N_h and $W_h = \frac{N_h}{N}$ so that $\left(0 < W_h < 1, \sum_1^H W_h = 1\right)$, and our problem is to suitably and unbiasedly estimate

$$\bar{Y} = \sum_1^H W_h \bar{Y}_h$$

writing y_{hi} the y-values for the unit i in the hth stratum of N_h individuals. Since we cannot identify the units in the respective strata and we cannot form the stratum-specific frames, we cannot technically choose a stratified sample. But, instead we may initially take a simple random sample without replacement (SRSWOR) of a size, say n'. From such a sample it is possible to identify the respective stratum-wise members, n'_h in numbers, $h = 1 \cdots, H$, such that $\sum_1^H n'_h = n'$. Letting $w_h = \frac{n'_h}{n'}$ as estimators of $W_h, h = 1, \cdots, H$ and introducing expectation, variance operators E_1, V_1 with respect to the above SRSWOR we may further take independent SRSWORs of sizes $n_h \left(2 \leq n_h < n'_h, h = 1, \cdots H, \sum_1^H n_h = n\right)$ out of these n'_h units of the respective strata, $h = 1, \cdots, H$. This is double sampling applicable to stratification. Let E_2, V_2 denote expectation, variance operators in respect of the SRSWOR independently strata-wise from the initial SRSWOR.

Let $\bar{y}'_h$ be the sample means of the n'_h-values of y_{hi} and $\bar{y}_h$ the sample means of the n_h-values in the second-phase samples.

Let
$$\bar{y}_{dst} = \sum_{h=1}^H w_h \, \bar{y}_h$$

be taken as the two-phase stratified sample estimator of $\bar{Y}$. We may then observe

$$E_2(\bar{y}_h) = \bar{y}'_h, \; E_1(w_h) = W_h$$

because $\underline{n}' = (n'_1, \cdots, n'_h, \cdots n'_H)$ has a generalized hypergeometric distribution with parameters n and $\underline{W} = (W_1, \cdots, W_h, \cdots, W_H)$.

So, $E = E_1 E_2$ and $V = E_1 V_2 + V_1 E_2$ the overall expectation, variance operators we may now apply to derive the following:

$$E(\bar{y}_{dst}) = E_1 \sum_h w_h \bar{y}'_h = E_1 \left[\frac{1}{n'} \sum_h \sum_1^{n'_h} y_{hi} \right]$$

$$= E_1(\bar{y}') = \bar{Y}, \; \text{writing} \; \bar{y}' \; \text{the sample mean}$$

of the n' initially drawn sample of the y-values.

Since $\quad E_2(\bar{y}_{dst}) = \bar{y}',\, V_1(E_2\bar{y}_{dst}) = V_1(\bar{y}') = \dfrac{N-n'}{Nn'}S^2,$

writing $\quad S^2 = \dfrac{1}{N-1}\sum_{1}^{H}\sum_{1}^{N_h}(y_{hi}-\bar{Y})^2.$

Now, $\quad V_2(\bar{y}_{dst}) = V_2\left[\sum_{1}^{H}\dfrac{n'_h}{n'}\left(\dfrac{1}{n_h}\sum_{1}^{n_h}y_{hi}\right)\right]$

$= \sum_{1}^{H}\dfrac{n'_h}{n'}\dfrac{n'_h}{n'}\left[\left(\dfrac{1}{n_h}-\dfrac{1}{n'_h}\right)\sum_{1}^{n'_h}(y_{hi}-\bar{y}'_h)^2\right]\bigg/(n'_h-1)$

$= \dfrac{1}{n'}\sum_{1}^{H}\dfrac{n'_h}{n'}\left(\dfrac{1}{\gamma_h}-1\right)(s'_h)^2,$

writing $\quad (s'_h)^2 = \dfrac{1}{(n'_h-1)}\sum_{1}^{n'_h}(y_{hi}-\bar{y}'_h)^2,\quad \gamma_h = \dfrac{n_h}{n'_h}.$

It follows, vide Chaudhuri (2010) and Cochran (1977) that

$$E_1[V_2(\bar{y}_{dst})] = \dfrac{1}{n'}\sum_{1}^{H}W_h\left(\dfrac{1}{\gamma_h}-1\right)S_h^2$$

So, $\quad V(\bar{y}_{dst}) = \left(\dfrac{1}{n'}-\dfrac{1}{N}\right)S^2 + \dfrac{1}{n'}\sum_{1}^{H}\left(\dfrac{1}{\gamma_h}-1\right)W_hS_h^2.$

Cochran (1977), writing $g' = \dfrac{N-n'}{N-1}$, by a very complicated analysis derived for $V(\bar{y}_{dst})$ the unbiased variance estimator as

$$v_1 = \dfrac{n'(N-1)}{(n'-1)N}\Bigg[\sum_h w_h s_h^2\left(\dfrac{1}{n'\gamma_h}-\dfrac{1}{N}\right) + \dfrac{g'}{n'}\sum_h w_h(\bar{y}_h-\bar{y}_{dst})^2$$
$$+ \dfrac{g'}{n'}\sum_h s_h^2\left(\dfrac{w_h}{N}-\dfrac{1}{n'\gamma_h}\right)^2\Bigg].$$

Chaudhuri (2010), however, derived the alternative unbiased estimator for $V(\bar{y}_{dst})$ as

$$v_2 = \dfrac{\left(\dfrac{1}{n'}-\dfrac{1}{N}\right)\dfrac{N}{N-1}}{1-\left(\dfrac{1}{n'}-\dfrac{1}{N}\right)\dfrac{N}{N-1}}\left[\left\{\sum w_h\left(\bar{y}_h^2-\dfrac{1}{n''_h}\right)\left(\dfrac{1}{\gamma_h}-1\right)s_h^2\right\}-(\hat{y}_{dst})^2\right]$$
$$+ \dfrac{1}{n'}\sum\left(\dfrac{1}{\gamma_h}-1\right)w_h s_h^2.$$

Next let us turn to Regression Estimators for Finite Population totals or means. Let us recall how Basu (1971)proves the non-existence of a uniformly minimum variance (UMV) unbiased estimator (UMVUE) among all unbiased estimators (UE) for a finite population total. Supposing if possible the UMVUE for Y may exist as $t_0 = t_0(s, \underline{Y})$ with $E_p(t_0) = Y \, \forall \, \underline{Y}$ and $\underline{A} = (a_1, \cdots, a_i, \cdots, a_N)$ be a vector in the space Ω of all possible $\underline{Y}$'s then, he considered the UE for Y as

$$t_A = t_A(s, \underline{Y}) = t_0(s, \underline{Y}) - t_0(s, \underline{A}) + A,$$

writing $A = \sum a_i$. In case one starts with the Horvitz and Thompson's (1952) UE for Y as

$$t_{HT} = \sum_{i \in s} \frac{y_i}{\pi_i}, \pi_i > 0 \, \forall \, i,$$

then, analogously to t_A, follows the UE for Y as

$$t_{GD} = \sum_{i \in s} \frac{y_i}{\pi_i} - \sum_{i \in s} \frac{a_i}{\pi_i} + \sum a_i$$

$$= \sum_{\in s} \frac{y_i - a_i}{\pi_i} + A.$$

With $\underline{A}$ pre-assigned and hence known, this t_{GD} is called a Generalized Difference Estimator for Y. If one has knowledge about the values x_i in $\underline{X} = (x_1, \cdots, x_i, \cdots, x_N)$ with $X = \sum_1^N x_i$, then taking $a_i = \beta x_i, i \in U$ with β unknown one may take the GDE (Generalized Difference Estimator) for Y as

$$t_\beta = t_\beta(s, \underline{Y}) = \sum_{i \in s} \frac{y_i}{\pi_i} + \beta \left(X - \sum_{i \in s} \frac{x_i}{\pi_i} \right)$$

which is not applicable when β is unknown. For an SRSWOR this t_β takes the form

$$t_\beta = t_\beta(s, \underline{Y}) = N \left[\bar{y} + \beta (\bar{X} - \bar{x}) \right]$$

writing $\bar{y}, \bar{x}$ for sample means of y, x and $\bar{X} = \frac{1}{N} \sum_1^N x_i$, the population mean of x. This estimator is called the Difference Estimator for Y based on SRSWOR.

Writing $y_i = \beta x_i + \epsilon_i, i \in U$ and treating ϵ_i as unknowable random errors in the linear regression of y on x, one may derive the least squares estimate of β as

$$b = \frac{\sum_{i \in s}(y_i - \bar{y})(x_i - \bar{x})}{\sum_{i \in s}(x_i - \bar{x})^2}.$$

Then,
$$t_b = N \left[\bar{y} + b(\bar{X} - \bar{x}) \right]$$

is regarded as the Linear Regression Estimator for Y based on an SRSWOR.

Writing
$$B = \frac{\sum_1^N (y_i - \bar{Y})(x_i - \bar{X})}{\sum_1^N (x_i - \bar{X})^2}$$

as the population analogue of b, the sample coefficient of regression of y on x, we may approximate for large samples the regression estimator t_b by

$$t_B = N\left[(\bar{y} - B\bar{x}) + B\bar{X}\right]$$
$$= N\left[\frac{1}{n}\sum_{i \in s}(y_i - Bx_i)\right] + BX.$$

This is helpful in deriving an approximate variance formula for t_b in large samples. Analogous to B one may note

$$R_N = \frac{\sum_1^N (y_i - \bar{Y})(x_i - \bar{X})}{\sqrt{\sum (y_i - \bar{Y})^2}\sqrt{\sum (x_i - \bar{X})^2}}$$

as the finite population correlation coefficient between y and x.

Writing
$$Sy^2 = \frac{1}{N-1}\sum_1^N (y_i - \bar{Y})^2, \quad S_x^2 = \frac{1}{N-1}\sum_1^N (x_i - \bar{X})^2$$
$$C_{yx} = \frac{1}{N-1}\sum_1^N (y_i - \bar{Y})(x_i - \bar{X}),$$

the finite population variances and covariance of y and x, we may note that

$$B = R_N \frac{S_y}{S_x} \quad \text{or} \quad B^2 = R_N^2 \frac{Sy^2}{S_x^2}, \quad C_{yx} = R_N S_y S_x.$$

So,
$$V(t_b) \simeq V(t_B) = N^2 \frac{N-n}{Nn} \frac{1}{N-1} \sum_{i=1}^N (y_i - Bx_i)^2$$
$$= N^2 \left(\frac{1}{n} - \frac{1}{N}\right) \frac{1}{N-1} \sum_1^N \left[(y_i - \bar{Y}) - B(x_i - \bar{X})\right]^2.$$

Since
$$\bar{Y} \simeq B\bar{X}$$

or
$$V(t_b) \simeq N^2 \left(\frac{1}{n} - \frac{1}{N}\right)\left[S_y^2 + B^2 S_x^2 - 2BC_{yx}\right]$$
$$= N^2 \left(\frac{1}{n} - \frac{1}{N}\right) S_y^2 \left[1 - R_N^2\right].$$

Since $BS_x = R_N S_y$ and $BC_{yx} = R_N^2 S_y^2$ for $\bar{Y}$, then we have the linear regression estimator based on SRSWOR of size n as

$$\bar{y}_{lr} = \frac{t_b}{N} = \bar{y} + b(\bar{X} - \bar{x});$$

hence,
$$V(\bar{y}_{lr}) \simeq \left(\frac{1}{n} - \frac{1}{N}\right) S_y^2 (1 - R_N^2).$$

An unbiased estimator for

$$\frac{1}{N-1} \sum_{1}^{N} \left[y_i + B(\bar{X} - x_i) - \bar{Y} \right]^2 = \frac{1}{N-1} \sum_{1}^{N} \left[(y_i - \bar{Y}) - B(x_i - \bar{X}) \right]^2$$

is
$$\frac{1}{n-1} \sum_{i \in s} \left[(y_i - \bar{y}) - B(x_i - \bar{x}) \right]^2.$$

This may be approximated by

$$s_e^2 = \frac{1}{n-1} \sum_{i \in s} \left[(y_i - \bar{y}) - b(x_i - \bar{x}) \right]^2$$
$$= s_y^2 + b^2 s_x^2 - 2b \hat{C}_{yx}$$

writing
$$s_y^2 = \frac{1}{n-1} \sum_{i \in s} (y_i - \bar{y}), \ s_x^2 = \frac{1}{n-1} \sum (x_i - \bar{x})^2$$

and
$$\hat{C}_{yx} = \frac{1}{n-1} \sum_{i \in s} (y_i - \bar{y})(x_i - \bar{x}).$$

Now writing
$$r = \frac{\sum (y_i - \bar{y})(x_i - \bar{x})}{\sqrt{\sum (y_i - \bar{y})^2} \sqrt{\sum (x_i - \bar{x})^2}},$$

the correlation coefficient between y and x from an SRSWOR of size n, we may note $\hat{C}_{yx} = r_{yx} s_y s_x$ and hence,

$$s_e^2 = s_y^2 + r^2 s_y^2 - 2r^2 s_y^2 = s_y^2 (1 - r^2).$$

So, an approximately unbiased estimator for

$$V(\bar{y}_{lr}) \quad \text{is} \quad \left(\frac{1}{n} - \frac{1}{N}\right) s_y^2 (1 - r^2)$$

Cochran (1977, pp 193–195) may be referred to for a more elegant presentation.

An important deficiency in the use of the regression estimator

$$\bar{y}_{lr} = \bar{y} + b(\bar{X} - \bar{x})$$

for $\bar{Y}$ based on an SRSWOR is that it involves $\bar{X}$, the population mean of the values of x_i for $i \in U$. If this $\bar{X}$ is not available and x_i-values like y_i-values, are also not available before the start of the sample survey, then Double Sampling may start playing its role. Double Sampling in Regression estimation in the special case of SRSWOR works as follows.

Let from U an SRSWOR of a size n' be drawn as s' and the values x_i for $i \in s'$ be observed yielding the initial sample mean $\bar{x}' = \frac{1}{n'}\sum_{i \in s'} x_i$. Then, from s' an SRSWOR of size $n(2 \leq n < n')$ is drawn as say, s and y_i-values are observed for i in s. Thus, (y_i, x_i)-values are now at hand for $i \in s$. Thus, (s, s') is now our double sample with values observed as

$$x_i, i \in s'; (y_i, x_i), i \in s, s \subset s'.$$

Then the regression estimator for $\bar{Y}$ on a double sample is

$$\bar{y}_{dlr} = \bar{y} + b(\bar{x}' - \bar{x}).$$

Here
$$\bar{y} = \frac{1}{n}\sum_{i \in s} y_i, \quad \bar{x} = \frac{1}{n}\sum_{i \in s} x_i, \quad \text{and}$$

$$b = \frac{\sum_{i \in s}(y_i - \bar{y})(x_i - \bar{x})}{\sum_{i \in s}(x_i - \bar{x})^2}.$$

By E_1, V_1 we shall denote the expectation, variance operators for the initial SRSWOR of size n', by E_2, V_2 those over the sub-sampling from s' by SRSWOR of size n and $E = E_1 E_2, V = E_1 V_2 + V_1 E_2$ those over the overall double sampling by SRSWOR as described above.

It follows that $E_2(\bar{y}) = \frac{1}{n'}\sum_{i \in s'} y_i = \bar{y}'$.

It will be convenient, following Cochran (1977), to approximate b by

$$B = \frac{\sum_1^N (y_i - \bar{Y})(x_i - \bar{X})}{\sum_1^N (x_i - \bar{X})} = C_{yx} S_x^2$$

to derive $V(\bar{y}_{dlr})$ and a suitable estimator for the latter.

Then
$$\bar{y}_{dlr} \simeq \bar{y} + B(\bar{x}' - \bar{x}) = \bar{y}'_{dlr}.$$
$$E_2(\bar{y}_{dlr}) \simeq \bar{y}'$$

and hence,
$$E(\bar{y}_{dlr}) \simeq \bar{Y}.$$

Also, writing
$$u_i = y_i - Bx_i,$$
$$s'^2_u = \frac{1}{(n'-1)}\sum_{i \in s'}(u_i - \bar{u}')^2,$$
$$\bar{u}' = \frac{1}{n'}\sum_{i \in s'} u_i,$$

it follows that
$$V_2(\bar{y}'_{dlr}) \simeq \left(\frac{1}{n} - \frac{1}{n'}\right) s'^2_u, \quad \text{vide Cochran (1977)}.$$

So, $V\left(\bar{y}'_{dlr}\right) \simeq V_1\left(\bar{y}'\right) + \left(\dfrac{1}{n} - \dfrac{1}{n'}\right) E_1 s'^2_u$

$= \left(\dfrac{1}{n'} - \dfrac{1}{N}\right) S_y^2 + \left(\dfrac{1}{n} - \dfrac{1}{n'}\right) \dfrac{1}{N-1} \sum_1^N \left[(y_i - \bar{Y}) - B(x_i - \bar{X})\right]^2$

$= \left(\dfrac{1}{n'} - \dfrac{1}{N}\right) S_y^2 + \left(\dfrac{1}{n} - \dfrac{1}{n'}\right) S_y^2 (1 - R_N^2)$

$= \dfrac{S_y^2(1-R_N^2)}{n} + \dfrac{S_y^2 R_N^2}{n'} - \dfrac{S_y^2}{N}$, vide Cochran (1977, p 339).

It is not easy to derive a reasonable estimator for this though Cochran (1977, pp 340 and 343) and Chaudhuri (2010, p 108) have presented two and one, respectively. This is left as an exercise.

The ratio estimation has been covered in detail in Chapter 3. In the context of Double Sampling with SRSWOR in both the phases we need to add the following.

Based on an SRSWOR of size n the ratio estimator for Y is

$$\hat{Y}_R = X\dfrac{\bar{y}}{\bar{x}} \quad \text{with usual notations.}$$

If we suppose $x_i, i \in U$ are not known to start with, a double sample in SRSWOR in both the phases may be obtained as discussed in the context of double sampling in linear regression. The Double sample ratio estimator for Y is

$$\hat{Y}_{dR} = X^* \hat{R}$$

writing $\quad X^* = N\bar{x} \quad$ and $\quad \hat{R} = \dfrac{\bar{y}}{\bar{x}} = \dfrac{N\bar{y}}{N\bar{x}} = \dfrac{\hat{Y}}{\hat{X}}$.

Let $Y^* = N\bar{y}'$. Further, following Chaudhuri (2010), let us write

$$e_1 = \dfrac{\hat{Y} - Y^*}{Y^*} \Rightarrow \hat{Y} = Y^*(1 + e_1),$$

$$e_2 = \dfrac{\hat{X} - X^*}{X^*} \Rightarrow \hat{X} = X^*(1 + e_2),$$

and

$$e = \dfrac{Y^* - Y}{Y} \Rightarrow Y^* = Y(1+e).$$

Then, $\quad \hat{Y}_{dR} = X^* \dfrac{Y^*(1+e_1)}{X^*(1+e_2)} = \dfrac{Y(1+e_1)(1+e)}{(1+e_2)}.$

Assuming n large enough relative to n', we suppose $|e_2| < 1$ and so we may expand
$$\frac{1}{1+e_2} \quad \text{as} \quad 1 - e_2 + e_2^2,$$
neglecting further terms in the Taylor series expansion we may take
$$E\left(\hat{Y}_{dR}\right) \simeq Y E_1 E_2 \left[(1+e_1)\left(1-e_2+e_2^2\right)(1+e)\right].$$
So,
$$E\left(\hat{Y}_{dR}\right) \simeq Y + Y E_1 \left[(1+e)\left\{\frac{V_2\left(\hat{X}\right)}{E_2^2\left(\hat{X}\right)} - \frac{\mathrm{cov}_2\left(\hat{X},\hat{Y}\right)}{E_2\left(\hat{X}\right) E_2 \hat{Y}}\right\}\right].$$

From this Chaudhuri (2010) takes as an estimator for the bias of $\hat{Y}_{dR}$ as an estimator of Y as

$$b\left(\hat{Y}_{dR}\right) = \hat{Y}\left[\frac{\hat{V}_2\left(\hat{X}\right)}{\left(\hat{X}\right)^2} - \frac{\mathrm{c\hat{o}v}_2\left(\hat{Y},\hat{X}\right)}{\hat{X}\hat{Y}}\right].$$

These results are too crude and need not be further pursued. The Mean Square Error of $\hat{Y}_{dR}$ about Y is approximately taken as

$$M\left(\hat{Y}_{dR}\right) = Y E_1 E_2 \left[e\left\{1+(e_1-e_2)\right\}\right]^2$$
$$= E_1 \left(Y^* - Y\right)^2 \left[\frac{V_2\left(\hat{Y}\right)}{(Y^*)^2} + \frac{V_2\left(\hat{X}\right)}{(X^*)^2} - \frac{2\mathrm{cov}_2\left(\hat{Y},\hat{X}\right)}{Y^* X^*}\right].$$

An estimator for this MSE Chaudhuri (2010) employed as

$$m\left(\hat{Y}_{dR}\right) = \hat{V}_1\left(Y^*\right)\left[\frac{\hat{V}_2\left(\hat{Y}\right)}{(Y^*)^2} + \frac{\hat{V}_2\left(\hat{X}\right)}{(X^*)^2} + \frac{\mathrm{cov}_2\left(\hat{Y},\hat{X}\right)}{Y^* X^*}\right]$$

with easy details for SRSWOR in both phases. Next let us consider Post-stratification.

Post-stratification has a position intermediate between stratification in a single phase and stratification in two phases.

Suppose the strata are defined in terms of the values of the variable of interest y lying in the disjoint ranges $(a_{h-1} \leq y < a_h), h = 1, \cdots, H$ with $H(>2)$ suitably chosen, respectively, specifying the H strata. Further let $N_h, \sum_1^H N_h = N$, the strata sizes and hence the strata proportions

$$W_h = \frac{N_h}{N}(0 < W_h < 1, \sum_1^H W_h = 1)$$

Exploring Improved Ways

be known. But what are crucially unknown are the individual population members with y-values in the respective strata. So, one cannot determine a strata-wise frame to draw a stratified sample. In this situation the strata are called Post-strata. Since the W_h's are supposedly known, they may be suitably utilized in estimating population mean

$$\bar{Y} = \sum_{h=1}^{H} W_h \bar{Y}_h$$

with the notations already explained.

Let an SRSWOR of a pre-assigned size n be drawn and surveyed. Gathering the n sampled y-values, the sampled units may then be assigned to the respective post-strata. Then let the numbers $n_h (0 \leq n_h \leq n, \sum_1^H n_h = n)$ of sampled units assignable to the respective post-strata be determined. Since for all the H post-strata the n_h-values need not be found positive, let us write

$$\bar{y}_h = \frac{1}{n_h} \sum_h y_i, \quad \sum_h \text{ denoting the } h\text{th post-stratum, if } n_h > 0$$

and $\quad \bar{y}_h = 0 \quad$ if $n_h = 0, h = 1, \cdots, H$.

Then,
$$\bar{y}_{pst} = \sum_1^N W_h \bar{y}_h \frac{I_h}{E(I_h)}$$

may be taken as an unbiased estimator for $\bar{Y}$, writing

$$I_h = 1, \text{ if } n_h > 0,$$
$$= 0, \text{ if } n_h = 0, h = 1, \cdots, H.$$

Now, $\quad E(I_h) = \text{Prob}(I_h = 1) = \text{Prob}(n_h > 0) = 1 - \dfrac{\binom{N-N_h}{n}}{\binom{N}{n}}.$

Doss, Hartley, and Somayajulu (1979) propose for $\bar{Y}$ the following alternative ratio-type estimator, namely,

$$\bar{y}'_{pst} = \frac{\sum_h W_h \bar{y}_h \frac{I_h}{E(I_n)}}{\sum W_h \frac{I_h}{E(I_h)}}.$$

They claim the following linear invariance property of $\bar{y}'_{pst}$ not shared by $\bar{y}_{pst}$.

Let $y = a + bx$. Then one gets $\bar{y}'_{pst} = a + b\bar{x}'_{pst}$ with obvious notation. But $\bar{y}_{pst} \neq a + b\bar{x}_{pst}$ because $a \sum W_h \frac{I_h}{E(I_h)} \neq a$.

4.6 CONTROLLED SAMPLING

Goodman and Kish (1950) introduced the concept and technique of Controlled sample selection. Sometimes, especially in sampling a population of units spread in a wide geographical region, a noteworthy number of units appear unsuitably located discouraging their coverage in an actual survey. Such units, if identified, are called Non-preferred units while the remaining ones are termed as Preferred units in the population. So, even if taking an SRSWOR of a given size n from a population of size $N\,(>2)$ is contemplated, it is deemed desirable to confine selection to only of the Preferred units. The Non-Preferred are avoided ones to the extent possible retaining the properties of the sample mean $\bar{y}$ being exactly unbiased for the population mean $\bar{Y}$ and its variance remaining intact as $V(\bar{y}) = \frac{N-n}{Nn}S^2, S^2 = \frac{1}{N-1}\sum_1^N (y_i - \bar{Y})^2$ for such a constrained sampling scheme, if duly implemented as indicated. Thus, we may enunciate the Definition. Controlled Sampling means selection avoiding Non-Preferred units to the extent possible but choosing from among the Preferred ones, yet retaining certain desirable properties of certain estimators based on a basic sampling scheme which is thus amended to keep the probability of choosing any Non-Preferred unit restricted to a small value $\alpha\,(0 \leq \alpha < 1)$.

The variance of $\bar{y}$ based on SRSWOR of size n is determined by the property of this sampling scheme by the linear restriction

$$\pi_{ij} = \sum_{s \ni i,j} p(s) = \frac{n(n-1)}{N(N-1)} \qquad (4.9.1)$$

as satisfied by this scheme, because for it,

$$p(s) = \frac{1}{\binom{N}{n}} \ \forall\, s \quad \text{with } p(s) > 0. \qquad (4.9.2)$$

But one may employ any sampling scheme with $p(s)$ subject to (4.9.1) though (4.9.2) need not be satisfied for it. For SRSWOR the support of the sample space, namely the number of samples with positive selection probability, is $\binom{N}{n}$. One may find a sampling design p subject to (4.9.1) with a support size much less than $\binom{N}{n}$. If one may lay out a sampling design p subject to (4.9.1) but in addition, $\sum_1 p(s) \leq \alpha, 0 < \alpha < 1$ with a pre-assigned α and samples s with $p(s) \geq 0, \sum_1$ denotes the sum over the Non-Preferred units only of size n, which is the number of units in it, and each is distinct. An immediate implication of this is that to derive a Controlled Sampling scheme from an SRSWOR scheme, the support size for the former has to be reduced suitably from that of the former which is $\binom{N}{n}$.

Sukhatme and Avadhani (1965) and Avadhani and Sukhatme (1965, 1966, 1968, 1972, 1973) developed and studied the problem of effecting such controls on SRSWOR. One classical but obvious way to reduce the support size of an

SRSWOR is to employ stratified sampling with independent SRSWOR in each stratum because

$$\prod_{h=1}^{H}\binom{N_h}{n_h} < \binom{N}{n} \text{ for } \sum_{1}^{H} N_h = N \text{ and } \sum_{1}^{H} n_h = n.$$

But it is more interesting to enforce Control on SRSWOR within each stratum. Gupta, Mandal, and Parsad (2012) in their book have described many necessary details.

For a systematic application of control on SRSWOR one needs to pay attention to the pioneering work of Chakrabarti (1963) who has described the use of incidence matrices in experimental designs in developing schemes for sampling from finite populations.

Chakrabarti (1963) and also Avadhani and Sukhatme (1973) propose the use of a Balanced Incomplete Block Design (BIBD) to construct the Controlled Sampling Scheme. A BIBD is a collection of b blocks of size k each of a set of v units such that each unit occurs in the blocks a common number r of times and each pair of units $i, j\, (i \neq j)$ also occurs a common number λ of times $\ni. bk = vr$ and $\lambda(v-1) = r(k-1)$. To see its link with an SRSWOR being treated here, let $v = N$ and $k = n$. Then one derives

$$r = b\frac{n}{N} = \frac{\lambda(N-1)}{(n-1)} \text{ or } \lambda = \frac{bn(n-1)}{N(N-1)}.$$

So, on constructing a BIBD we may regard b as the total number of samples giving the Support size and on choosing one of them at random with probability $\frac{1}{b}$ one may ensure the observance of the requirement

$$\pi_{ij} = \frac{n(n-1)}{N(N-1)} = \frac{\lambda}{b}$$

for the Controlled Sampling scheme for which the support size is b which is much less than $\binom{N}{n}$. On identifying the Preferred and Non-preferred samples suitably, the samples in the space of support b the problem of constructing a Controlled sample corresponding to SRSWOR of size n from a population of size N is easy to solve.

But a problem persists because a BIBD does not exist in certain cases when $b \not< \binom{N}{n}$. For example, with $v = N = 8$ and $k = n = 3, \not\exists$ a BIBD with $b < \binom{8}{3} = 56$.

Wynn (1977) and Foody and Hedayat (1977) prescribe taking repeated blocks when a BIBD does not exist with $b < \binom{v}{k}$. Gupta et al. (2012) give a table to show how with $v = 8$ and $k = 3$ on taking $b = 22\,(< 56 = \binom{8}{3})$, and using repeated blocks a controlled sample may be suitably chosen. But with larger values of N and n a choice may be difficult to make. Avadhani and Sukhatme (1973) suggest the following course: Divide N into groups of sizes $N_1, N_2, \cdots, N_G = N$; choose integer $n'_i \cdots \ni n_1 < n'_i < N_i, i = 1, \cdots, G. \ni .$

A BIBD exists for each $v_i = n'_i, b_i, r_i, n_i, \lambda_i, i = 1, \cdots G$; $\sum_{i=1}^{G} n_i = n$ and $n_i \alpha N_i \,\forall i$; then take SRSWOR of size n'_i from N_i for each $i = i, \cdots G$; identify n_i Preferred units out of each group of n'_i units and use the above BIBDs to take controlled samples from each group and hence from the entire population.

A path-breaking way out for constructing Controlled Samples covering the original SRSWOR and original varying probability sampling (VPS) schemes has been provided by Rao and Nigam (1990) recommending application of Linear Programming Techniques utilizing the Simplex Method. Gupta, Nigam, and Kumar (1982) used BIBD along with the HTE, namely,

$$t_{HT} = \sum_{i \in s} \frac{y_i}{\pi_i}$$

for s in the support S of the design p providing a solution using a BIBD such that $\pi_i = np_i$ is realized with p_i as the normed size-measures at hand only provided n, N and p_i's are such that

$$\sum_{i \in s} p_i > \frac{n-1}{N-1} \forall \, s \in S.$$

Hedayat, Lin, and Stufken (1989) applied a new "Method of emptying boxes" to construct a Controlled VPS design yielding $\pi_i \alpha p_i, \pi_{ij} > 0$ and $\pi_i \pi_j \geq \pi_{ij} \forall \, i \neq j$, yielding a uniformly non-negative Yates and Grundy (1953) unbiased variance-estimator for the t_{HT}. However, Nigam et al. (1984) employed BIBD to obtain a controlled VPS scheme ensuring t_{HT} to have $\pi_i \alpha p_i$ and $c \, \pi_i \pi_j \leq \pi_{ij} \leq \pi_i \pi_j \forall \, i \neq j$ with c in (0,1) that provides a stable Yates-Grundy variance-estimator for t_{HT}. Wynn (1977) added a procedure that ensures moreover that the support size $\leq N_{c_2}$, but his method is not fully constructive to ease the derivation of a controlled scheme with practical applicability.

Rao and Nigam (1990) address the linear programming problem of minimizing Objective Function $\phi = \sum_{s \in S_1} p(s), S_1 \equiv$ the set of Non-preferred samples with respect to the variables $\{p(s) | s \in S\}$ where S is the Support of all samples if two or more units are subject to the constraints.

i. $\sum_{s \ni i,j} p(s) = \pi_{ij} \, (i < j = 1, \cdots N)$ and
ii. $p(s) \geq 0 \, \forall \, s \in S$

such that

$$\sum_{\substack{j=1 \\ (j \neq i)}}^{N} \pi_{ij} = (n-1) \pi_i = n(n-1) p_i$$

with notations as usual.

In order to have a concrete region of feasible solutions, Rao and Nigam (1990) start with π_{ij}'s given by a known sampling scheme due to Sampford (1967) described in detail by Chaudhuri and Vos (1988, pp 217–218) having the desirable properties (1) $0 < \pi_{ij} \leq \pi_i \pi_j = n^2 p_i p_j$ and (2) $V(t_{HT}) \leq V(t_{HH})$, t_{HH} denoting Hansen and Hurwitz's (1943) estimator based on PPSWR scheme in n draws.

Rao and Nigam (1990) also finds a controlled sampling scheme on minimizing another objective function $\phi = \sum_{s \in S_1} p(s) w(s)$ allowing specified weight functions $w(s) \geq o$. They also got optimal solutions minimizing the same objective functions but on revising the constraints as

i. $\sum_{s \ni i} p(s) = n p_i \ (i = 1, \cdots, N)$
ii. $C(n p_i)(n p_j) \leq \sum_{s \ni i,j} p(s) \leq (n p_i)(n p_j) \ (i < j = 1, \cdots, N)$
iii. $p(s) \geq 0 \ \forall \ s \in S$.

Rao and Nigam (1990) have given us a solution for deriving a Controlled Version of the sampling strategy involving the Lahiri, Midzuno, Sen (1951; 1952; 1953) sampling scheme and the ratio estimator

$$t_R = X \frac{\sum_{i \in s} y_i}{\sum_{i \in s} x_i}$$

employing Chakrabarti's (1963) BIBD approach examining the availability of uniformly non-negative Unbiased variance estimator for this t_R.

5 Modeling

Introduction. Super-population modeling. Prediction approach. Model-assisted approach. Bayesian methods. Spatial smoothing. Sampling on successive occasions: Panel rotations. Non-response and not-at-homes.

5.1 INTRODUCTION

The crucial message we have thrashed out for the benefit of the readers that in the finite population set-up we are concerned with a best strategy or a best estimator for a finite population total, given a sampling design, does not in fact exist. As a way out the finite population vector $\underline{Y} = (y_1, \cdots, y_i \cdots, y_N)$ is supposed to be a random vector. Its probability distribution defines a population called the Super-Population relative to the finite survey population $U = (1, \cdots, i, \cdots, N)$ of the labels identifying the units or individuals of the population. This distribution is not needed to be strictly specific. A class of distributions is enough, and this class defines a model. Expectation, variance, covariance, etc., are defined for such classes. We may then be satisfied on controlling model-expectation of sampling design-based variances of estimators in search of optimal strategies or optimal estimators, given specific sampling designs.

A competing approach is to look for predictors of finite population parameters regarding the latter as random variables whose expectations may be estimated. This approach starts with sample values given at hand and needs to estimate the conditional expectation of the totals of the un-sampled values given the values for the sampled units. Unfortunately, super-population modeling yields optimal strategies that involve unknowable parameters. The Prediction approach also needs prying into the unknown characteristics of the populations. Hence, both approaches are doomed to failure unless aided by face-saving devices. Asymptotics provide a much-sought-after relief. A Bayesian approach is more straightforward but is not greatly successful in survey sampling. The most fruitful approach seems to be the model-assisted approach where a simple model is postulated suggesting either a model-design-based estimator or a predictor which owing to inapplicability because of unknown model parameters in it is modified into simpler forms bereft of the unknowable parameters turns out robust and amenable to design-based analyses.

When a population is to be surveyed more than once, rotating panels retaining some sampled elements but replacing others by new entrants and dropping others is often needed to study changes over time. To examine the observational errors in surveys is another topic to engage an investigator's attention. To tackle non-response intentional or otherwise is another requirement. Here use of models often proves effective. In examining changes in populations across time there are two approaches. In one, surveys are conducted from time to time keeping some units common and allowing others to change and study the changes in values. In the other, over time points different estimators are developed making use of modeled inter-relationships among them over time in effecting improvements in estimation of the current and future occasions making use of the past. The Kalman filter technique in Time Series based on a Bayesian approach is elegantly applicable in sampling on successive occasions.In negotiating with non-responses there are two approaches: (1) Weighting adjustments applied when for a sampled person no response at all is available and (2) Imputation Techniques applicable when data on sampled persons are gathered on some of the items but not on certain others.

5.2 SUPER-POPULATION MODELING

$U = (1, \cdots, i, \cdots, N)$ is a finite population of N elements.

y, a real variable, y_i is its value on i in U.
$\underline{Y} = (y_1, \cdots, y_i, \cdots, y_N)$ is the vector of y_i's, i $\in U$.

Since for $Y = \sum_1^N y_i$ and $\bar{Y} = \frac{Y}{N}$ there do not exist UMV unbiased estimators for useful sampling designs, and the other criteria for discrimination among competing estimators like completeness, admissibility, sufficiency, minimal sufficiency, etc., do not prove effectively selective in terms of the design-based approach of inference in survey sampling which is a Fixed Population approach, the following Model-based approach is of necessity introduced. Most importantly, $\underline{Y}$ is now treated as a random vector each $y_i, i \in U$ being thus a random variable. The probability distribution of $\underline{Y}$ defines then a Population. This population is regarded as a Super-Population. The probability distribution of $\underline{Y}$, however, is not required to be too specific. It is enough to talk about a class of probability distributions for which low-order moments like mean, variance, covariance, etc., are only required to exist as finite quantities. Such a class is called a Model. Indicating such a class is called Modeling or Super-population modeling.

Let $t = t(s, \underline{Y})$ denote a Design-unbiased estimator for Y so that

$$Y = E_p(t) = \sum_s p(s) t(s, \underline{Y}).$$

Let $\underline{Y} = (y_1, \cdots, y_i, \cdots, y_N)$ be so modeled that

i. y_i's are independently distributed random variables such that
ii. the model-based means $\mu_i = E_m(y_i)$ are finite for every i in

$$U = (1, \cdots, N), \mu = \sum_1^N \mu_i.$$

Modeling

iii. The finite variances are $\sigma_i^2 = E_m \left(y_i - \mu_i\right)^2 > 0 \ \forall \ i \in U$.

Let the model-based expectation and variance operators E_m, V_m be such that the design-based expectation, variance operators are such that

$$E_m E_p = E_p E_m = E,$$

the overall expectation operator and the overall variance operator V be such that

$$V = E_p V_m + V_p E_m = E_m V_p + V_m E_p.$$

Thus, E_p commutes with E_m. Then the following will follow:

Let
$$t_{HT} = \sum_{i \in s} \frac{y_i}{\pi_i} = t_{HT}\left(s, \underline{Y}\right).$$

Then,
$$E_p\left(t_{HT}\right) = Y.$$

Let
$$t = t_{HT} + h, \text{ i.e., } t\left(s, \underline{Y}\right) = t_{HT}\left(s, \underline{Y}\right) + h\left(s, \underline{Y}\right).$$

Then,
$$0 = E_p\left(h\right) = \sum_s p\left(s\right) h\left(s, \underline{Y}\right)$$
$$= \sum_{s \ni i} p\left(s\right) h\left(s, \underline{Y}\right) + \sum_{s \not\ni i} p\left(s\right) h\left(s, \underline{Y}\right),$$

i.e.,
$$\sum_{s \ni i} p\left(s\right) h\left(s, \underline{Y}\right) = -\sum_{s \not\ni i} p\left(s\right) h\left(s, \underline{Y}\right).$$

Then, $E_m V_p\left(t\right) = E_m E_p\left(t - Y\right)^2$
$$= E_m E_p \left[\left(t - E_m\left(t\right)\right) + \left(E_m\left(t\right) - E_m Y\right) - \left(Y - E_m Y\right)\right]^2$$
$$= E_p \left[\ E_m\left(t - E_m\left(t\right)\right)^2 + \left(E_m\left(t\right) - E_m Y\right)^2 \right.$$
$$+ E_m\left(Y - E_m Y\right)^2$$
$$+ 2\left(E_m\left(t\right) - E_m Y\right) E_m\left(t - E_m\left(t\right)\right)$$
$$- 2 E_m\left(Y - E_m Y\right)\left(t - E_m\left(t\right)\right)$$
$$\left. - 2\left(E_m\left(t\right) - E_m Y\right) E_m\left(Y - E_m Y\right)\ \right]$$
$$= E_p V_m\left(t\right) + E_p \Delta_m^2\left(t\right) + V_m\left(Y\right)$$
$$- 2 E_m\left(Y - E_m Y\right) \left(E_p\left(t\right) - E_m E_p\left(t\right)\right)$$
$$= E_p V_m\left(t\right) + E_p \Delta_m^2\left(t\right) - V_m\left(Y\right)$$

$$\Delta_m(t) = E_m(t - E_m(t)).$$

Now,
$$E_p V_m(t) = E_p V_m(t_{HT} + h)$$
$$= E_p V_m(t_{HT}) + E_p V_m(h) + 2 E_p C_m(t_{HT}, h)$$

writing C_m as the model-based covariance-operator.

Now,
$$E_p C_m(t_{HT}, h) = E_p E_m \left[\left(\sum_{i \in s} \frac{y_i - \mu_i}{\pi_i} \right) h(s, \underline{Y}) \right]$$
$$= E_m \left[\sum_{i=1}^{N} \left(\frac{y_i - \mu_i}{\pi_i} \right) \sum_{s \ni i} p(s) h(s, \underline{Y}) \right]$$
$$= E_m \sum_{1}^{N} \left(\frac{y_i - \mu_i}{\pi_i} \right) \left[- \sum_{s \not\ni i} p(s) h(s, \underline{Y}) \right]$$
$$= - \sum_{1}^{N} \left[E_m(y_i - \mu_i) \frac{1}{\pi_i} \right] \sum_{s \not\ni i} p(s) h(s, \underline{Y}) = 0$$

because $\sum_{s \not\ni i} p(s) h(s, \underline{Y})$, does not contain y_i but may contain y_j's ($j \neq i$) that are independent of y_i.

So,
$$E_m V_p(t) = E_p V_m(t_{HT}) + E_p V_m(h) + E_p \Delta_m^2(t) - V_m(Y)$$
$$= E_p \sum_{i \in s} \frac{\sigma_i^2}{\pi_i^2} + E_p V_m(h) + E_p \Delta_m^2(t) - \sum_{1}^{N} \sigma_i^2$$
$$= \sum_{1}^{N} \Delta_i^2 \left(\frac{1}{\pi_i} - 1 \right) + E_p V_m(h) + E_p \Delta_m^2(t).$$

Let a particular t be
$$t_\mu = \sum_{i \in s} \frac{y_i - \mu_i}{\pi_i} + \mu$$
$$= t_{HT}(s, Y) + h_\mu(s, \underline{Y}), \text{ writing } h_\mu(s, \underline{Y}) = -\sum_{i \in s} \frac{\mu_i}{\pi_i} + \mu$$
$$= h_\mu.$$

Then
$$V_m(h_\mu) = 0, \quad \Delta_m(t_\mu) = E_m(t_\mu) - \mu = 0.$$

So,
$$E_m V_p(t_\mu) = \sum_{1}^{N} \sigma_i^2 \left(\frac{1}{\pi_i} - 1 \right).$$

So,
$$E_m V_p(t) \geq E_m V_p(t_\mu) = \sum_{1}^{N} \sigma_i^2 \left(\frac{1}{\pi_i} - 1 \right).$$

Modeling

So, among all p unbiased t's, the optimal one is

$$t_\mu = \sum_{i \in s} \frac{y_i - \mu_i}{\pi_i} + \mu.$$

However, μ_i is usually unknown. So t_μ is not usable. In Section 5.4 we shall discuss this issue.

Now, Cauchy inequality gives us

$$\left(\sum_1^N \pi_i\right)\left(\sum_1^N \frac{\sigma_i^2}{\pi_i}\right) \geq \left(\sum \sigma_i\right)^2 \quad (5.1)$$

Letting $\sum_1^N \pi_i = E_p v(s) = v$ and $\sum_1^N \sigma_i = \sigma$, it follows that

$$E_m V_p(t) \geq \frac{\sigma^2}{v} - \sum \sigma_i^2$$

if $v(s) = n \; \forall \; s$ with $p(s) > 0$, then (5.1) reduces to an equality if

$$\frac{\sigma_i}{\sqrt{\pi_i}} \propto \sqrt{\pi_i} \text{ or } \pi_i \propto \sigma_i \text{ giving } \pi_i = \frac{n\sigma_i}{\sigma}.$$

If $\mu_i = \beta x_i$, with β unknown but x_i positive and known and $\sigma_i^2 \propto x_i^2$, then optimal π_i is given by

$$\pi_i = \frac{nx_i}{X}. \quad (5.2)$$

A design subject to (5.2) is called a πps or IPPS design. In this case $\pi_i \propto \mu_i \propto \sigma_i$.

Godambe and Thompson (1977) is the principal reference for the work in this section.

5.3 PREDICTION APPROACH

Brewer (1963) and Royall (1970) pioneer the third inferential approach in the field of Survey Sampling besides the two covered already, namely the Classical-Design-based and the Model-Design-based super-population-modeling approaches.

When on the finite survey population $U = (1, \cdots i, \cdots N)$ the Vector $\underline{Y} = (y_i, \cdots, y_i, \cdots, y_N)$ of real numbers y_i, $i \in U$ is defined and a sample s from U is surveyed gathering the survey data

$$d = (s, y_i | i \in s) = ((i_1, y_{i1}) \cdots, (i_n, y_{in})),$$

say, with $s = (i_1, \cdots, i_n)$, at hand with the objective of estimating $Y = \sum_1^N y_i$, the following idea may be conceived following Brewer (1963) and Royall (1970).

Letting
$$Y = \sum_{i=1}^{N} y_i = \sum_{i \in s} y_i + \sum_{i \notin s} y_i,$$

the portion $\sum_{i \in s} y_i$ of Y being already ascertained, a contemplated estimator

$$t = t(s, \underline{Y}) = \left(\sum_{i \in s} y_i \right) + \left(t - \sum_{i \in s} y_i \right)$$

may be adjudged to come close to Y only if $\left(t - \sum_{i \in s}\right)$ turns up near $\sum_{i \notin s} y_i$. But $t - \sum_{i \in s} y_i = t(s, \underline{Y}) - \sum_{i \in s} y_i$, by definition of an estimator for Y involves only y_i for $i \in s$, while $\sum_{i \notin s} y_i$ involves no such y_i, for $i \in s$. So, logically $\left(t - \sum_{i \in s} y_i\right)$ cannot be claimed to be close to $\sum_{i \notin s} y_i$, unless we postulate that the entries in $\underline{Y} = (y_i, \cdots, y_N)$ are mutually inter-related.

A convenient way to deal with $\underline{Y}$ with its entries inter-related is to postulate that $\underline{Y}$ is a random Vector suitably distributed. Consequently, $Y = \sum_{i}^{N} y_i$ is also a random variable and hence it cannot be estimated. One may only predict its value conditionally on the observable values of the sampled co-ordinates $y_i, i \in s$ given as surveyed. This is the essence of the Prediction approach in survey sampling, as in the concept of the super-population modeling here also the probability distribution of $\underline{Y}$ need not be too pin-pointed. It is enough to spell out for $\underline{Y}$ a class of probability distributions as a Model. Let us suppose (i) the entries of $\underline{Y}$ are independently distributed, (ii) the model-based moments namely the means $\mu_i = E_m(y_i)$ exist as certain finite numbers, and (iii) the variances $E_m(y_i - \mu_i)^2 = \sigma_i^2 (> 0 \forall i \in U)$ are each finite. We may lay down further stipulations as and when needed.

In particular, let $\mu_i = \beta x_i$, with β as an unknown constant but x_i's be known finite numbers with $X = \sum_{1}^{N} x_i$. Then, to start with, let us need $E_m(t) = E_m(Y) \forall s$, with $p(s) > 0$, called the Condition of Model-unbiasedness of the predictor t for Y.

Thus, for
$$t = \sum_{i \in s} y_i + \left(t - \sum_{i \in s} y_i \right)$$

we need
$$E_m \left[\left(t - \sum_{i \in s} y_i \right) + \sum_{i \in s} y_i \bigg| (s, y_i, i \in s) \right]$$
$$= E_m[Y | s, y_i, i \in s] = \sum_{i \in s} y_i + \beta \sum_{i \notin s} x_i$$

justifying the choice
$$t = \sum_{i \in s} y_i + \hat{\beta} \sum_{i \notin s} x_i$$

Modeling

with $\hat{\beta}$ as a function of $s, y_i | i \in s$ subject to

$$E_m\left(\hat{\beta}\right) = \beta \ \forall \ (s, y_i | i \in s).$$

Now let us impose the simplifying linearity restriction on $\hat{\beta}$, namely,

$$\hat{\beta} = \sum_{i \in s} l_i y_i$$

demanding the choice of l_i's subject to

$$E_m\left(\sum_{i \in s} l_i y_i\right) = \beta \ \forall \ s, p(s) > 0$$

or

$$\beta \sum_{i \in s} l_i x_i = \beta \ \forall \ s, p(s) > 0.$$

Thus, l_i's are subject to

$$\sum_{i \in s} l_i x_i = 1 \ \forall \ s, p(s) > 0. \tag{5.3}$$

Then for an appropriate choice of l_i's and hence $\hat{\beta} = \sum_{i \in s} l_i y_i$ let us demand $E_m\left[(t-Y)^2 | s, y_i, i \in s, p(s) > 0\right]$ be the minimum for the choice of l_i's subject to (5.3). Thus, we need

$$t_m\left[(t-Y)^2 | s, y_i, i \in s, p(s) > 0\right]$$

$$= E_m\left[\left(\sum_{i \in s} y_i\right) + \hat{\beta} \sum_{i \notin s} x_i - \sum_{i \in s} y_i - \sum_{i \notin s} y_i\right]^2$$

$$= E_m\left[\left(\hat{\beta} - \beta\right) \sum_{i \notin s} x_i - \sum_{i \notin s} (y_i - \beta x_i)\right]^2$$

$$= \left(\sum_{i \notin s} x_i\right)^2 E_m\left(\hat{\beta} - \beta\right)^2 + \sum_{i \notin s} \sigma_i^2$$

because $\hat{\beta} = \sum_{i \in s} l_i y_i$ is independent of all y_i's for $i \notin s$.

Now we need to minimize $E_m\left(\hat{\beta} - \beta\right)^2 = E_m\left[\sum_{i \in s} l_i (y_i - \beta x_i)\right]^2$ with l_i's subject to (5.3) is $\sum_{i \in s} l_i^2 \sigma_i^2$ with respect to l_i's subject to (5.3).

We are to solve the equations

$$\frac{\delta}{\delta l_i}\left[\sum_{i\in s} l_i^2 \sigma_i^2 - \lambda\left(\sum_{i\in s} l_i x_i - 1\right)\right] = 0$$

or
$$2 l_i \sigma_i^2 = \lambda x_i \quad \text{for} \quad i \in s, \tag{5.4}$$

here λ denoting a Lagrangian undetermined multiplier.

Then, (5.4) gives
$$l_i = \frac{\lambda}{2} \frac{x_i}{\sigma_i^2}$$

$$l_i x_i = \frac{\lambda}{2} \frac{x_i^2}{\delta_i^2}, \quad \text{whence}$$

$$1 = \sum_{i\in s} l_i x_i = \frac{\lambda}{2} \sum_{i\in s} \frac{x_i}{\sigma_i^2}, \quad \text{giving}$$

$$\frac{\lambda}{2} = \frac{1}{\sum_{i\in s}\frac{x_i^2}{\sigma_i^2}} \quad \text{or} \quad l_i = \frac{\frac{x_i}{\sigma_i^2}}{\sum_{i\in s}\frac{x_i^2}{\sigma_i^2}}, \quad \text{using (5.4)}.$$

Hence, the optimal predictor for Y is

$$t_{opt} = \sum_{i\in s} y_i + \left(\sum_{i\notin s} x_i\right) \frac{\sum_{i\in s}\frac{y_i x_i}{\sigma_i^2}}{\sum_{i\in s}\frac{x_i^2}{\sigma_i^2}}.$$

In particular, if we impose the simplifying assumption that $\sigma_i^2 = \sigma^2 x_i \,\forall\, i$, then, t_{opt} reduces to the Ratio predictor

$$t_R = \sum_{i\in s} y_i + \left(\frac{\sum_{i\in s} y_i}{\sum_{i\in s} x_i}\right)\left(\sum_{i\notin s} x_i\right) = X\frac{\sum_{i\in s} y_i}{\sum_{i\in s} x_i}$$

$$= X\frac{\bar{y}}{\bar{x}}, \text{ writing } \bar{y}, \bar{x} \text{ for the sample means of } y- \text{ and } x\text{-values.}$$

Clearly, $E_m(t_R) = \beta X = E_m(Y)$.
Thus, t_R is a model-unbiased predictor of Y.

Also,
$$E_m(t_R - Y)^2 = E_m\left[\frac{X}{\bar{x}}\left(\beta\bar{x} + \frac{1}{n}\sum_{i\in s}\epsilon_i\right) - \beta X - \sum_1^N \epsilon_i\right]^2$$

writing $y_i = \beta x_i + \epsilon_i, i \in U$

under the postulated model. Of course, $E_m(\epsilon_i) = 0, V_m(\epsilon_i) = \sigma_i^2 = \sigma^2 x_i$ with $\sigma\,(>0)$ and β as unknowns, x_i's known, and ϵ_i's as independent random variables, $i \in U$.

Modeling

So, $E_m (t_{opt} - Y)^2 = V_m (t_{opt} - Y)$ because $E_m (t_{opt} - Y) = 0$ is a measure of error of t_{opt} about Y.

Now, $E_m (t_{opt} - Y)^2 = E_m \left(X \dfrac{\bar{y}}{\bar{x}} - \sum_1^N y_i \right)^2$,

$$= E_m \left[\dfrac{X}{\bar{x}} \left(\beta \bar{x} + \dfrac{\sum_{i \in s} \in_i}{n} \right) - \left(\beta X + \sum_1^N \in_i \right) \right]^2$$

$$= E_m \left[\dfrac{X}{\sum_{i \in s} x_i} \sum_{i \in s} \in_i - \sum_{i \in s} \in_i - \sum_{i \notin s} \in_i \right]^2$$

$$= E_m \left[\left(\sum_{i \in s} \in_i \right) \left(\dfrac{X}{\sum_{i \in s} x_i} - 1 \right) - \sum_{i \notin s} \in_i \right]^2$$

$$= \left(\dfrac{X - \sum_{i \in s} x_i}{\sum_{i \in s} x_i} \right)^2 \sigma^2 \left(\sum_{i \in s} x_i \right) + \sigma^2 \sum_{i \notin s} x_i$$

$$= \left(\sum_{i \notin s} x_i \right)^2 \dfrac{\sigma^2}{\sum_{i \in s} x_i} + \sigma^2 \sum_{i \notin s} x_i$$

$$= \sigma^2 \left(\sum_{i \notin s} x_i \right) \left[\dfrac{\sum_{i \notin s} x_i}{\sum_{i \in s} x_i} + 1 \right] = \sigma^2 \left(\sum_{i \notin s} x_i \right) \dfrac{X}{\sum_{i \in s} x_i}$$

$$= \sigma^2 \dfrac{N\bar{X}}{n\bar{x}} (N - n) \bar{X}_c, \text{ where } \bar{X}_c = \dfrac{1}{N - n} \left(\sum_{i \notin s} x_i \right)$$

$$= \sigma^2 \dfrac{N^2 (1 - f)}{n} \dfrac{\bar{X} \bar{X}_c}{\bar{x}}, \text{ writing } f = \dfrac{n}{N}.$$

$= MSE (t_{opt})$, the Mean Square Error of t_{opt} about Y.

In order to derive an unbiased estimator of $MSE (t_{opt})$ let us note the following.

Let
$$e_i = y_i - \hat{\beta} x_i = y_i - x_i \sum_{i \in s} l_i y_i$$

$$= y_i - x_i \dfrac{\sum_{i \in s} y_i x_i / \sigma_i^2}{\sum_{i \in s} x_i^2 / \sigma_i^2} = y_i - x_i \dfrac{\bar{y}}{\bar{x}}.$$

Then, consider

$$E_m \sum_{i \in s} \frac{e_i^2}{x_i} = E_m \left[\sum_{i \in s} \frac{1}{x_i} \left(y_i - x_i \frac{\bar{y}}{\bar{x}} \right)^2 \right]$$

$$= E_m \left[\sum_{i \in s} \frac{1}{x_i} \left(\beta x_i + e_i - \beta x_i - x_i \frac{\sum_{i \in s} e_i}{\sum_{i \in s} x_i} \right)^2 \right]$$

$$= \sum_{i \in s} \frac{1}{x_i} \left[\sigma^2 x_i + x_i^2 \frac{\sigma^2}{\sum_{i \in s} x_i} - 2 x_i \sigma^2 \right]$$

$$= (n-1) \sigma^2.$$

An unbiased estimator for σ^2 is

$$\hat{\sigma}^2 = \frac{1}{n-1} \sum_{i \in s} \frac{e_i^2}{x_i} = \frac{1}{n-1} \sum_{i \in s} \frac{1}{x_i} \left(y_i - \frac{\bar{y}}{\bar{x}} x_i \right)^2.$$

An unbiased estimator for $MSE(t_{opt})$ is

$$m(t_{opt}) = \frac{N^2 (1-f)}{n} \frac{\bar{X} \bar{X}_c}{\bar{x}} \hat{\sigma}^2.$$

5.4 MODEL-ASSISTED APPROACH

The Design-Model-based Super-population modeling approach gives us the optimal estimator for Y as

$$t_\mu = \sum_{i \in s} \frac{y_i - \mu_i}{\pi_i} + \mu$$

based on a suitable sampling design. But μ_i is usually unknown, and consequently, t_μ cannot be employed to estimate Y. Hence, it cannot be genuinely called an estimator at all.

In a specific situation one may write $\mu_i = \beta x_i$ with β unknown but x_i known and positive for every i in U giving t_μ as

$$t_\beta = \sum_{i \in s} \frac{y_i}{\pi_i} + \beta \left(X - \sum_{i \in s} \frac{x_i}{\pi_i} \right).$$

A postulated model brings us just here. To derive a genuine estimator out of this t_β we have now to employ an estimator for β as say

$$b_Q = \frac{\sum_{i \in s} y_i x_i Q_i}{\sum_{i \in s} x_i^2 Q_i}$$

with a suitable choice of Q_i as a positive constant free of the elements of $\underline{Y}$. This is because β should be recognized as a coefficient of regression of the variable y on the variable x. The model only assists us in arriving at the form of

$$t_g = \sum_{i \in s} \frac{y_i}{\pi_i} + b_Q \left(X - \sum_{i \in s} \frac{x_i}{\pi_i} \right),$$

called the Generalized regression or the Greg estimator or also the Greg predictor for Y, because (i) forgetting the model we may treat Y as a constant and t_g as an estimator for it or (ii) recognizing the model treating $\underline{Y}$ as a random vector and Y as a random variable, t_g may be treated as a model-based predictor for Y. But at this stage it is important to study the properties of t_g from the consideration of sampling design only now onwards. Hence, the approach to be followed now is recognized as a Model-Assisted approach.

Since t_g involves the ratio of two random variables according to the probability distribution p of s and also its product by another such random variable $\sum_{i \in s} \frac{x_i}{\pi_i}$, it is difficult to study the design-based properties of t_g according to its exact probability distribution in terms of p. But as a saving grace we may appeal to the asymptotic distribution of t_g following the approach of Brewer (1979). Brewer considers an infinite sequence of finite populations and the corresponding sequences of vectors of real values defined on them and of the infinite sequences of probability distributions starting respectively with the entities like $U, \underline{Y}$, and p. Then he considers limiting values of the expectations of various functions based on the sequences of samples from the sequences of finite populations and the vector-valued real numbers. Then, applying Slutzky's [cf. Cramer (1946)] theorem concerning rational functions of random variables like t_g, he could define concepts of Asymptotic Design-Unbiasedness (ADU) and Asymptotic Design Consistency. Operationally, he could calculate asymptotic design-expectations of ratios and products of s-based random variables as ratios and products of the corresponding expectations of those variables at ease.

Thus, honoring Brewer's (1979) theoretical approach, we may calculate the Asymptotic Design expectation of t_g as

$$\lim E_p(t_g) = E_p \left(\sum_{i \in s} \frac{y_i}{\pi_i} \right) + \frac{\lim E_p \left(\sum_{i \in s} \frac{y_i x_i Q_i \pi_i}{\pi_i} \right)}{\lim E_p \left(\sum_{i \in s} \frac{x_i^2 Q_i \pi_i}{\pi_i} \right)} \left(X - E_p \left(\sum_{i \in s} \frac{x_i}{\pi_i} \right) \right) = Y.$$

Thus, t_g is an asymptotically design unbiased estimator for Y, whatever $Q_i \, (> 0)$ may be chosen. In order to study further a few design-based properties of t_g, let us proceed as follows.

Let us note that
$$\lim E_p \left(\sum_{i \in s} y_i x_i Q_i \right) = \sum_{1}^{N} y_i x_i Q_i \pi_i$$

and
$$\lim E_p \left(\sum_{i \in s} x_i^2 Q_i \right) = \sum_{1}^{N} x_i^2 Q_i \pi_i.$$

Then, let us write
$$t_G = \sum_{i \in s} \frac{y_i}{\pi_i} + B_Q \left(X - \sum_{i \in s} \frac{x_i}{\pi_i} \right)$$

writing
$$B_Q = \frac{\sum_1^N y_i x_i Q_i \pi_i}{\sum_1^N x_i^2 Q_i \pi_i} = \lim E_p b_Q.$$

Then, let further
$$t_G = \sum_{i \in s} \frac{y_i - B_Q x_i}{\pi_i} + B_Q X$$
$$= \sum_{i \in s} \frac{E_i}{\pi_i} + B_Q X,$$

writing
$$E_i = (y_i - B_Q x_i), i \in U.$$

Approximating t_g by t_G, assuming a large sample-size, one may take

$$V_p(t_g) \simeq V_p(t_G) = V_p \left(\sum_{i \in s} \frac{E_i}{\pi_i} + B_Q X \right) = V_p \left(\sum_{i \in s} \frac{E_i}{\pi_i} \right)$$
$$= \sum_1^N E_i^2 \frac{1 - \pi_i}{\pi_i} + \sum \sum_{i \neq j} E_i E_j \frac{\pi_{ij} - \pi_i \pi_j}{\pi_i \pi_j}$$
$$= \sum \sum_{i < j} (\pi_i \pi_j - \pi_{ij}) \left(\frac{E_i}{\pi_i} - \frac{E_j}{\pi_j} \right)^2 + \sum \alpha_i \frac{E_i^2}{\pi_i}$$

with
$$\alpha_i = 1 + \frac{1}{\pi_i} \sum_{j \neq i}^N \pi_{ij} - \sum_1^N \pi_i.$$

Writing $e_i = y_i - b_Q x_i$ as an estimator for $V_p(t_g)$
one may take
$$m(t_g) = \sum_{i \in s} e_i^2 \frac{1 - \pi_i}{\pi_i} \frac{I_{si}}{\pi_i} + \sum \sum_{i \neq j \in s} e_i e_j (\pi_{ij} - \pi_i \pi_j) \frac{I_{sij}}{\pi_{ij}}$$

provided $\pi_{ij} > 0 \ \forall \ i \neq j,$ and one may also employ
$$m(t_g) = \sum_{i < j \in s} (\pi_i \pi_j - \pi_{ij}) \left(\frac{e_i}{\pi_i} - \frac{e_j}{\pi_j} \right)^2 + \sum_{i \in s} \frac{e_i^2 I_{si}}{\pi_i \pi_i}.$$

Alternatively, writing
$$g_{si} = 1 + \left(X - \sum_{i \in s} \frac{x_i}{\pi_i} \right) \frac{x_i Q_i \pi_i}{\sum_{i \in s} x_i^2 Q_i},$$

we get $t_g = \sum_{i \in s} \frac{y_i}{\pi_i} g_{si}$, alternative approximate estimators of $V_p(t_g)$ may be written down as $m'(t_g) = m(t_g) \Big|_{e_i = g_{si} e_i}.$

Modeling

At this stage we may illustrate suitable choices of $Q_i, i \in U$. We have $y_i = \beta x_i + \epsilon_i, V_m(\epsilon_i) = \sigma_i^2, i \in U$. On taking $\sigma_i^2 = \sigma^2 x_i$ or $\sigma^2 x_i^2, \sigma^2 x_i^g U$, etc., it seems reasonable to take $Q_i \propto \frac{1}{\sigma_i^2}$, and hence $Q_i = \frac{1}{x_i}, \frac{1}{x_i^2}, \frac{1}{x_i^g}$, etc. Two more choices of Q_i we shall presently illustrate.

The prediction approach has given us an optimal linear predictor for Y as

$$t_\beta = \sum_{i \in s} y_i + \frac{\sum_{i \in s} y_i x_i / \sigma_i^2}{\sum_{i \in s} x_i^2 / \sigma_i^2} \left(\sum_{i \notin s} x_i \right).$$

But as σ_i^2 is unknown, Brewer (1979) suggests replacing $\frac{1}{\sigma_i^2}$ in t_β by w_i, taking w_i's as certain assignable design-weights so as to ensure

$$t_w = \sum_{i \in s} y_i + \left(\frac{\sum_{i \in s} y_i x_i w_i}{\sum_{i \in s} x_i^2 w_i} \right) \sum_{i \notin s} x_i$$

to have the Asymptotic Design-unbiased property, namely, $\lim E_p(t_w) = Y$. This leads to

$$\lim E_p(t_w) = \sum_1^N y_i \pi_i + \frac{\sum_1^N y_i x_i w_i \pi_i}{\sum_1^N x_i^2 w_i \pi_i} \left(X - \sum_1^N x_i \pi_i \right).$$

Then, on taking $w_i = \frac{1-\pi_i}{\pi_i x_i}$, Brewer (1979) checks this to match Y and hence is derived the Brewer's (1979) ADU predictor for Y as

$$t_B = \sum_{i \in s} y_i + \frac{\sum_{i \in s} y_i \frac{1-\pi_i}{\pi_i}}{\sum_{i \in s} x_i \frac{1-\pi_i}{\pi_i}} \left(\sum_{i \notin s} x_i \right).$$

It is a simple exercise to check algebraically that this t_B matches a Greg predictor on taking $Q_i = \frac{1-\pi_i}{\pi_i x_i}$ in the latter. Again on choosing $Q_i = \frac{1}{\pi_i x_i}$ in t_g one may derive the Greg predictor t_g for Y as a general ratio predictor:

$$t_R = X \left(\sum_{i \in s} \frac{y_i}{\pi_i} \Big/ \sum_{i \in s} \frac{x_i}{\pi_i} \right).$$

5.5 BAYESIAN METHODS

We have already seen that in the context of survey sampling for the population $U = (1, \cdots, i \cdots, N)$ on which is defined a Vector $\underline{Y} = (y_1, \cdots, y_i, \cdots, y_N)$ with y_i as the value of the ith member of U on a real variable, on taking a sample $s = (i_1, \cdots, i_n)$ as an ordered sequence from U and surveying to generate the sampled observations $(y_{i1}, \cdots, y_{in})$, data at hand are $d = ((i_1, y_{i1}), \cdots, (i_n, y_{in})) = (s, \underline{y})$, say, the Probability of observing d is $P_{\underline{Y}}(d) = p(s) I_{\underline{Y}}(d)$.

Here p is supposed to be a Non-Informative sampling design and $I_{\underline{Y}}(d) = 1/0$ according to $\underline{Y} \in \Omega_d$ or $\underline{Y} \notin \Omega_d$, respectively. Here Ω_d is the part of the parametric space $\Omega = \{\underline{Y} | -\alpha < a_i \leq y_i < b_i < +\alpha, i \in U\}$ that is consistent with d, i.e.,

$$\Omega_d = \{\underline{Y} | \underline{Y} \in \Omega \text{ and } y_i\text{'s for } i \in s \text{ as are observed in } d\}.$$

It is customary to regard $P_{\underline{Y}}(d)$ formally as the same as the likelihood of $\underline{Y}$ given the data d and write it as

$$L(\underline{Y}|d) = P_{\underline{Y}}(d) = p(s) I_{\underline{Y}}(d).$$

So, $\qquad L(\underline{Y}|d) = p(s) \quad \text{for every} \quad \underline{Y} \in \Omega_d$

and $\qquad L(\underline{Y}|d) = o \quad \text{for every} \quad \underline{Y} \notin \Omega_d.$

Thus, for a Non-Informative design the likelihood is a mere constant—it is non-zero when a $\underline{Y}$ is supported by the data at hand and is zero for every other $\underline{Y}$ which is inconsistent with the data. Out of $\underline{Y}$ the elements y_i for $i \in s$ are as observed and the likelihood says nothing about the y's for the unsampled units of U. So, the likelihood has no discriminating capabilities concerning diverse vector points $\underline{Y}$ in Ω to provide differential supports when the data d are at hand. Godambe (1966) was the first researcher to point out this sterility of the Likelihood in Survey sampling. Basu (1969), however, was the first researcher to be delighted with this flatness in the likelihood and aridity in its inference-yielding potentials. Basu pointed out that introducing an operationally effective "prior" for $\underline{Y}$ and combining that with this simple likelihood to produce a posterior for $\underline{Y}$ given the data d, the likelihood, and the prior so that on choosing a simple loss function, e.g., the square error loss function, a suitable inference may be accomplished for a parametric function $\theta(\underline{Y})$, say. To view this formally, let $q(\underline{Y})$ be taken as a prior for $\underline{Y}$ leading to the posterior

$$q*(\underline{Y}|d) = \frac{q(\underline{Y}) L_d(\underline{Y})}{\int_\Omega q(\underline{Y}) L_d(\underline{Y}) d\mu}.$$

Here the integration is with respect to a measure μ for which $q(.)$ is the density.

Since we are restricting to non-informative designs only, the likelihood involves no unobserved y's, it follows that the posterior simplifies to

$$q^*(\underline{Y}|d) = \frac{q(\underline{Y})}{\int_\Omega q(\underline{Y}) d\mu} = q(\underline{Y}) t(d), \text{ say, for } \underline{Y} \in \Omega_d$$

but $\qquad q^*(\underline{Y}|d) = 0 \ \forall \ \underline{Y} \notin \Omega_d.$

This posterior initiated by Godambe (1966, 1969) was also studied by Basu (1969). Obviously it does not involve a sampling design. With Bayesian inference which is yielded by a posterior alone, it does not matter how the sample observations are gathered though the strength or the inference depends of course on the quality of the data which form its basis. How they are gathered does not influence the inference so long as it is realized from the posterior.

If we intend to estimate the finite population total Y allowing a square error loss, then an estimator $f(d)$ that minimizes the posterior risk $E_{q^*}[Y - f(d)]^2$ is just the posterior expectation of Y given d, namely $E_{q^*}(Y|d)$.

As long as an analytically simple prior may be employed, this solution is very easy to derive. This $E_{q^*}(Y|d)$ is called the Bayes estimator for Y. This estimator also minimizes the prior risk which is

$$E_{q(d)} E_{q^*} \left[Y - E_{q^*}(Y|d) \right]^2$$

writing $q(d)$ as the marginal prior for y's in d. This also suggests how a sampling design p is to be employed in Bayes estimation. This is because $E_p \left[E_{q(d)} E_{q^*}(Y - E_{q^*}(Y|d)) \right]^2$ is required to be minimized. If there exists a sample s_o, say such that for the data

$$d_o = (s_o, \underline{y}, i \in s_o), \quad E_{q(d)} E_{q^*}(Y - E_{q^*}(Y|d))^2$$

is the minimum over the variation in d, then p should be such that

$$p(s_o) = 1 \quad \text{and} \quad p(s) = 0 \ \forall \ s \neq s_o.$$

So, a Bayesian approach may also yield an optimal sampling design which obviously as above is a purposive design. The above analysis though apparently simple, nevertheless is quite impracticable. Bayesian methods as a matter of fact have not yet turned out quite effective or popular except in specific situations like Small Area Estimation, a topic we shall take up in Chapter 7.

Ericson (1969) is an eminent votary of the efficacy of Bayesian inference in finite population survey sampling. He recommends postulating exchangeable priors to be combined with the sterile likelihood in survey sampling to derive appropriate posterior to facilitate Bayes estimation for finite population totals and means. An exchangeable prior for a vector of variable co-ordinates is one which is invariant over the permutation of these co-ordinates. Ericson (1969) hereby supports the use of SRSWOR in his Bayesian approach in Survey sampling. Scott (1977) considered a situation when a vector $\underline{X} = (x_{ij}, \cdots, x_i \cdots, x_N)$ of real numbers is known fully or at least partially, allowing him to contend with a likelihood $L_d(\underline{Y}|\underline{X})$, postulate a prior $q(\underline{Y}|\underline{X})$, both involving $\underline{X}$ so as to derive a posterior

$$q^*_d(\underline{Y}|\underline{X}) = \frac{L_d(\underline{Y}|\underline{X}) q(\underline{Y}|\underline{X})}{\int_{\Omega_d} L_d(\underline{Y}|\underline{X}) q(\underline{Y}|\underline{X}) d_\mu(\underline{Y}|\underline{X})} \quad \forall \ \underline{Y} \in \Omega_d.$$

With this posterior Scott(1977) could afford to justify the use of a sampling design involving $\underline{X}$ as size-measures and thereby recommend varying probability sampling design in employing Bayes estimation of the population total Y.

Godambe (1968) introduced the concept of Bayesian Sufficiency in the following way. Let ψ be a class of priors q. If for an estimated parametric function $\Upsilon(\underline{Y})$ the posterior based on data d and every prior q in the class ψ depends on d only through a statistic $t = t(d)$, then t is called a Bayesian Sufficient Statistic for Υ for every prior in ψ. In order to show the importance of this concept Godambe (1968) and Godambe and Thompson (1977) postulated a product-measure prior of the form

$$q(\underline{Y}) = \Pi_{i=1}^{N} q_i(Y_i)$$

taking q_i as a prior for $Y_i, i \in U$, and in addition, restricted to well-defined class of Invariant parametric functions and Invariant statistics and showed that $\left(s, \sum_{i \in s} y_i\right)$ is a Bayes Sufficient Statistic for Y and so also are

$$\frac{N}{n} \sum_{i \in s} (y_i - a_i) + \sum_{1}^{N} a_i \text{ and } \left(\sum_{1}^{N} a_i\right) \left(\sum_{i \in s} y_i \Big/ \sum_{i \in s} a_i\right)$$

for Y with a_i's as known constants for $i \in U$. Chaudhuri (1977) showed that for product-measure priors, among invariant statistics,

$$\left(s, \bar{y} = \frac{1}{n} \sum_{i \in s} y_i, s_y^2 = \frac{1}{(n-1)} \sum_{i \in s} (y_1 - \bar{y})^2\right)$$

is Bayes Sufficient for $S_y^2 = \frac{1}{(N-1)} \sum_{1}^{N} (y_i - \bar{Y})^2$ and also for $(\bar{Y}, S_y^2)$.

Chaudhuri (1977), in addition, adopting Hall, Wijsman, and Ghosh's (1965) concept of Invariantly Sufficient Statistics and Cochran's (1977) idea of Finite Consistency, namely a statistic coinciding with the parameter it seeks to estimate when s coincides with U itself, proved that for (i) S_y^2 and (ii) $\frac{S_y}{\bar{Y}}$ respectively (i)' s_y^2 and (ii)' $\frac{s_y}{\bar{y}}$ are unique Cochran Consistent Invariantly Bayes Sufficient Statistics among origin- and scale-invariant statistics.

Ericson (1969) considering a mixture of product-measure priors

$$q(\underline{Y}) = \int_{\underline{\alpha}} \prod_{1}^{N} q_i(y_i | \underline{\alpha}) \, d\mu(\underline{\alpha})$$

with $\underline{\alpha} = (\alpha_1, \cdots \alpha_N)$ as a vector of real constants and a corresponding posterior of the form

$$q^*(\underline{Y}|d) \propto \int_{\underline{\alpha}} \left[\prod_{i \notin s} q_i(y_i|\underline{\alpha})\right] d\mu(\underline{\alpha}|d'),$$

with d' as the data based on y_i's,$i \notin s$, derived several interesting results, as are quoted in Chaudhuri and Vos's (1988) text. We omit because of complicated theories underlying them as detailed by Ericson (1969, 1970). More Bayesian methods will be narrated in Chapters 6 and 7.

5.6 SPATIAL SMOOTHING

An alternative to the prediction approach we discussed earlier known as Kriging in the context of geostatistics considers how to foretell about something we may find in a new site or on an average in a number of sites in a big territory in terms of our findings already at hand in a number of sites we have covered in our investigations. This is spatial prediction. Let us elaborate.

It is a study of covariance functions or equivalently of variogram. Let y_{t+h} and y_t denote the variables at a site away with a displacement by a distance h, and that at an original site t, their covariance is denoted by $C(h) = Cov(t_{t+h}, y_t)$.

Let at n different sites $t_1, \cdots, t_n$ the sampled values observed be, respectively, $y_1, \cdots, y_n$ and variate value y_o at a new site t_o is to be predicted in terms of them. We follow Thompson (1992) to predict y_o by $\hat{y}_o$ such that

$$E(\hat{y}_o) = E(y_o) \quad \text{and} \quad \hat{y}_o = \sum_1^n a_i y_i$$

with the unknown a_i's to be determined so as to minimize $E(\hat{y}_0 - y_o)^2$.

The solution is obtained using Lagrangian undetermined multiplier m so as to achieve the minization under the constraint specified above as

$$\underline{f} = G^{-1} \underline{h}$$

taking $\underline{h} = \begin{bmatrix} C_{10} \\ \vdots \\ C_{n0} \\ 1 \end{bmatrix}$, $\quad \begin{array}{l} C_{i0} = \text{Cov}(y_i, y_0) \\ C_{ij} = \text{Cov}(y_i, y_j) \\ C_{00} = V(y_0). \end{array}$

$$\underline{f} = \begin{bmatrix} a_1 \\ \vdots \\ a_n \\ m \end{bmatrix} \quad \text{and} \quad G = \begin{bmatrix} C_{11} & C_{12} & \cdots & C_{1n} & 1 \\ \vdots & \vdots & & \vdots & \vdots \\ C_{n1} & C_{n2} & \cdots & C_{nn} & 1 \\ 1 & 1 & \cdots & 1 & 0 \end{bmatrix}$$

as an $(n+1) \times (n+1)$ matrix is assumed to be non-singular. The covariances of y's at sites at distances d apart are estimated by

$$\hat{C}(d) = \frac{1}{n_d} \sum_d (y_{ti} - \bar{y})(y_{tj} - \bar{y})$$

taking $\sum_d$ as the sum over sites at an average distance approximately d apart and the $\bar{y}$ as their mean. The covariance function $C(x)$ is then usually smoothened by fitting the curve

$$C(x) = a \ e^{-bx}$$

with the parameters a and b estimated by least squares analysis.

An alternative approach is to study the Variogram considering

$$\text{Var}(y_{t+h} - y_t) = 2\gamma(h)$$

calling $\gamma(h)$ the Semivariogram. To see the equivalence of the two approaches

$$\gamma(h) = C(o) - C(h)$$

is to be noted. For theory and application, refer to Thompson (1992).

5.7 SAMPLING ON SUCCESSIVE OCCASIONS: PANEL ROTATION

It is a common practice in Survey Sampling that a population needs to be repeatedly surveyed over a considerable period of time, say a few years at a stretch or with occasional gaps and recesses. The population need not remain intact across the time points, but in spite of changes like some units dropping out, some newly arriving, some particularly in business surveys changing their brands or changing their trade items, some getting amalgamated with others, some splitting up, and component parts changing proprietors and the like, essentially the population retains their major characteristics in common. In such situations it may be beneficial to draw samples on successive occasions, retaining some units in common over occasions, dropping some others, and taking a number of fresh units. How to implement such practices in a scientific manner and with what consequences and benefits positive and negative need to be studied in certain detail.

The theory of Sampling on successive occasions has been developing through diverse routes. Let us briefly narrate one by one. Jessen (1942), to our knowledge and belief, was the pioneering researcher in this area. Virtually an infinite population set-up was envisaged, and it was postulated to remain intact in its composition of its constituent units. He took a simple random sample, no matter whether with or without replacement on the first occasion, to be called the previous occasion. On a next occasion called the Current occasion when one sample was taken as a sub-sample at random from the earlier sample, this was to be called a matched sample for which both the previous values are known and the current values are freshly gathered, and in addition another current simple random sample is taken from the population independently of the initial simple random sample. Using the known previous sample total, say, $\sum_{i \in s_1} x_i$, and the previous and current sample values x_i and y_i in the matched sample s_m from s_1 an estimator for the current population total is obtained on applying the method of Double Sampling. Next using the y-values in the current independently drawn random sample, another estimator for the current population total Y is determined. These two estimators for Y are next combined employing appropriate weights so as to employ a convex function of these two estimators for Y to obtain a final estimator for the current total Y. The problems Jessen (1942) addressed were to choose appropriate matching sample size as a proportion of the previous sample size and suitable weights to combine the two current estimators for Y, first conditional on a given matching sample fraction and then over the variation of this fraction. Yates (1949), Patterson (1950), Tikkiwal (1951), and Eckler (1955) essentially continued with this approach but analyzed data gathered on more than two occasions. They introduced the concept of lag correlation coefficient between observations on the same units but measured on different occasions with succeeding lags and modeled these lag correlations as powers of the correlation coefficient lagging by unity.

In the next development a finite population setup was enforced, and previous and matched samples were taken without replacement. The current sample was taken either from the entire finite population supposed to remain unchanged on the two consecutive occasions or from the remainder on dropping either the initial sample or from the initial sample or from the population U minus S_m. Gurney and Daly (1965) introduced the concept of elementary estimates which involve for each sampled unit values pertaining to a single occasion only and of composite estimators composed of various elementary estimates. Kulldorff (1963) and Sen (1973) allowed the previous sample size to differ from the current sample size. Ravindra Singh (1972) took the current sample s_2 from $U - s_m$, while Kulldorff (1963) from $U - s_1$. Rao and Graham (1964) studied rotation sampling permitting sampled units to be retained or dropped for some occasions and to reappear after specific gaps. Pathak and Rao (1967) improved on Jessen's (1942) estimator by Rao-Blackwellization. Olkin (1958) and Goswamy and Sukhatme (1965) considered multiple auxiliary variables and used ratio and ratio-type estimators. Raj (1965) used the PPSWR method to choose s_1 and s_2 in independent manners but chose s_m by the SRSWOR method. Letting z_i be the positive size measure available on ith unit Raj's (1965) estimator for Y is of the form

$$e = \phi \Big[\frac{1}{m} \sum_{s_m} \Big(\frac{y_i - x_i}{z_i} \Big) + \frac{1}{n_1} \sum_{r=1}^{n_1} \frac{x_r}{z_r} \Big] + (1 - \phi) \frac{1}{n_2} \sum_{1}^{n} \frac{y_r}{z_r}.$$

Ghangurde and Rao (1969) choose s_1 and s_2 independently from U by the Rao, Hartley, and Cochran (1962) method but s_m from s_1 by the SRSWOR method. Chotai (1974) chooses s_m also by the RHC method. Avadhani and Sukhatme (1970) choose s_1 from U by SRSWOR, s_2 from $U - s_1$ by SRSWOR, using $x_i, i \in s_1$ as size-measures they take s_m by RHC method. Their second method chooses s_1, s_m from s_1 and s_2 from $U - s_1$ each by SRSWOR but using y_i in s_m and x_i in s_1 employ a double sample ratio estimator. Singh and Singh (1965), Singh (1968), Abraham, Khosla, and Kathuria (1969), Singh and Kathuria (1969), Singh and Singh (1973), and Kathuria (1975) use SRSWOR but in two or more stages and with and without stratification.

Chaudhuri and Arnab (1977) showed better strategies than Ghangurde and Rao's (1969) and Chotai's (1974), at least conditionally and also under super-population modeling in the following ways. Letting

$p_i \left(0 < p_i < 1 \ \forall \ i, \sum_1^N p_i = 1\right)$ denote available normed size-measures and Lahiri (1951), Midzuno (1952), Sen's (1953) and sampling scheme (LMS scheme in brief) for which on the first draw a unit is chosen with probability proportional to p_i and from the remaining population units an SRSWOR is taken and added to the unit chosen first. If the overall sample is of size n from a population of size N such that

$$\frac{1}{n} > p_i > \frac{n-1}{n}\frac{1}{N-1} \ \forall \ i,$$

then this sampling scheme is called Midzuno's ΠPS or IPPS scheme or LMS ΠPS scheme. Chaudhuri and Arnab (1977) take s_1, s_2 by LMS ΠPS and S_m by SRSWOR and alternatively by LMS ΠPS scheme and again s_1, s_2 by RHC scheme but s_m by LMS ΠPS scheme. Further, taking s_1, s_2 by SRSWOR but s_m with probability proportional to x_i-values in s_1 and alternatively s_m by LMS ΠPS method, Chaudhuri and Arnab (1982) show improvements over Avadhani and Sukhatme's (1970) method of sampling on two successive occasions. Following Gurney and Daly's (1965), Singh's (1968), and Raj's (1965) schemes, Chaudhuri and Arnab (1979) could improve upon the works by these predecessors considering the estimator

$$e = \frac{a}{m}\sum_{s_m}\frac{y_i}{z_i} + \frac{b}{m}\sum_{s_m}\frac{x_i}{z_i} + \frac{c}{n}\sum_1^n\frac{x_r}{z_r} + \frac{d}{n}\sum_1^n\frac{y_r}{z_r}$$

with a,b,c,d suitably chosen. Such an estimator they also employed to improve upon Chotai's (1974). Many details given by Chaudhuri (1988) are stressed to be consulted. This text gives several accounts on the subject on successive occasions by Rao and Bellhouse (1978), Tripathi and Srivastava (1979), Arnab (1981), Ravindra Singh (1972, 1980), and Chaudhuri (1985). All these works when dealing with sampling only on two contemporary occasions assume the population to remain unchanged in size and composition. Chaudhuri (2012) considered Panel Rotation Sampling. In periodic surveys a Panel means the set of sampled units "reporting" for a specific period. A fixed panel is good to estimate changes. A rotating panel is useful to mitigate respondent burden and to achieve efficient estimation for a current total or average. Rotation means organizing a process to keep some elements common in samples at different times, eliminating some sampled earlier when drawing subsequent fresh elements into a current sample. This is sampling on successive occasions with partial replacement of units and inducting fresh ones into a current sample. Fixed Panel and Rotating Panel are two well-known concepts in sampling in periodic surveys. Some units called Certainty Units need to be retained in every sample. Among the Non-certainty units those that are meant to be selected on certain specific occasions but may be discarded on others constitute the Rotating Panels. Panel Rotation sampling may be executed in various ways. "One-level rotation sampling" envisages each unit in a Panel should report for a time period t if it is retained for survey in the period t itself. Also, there may be "semi one-level rotation" or "multi-level rotation." In the former, a unit in a sample at every time period, reports for only one time period but out of those sampled and surveyed a fraction is retained and the complementary fraction is dropped for the next period but a same number of new units are to replace the latter. Alternatively, a unit in a panel in a selected sample may be required not only to report for the current period t but also may be for earlier periods $t-1, t-2$ etc., leading to a multi-level rotation case. In the monthly Canadian Labor Force Survey (LFS) one-sixth of the sample households is replaced totally every month. The sample is composed of six panels, each panel remaining to report for six consecutive monthly enumerative reporting before rotating out of the sample. The monthly Current Population Survey (CPS) by the US Bureau of the Census a 4-8-4 rotation scheme is followed. Here a specified panel in the sample remains in the fray for four consecutive months, drops out for the next eight months, and returns for the next four months to complete a cycle. Several such cycles of panels are constituted.

Chaudhuri (2012) considered how to work out an optimal Matching Sampling Fraction (MSF), namely how many units of the previously drawn sample are to be retained on the current occasion in designing a varying probability sampling scheme. He noted, vide Chaudhuri (1977, 1978), Chaudhuri and Arnab (1978), Chaudhuri and Mukhopadhyay (1978), and Lanke (1975), among others, that the Horvitz and Thompson's Estimator (HTE) has variance formulae in general not explicitly showing their behaviors with the size of the sample. Consequently, in dealing with the HTE in panel rotation sampling determining an appropriate MSF is not easy. Chaudhuri (2012) considers finding an MSF on employing the Rao, Hartley, Cochran (RHC) strategies in panel rotation sampling. Briefly he does the following.

Suppose in two nearby time-points a finite population remains intact in composition and size N and normed size-measures

$$p_i\left(0 < p_i < 1, \sum_1^N p_i = 1\right)$$

are available. To choose the sample of size n_1 on the first occasion the population is randomly split up into n_1 groups of sizes $N_1, \cdots, \cdots, N_n$ such that $\sum_{n_1} N_i = N$, writing $\sum_{n_1}$ as the sum over the n_1 groups. Letting $p_{i1}, \cdots, p_{iN_i}$ be the p_i-values for the units falling in the ith group and $Q_i = p_{i1} + \cdots + p_{iN_i}$, from the ith group one unit, say, i_j is chosen with probability $\frac{p_{ij}}{Q_i}$. Then, $t_1 = \sum_{n_1} x_i \frac{Q_i}{P_i}$, writing x_i, p_i for the x- and p-values for the units in the ith group, is RHC's unbiased estimator for $X = \sum_1^N x_i$. Next on the current occasion an independent RHC sample likewise is drawn with n_1, p_i, Q_i changed to n_2, p'_i, Q'_i; also a matched RHC sub-sampling from n_1 of size m is taken with p_i, Q_i changed to p'_i, Q'_i, and a generalized regression (greg) estimator is taken for Y as

$$t_G = \sum_m y'_i \frac{Q''_i}{p''_i} + b_R \left(t_1 - \sum_m x'_i \frac{Q''_i}{p''_i} \right),$$

with
$$y'_i = y_i \frac{Q'_i}{p'_i}, x'_i = x_i \frac{Q_i}{p_i},$$

$$b_R = \frac{\sum_{i \in s_m} y'_i x'_i R_i}{\sum_{i \in s_m} (x'_i)^2 R_i}, R_i > 0.$$

This R_i we choose as

$$R_i = \frac{1 - \frac{p''_i}{Q''_i}}{p''_i / (Q''_i)}.$$

Based on s_2 the RHC estimator is

$$t_2 = \sum_{n_2} y_i \frac{Q'_i}{p'_i}.$$

Finally t_2 and t_G are combined with suitable weights. Chaudhuri (2012) has given a choice for MSF, namely $\frac{m}{n_1}$. Chaudhuri (2013) gave a different choice of MSF in the following alternative presented in brief. The finite populations U_1 and U_2 of sizes N_1 and N_2 in two close-by consecutive occasions with respective totals X and Y of the same variable but changed from x to y over the time lapse are considered. Normed size-measures

$$p_i \left(0 < p_i < 1, \sum_1^{N_1} p_i = 1 \right) \quad \text{and} \quad q_i \left(0 < q_i < 1, \sum_1^{N_2} q_I = 1 \right)$$

are supposed to be at hand. From U_1 a sample s_1 of n_1 distinct units with positive inclusion probabilities π_i, π_{ij} of units and paired units is chosen on the first occasion to produce the HTE, namely

$$e_1 = \sum_{i \in s_1} \frac{x_i}{\pi_i}.$$

From s_1 a matched sub-sample s_m of size m is drawn with positive inclusion-probabilities π'_i, π'_{ij} for $i, (i,j)$ in s_1. From U_2 then independently of s_1 a current sample s_2 of size n_2 with positive inclusion probabilities π''_i, π''_{ij} for $i, (i,j)$ in U_2 is also drawn.

Let
$$x'_i = \frac{x_i}{\pi_i}, y'_i = \frac{y_i}{\pi_i} \quad \text{and} \quad y'_i = \beta x'_i + \in_i, i \in s_1,$$

Modeling

vide (6,4.4), (6.4.5), on p 226 of Sarndal, Swesson, and Wretman (CSW, 1992). Here β is an unknown constant and ϵ_i's are distributed independently with zero means and variances σ_i^2 for $i \in s_1$. Following CSW (1976) Chaudhuri (2013) estimates β by

$$b_Q = \sum_{i \in s_m} y_i' x_i' Q_i \Big/ \sum_{i \in s_m} (x_i')^2 Q_i$$

with Q_i as one of

$$\frac{1}{x_i'}, \frac{1}{(x_i')^2}, \frac{1}{\pi_i' x_i'}, \frac{1-\pi_i'}{\pi_i' x_i'}.$$

Then the greg estimator for Y is

$$t_1 = \sum_{i \in s_m} \frac{y_i'}{\pi_i'} + b_Q \left(\sum_{i \in s_1} x_i' - \sum_{i \in s_m} \frac{x_i'}{\pi_i'} \right).$$

Writing
$$B_Q = \left(\sum_{i \in s_1} y_i' x_i' Q_i \pi_i' \right) \Big/ \left(\sum_{i \in s_1} (x_i')^2 Q_i \pi_i' \right),$$
$$E_i' = y_i' - B_Q x_i',$$
and
$$e_i = y_i' - b_Q x_i',$$

an approximate formula for $V(t_1)$ is found and

$$v(t_1) = \sum_{i<j \in s_m} \sum \left(\frac{\pi_i' \pi_j' - \pi_{ij}'}{\pi_{ij}'} \right) \left(\frac{e_i'}{\pi_i'} - \frac{e_j'}{\pi_j'} \right)^2$$
$$+ \sum_{i<j \in s_m} \left(\frac{\pi_i \pi_j - \pi_{ij}}{\pi_{ij}} \right) \left(\frac{y_i}{\pi_i} - \frac{y_j}{\pi_j} \right)^2 \frac{1}{\pi_{ij}'}.$$

Also for $t_2 = \sum_{i \in s_2} \frac{y_i}{\pi_i''}$ a variance estimator is employed as

$$v(t_2) = \sum_{i<j \in s_2} \left(\frac{\pi_i'' \pi_j'' - \pi_{ij}''}{\pi_{ij}''} \right) \left(\frac{y_i}{\pi_i''} - \frac{y_j}{\pi_j''} \right)^2.$$

Then he takes $\bar{t} = \dfrac{t_1/v(t_1) + t_2/v(t_2)}{\frac{1}{v(t_1)} + \frac{1}{v(t_2)}}$

with a variance estimator

$$v = \frac{v(t_1) v(t_2)}{v(t_1) + v(t_2)}.$$

Then he postulates a model with $v_m(\epsilon_i) = \sigma^2 x_i'$, takes $Q_i = \frac{1}{x_i'}$, $\pi_i' = m\theta_i$, with $0 < \theta_i < 1 \,\forall\, i \in s_1$, $\sum_{i \in s_1} \theta_i = 1$ and notes $E_m V_C(t_1) = \frac{A}{m} - B$, V_C denoting variance over the matched sampling with s_1 fixed. Thus, he observes that m should be taken as large as possible with $m < n_1$ consistent with the budget approved.

For more up-to-date accounts of rotation sampling one should consult Park, Choi, and Kim (2007) and some of the research papers cited therein. As the author has no contribution to the type of material covered therein, no details thereupon are touched on here.

5.8 NON-RESPONSE AND NOT-AT-HOMES

So far we have continued to assume that once a population $U = (1, \cdots, i, \cdots, N)$ of size N with its respective units is defined on taking a sample s from it in any way we like, we may get hold of them and gather the intended variate-values y_i for each of the respective members i in s and thus accomplish a survey. But often all the selected units may not be found to give us the values we need. Some households (hh) may be closed at the time of the visits by the investigators. Some in the sampled households may refuse to respond to our queries. In case a chosen sample is a crop field, a farmer owning it or cultivating it may not be visible to answer our queries or permit us to measure the crop area or crop production or tell us where the produce has been transported to fetch a market or report to us the sale proceeds of the marketed crops, etc.

A general name given to such phenomena is Non-Response. Obviously survey data vitiated by non-response cannot help us to derive an unbiased estimator for a parameter we mostly need in survey sampling.

Let us discuss briefly a theory for rectification with a simplistic illustration. Let from a finite population $U = (1, \cdots, i, \cdots, N)$ of a known size an SRSWOR of size n be drawn with an intention to estimate the population mean $\bar{Y} = \frac{1}{N} \sum_1^N y_i$ of a variable y with values y_i for respective units i in U. Suppose y_i-values are gathered only from a sub-sample s_1 of size n_1 out of the entire sample s of size $n\,(> n_1)$, but no y_i-values are gathered from the remaining $n_2 = n - n_1$ units of the complement $s_2 = s - s_1$ of the full sample s. Then, if the sub-sample s_1 may be supposed to be an SRSWOR of size n_1 from the initial SRSWOR s of size n from U, then this s_1 is also just an SRSWOR of size n_1 from U. In such a case the mean $\bar{y}_1 = \frac{1}{n_1} \sum_{i \in s_1} y_i$ is an unbiased estimator for $\bar{y} = \frac{1}{n} \sum_{i \in s} y_i$ and hence is also an unbiased estimator for $\bar{Y} = \frac{\sum y_i}{N}$.

Consequently, $\quad V(\bar{y}_1) = \dfrac{N - n_1}{N n_1} S^2 = \left(\dfrac{1}{n_1} - \dfrac{1}{N}\right) S^2.$

Clearly, $V(\bar{y}_1) > V(\bar{y})$. Thus in this simplistic situation no bias ensues in estimation but only the efficiency declines. It is reasonable to assume that the non-respondents differ from the respondents in respect of their essential characteristic features. Hence, it is fair to anticipate bias in estimating $\bar{Y}$ by $\bar{y}_1$ rather than $\bar{y}$ which is here unavailable. To try to cut this possibilities of bias short the following procedure is worth attending to.

Let by an additional effort of persuasion, out of the initial non-responding sub-sample s_2 in s of size n_2 an SRSWOR of r units out of these n_2 units the y_i-values be available. Let $\bar{y}_{2r}$ be the mean of these r units of the SRSWOR from s_2. Then, consider the new estimator

$$\bar{y}' = \left(\frac{n_1}{n}\right)\bar{y}_1 + \left(\frac{n_2}{n}\right)\bar{y}_{2r}.$$

Then $$E_C(\bar{y}_{2r}) = \bar{y}_2 = \frac{1}{n_2}\sum_{i \in s_2} y_i\ ;$$

here E_C denoting the expectation operator conditionally on the samples s, s_1, and s_2 being at hand as described. Then,

$$E_C(\bar{y}') = \frac{n_1\bar{y}_1 + n_2\bar{y}_2}{n} = \bar{y} = \frac{1}{n}\sum_{i \in s} y_i.$$

Hence, $E(\bar{y}') = E(\bar{y}) = \bar{Y}$. Thus, $\bar{y}'$ is unbiased for $\bar{Y}$.

Now, applying the principle of double sampling we may derive, on writing $W_1 = \frac{N_1}{N}$ and $W_2 = \frac{N_2}{N} = 1 - W_1$ on defining N_1 to be the number of units in U who are anticipated to be the habitual respondents and $N_2 = N - N_1$ is the number of units unlikely to respond on the first attempt, some of the results are given below.

Further, let

$$S_1^2 = \frac{1}{(N_1 - 1)}\sum_1^{N_1}(y_i - \bar{Y}_1)^2,$$

$$S_2^2 = \frac{1}{(N_2 - 1)}\sum_{N_1+1}^{N}(y_i - \bar{Y}_2)^2, S^2 = \frac{1}{N}\sum_1^{N}(y_i - \bar{Y})^2,$$

$$\bar{Y}_1 = \frac{1}{N_1}\sum_1^{N_1} y_i, \bar{Y}_2 = \frac{1}{N_2}\sum_{N_1+1}^{N} y_i, \bar{Y} = \frac{1}{N}\sum_1^{N} y_i.$$

Again, we may write E_C, V_C as conditional expectation, variance operators, given the initial sample s and $y_i, i \in s$ held fixed, E, V the expectation, variance operators over selecting and surveying the initial sample s and E_o, V_o the overall expectation, variance operators implying $E_o = EE_C$ and $V_o =$

$EV_C + VE_C$. Then, we get

$$V\left(E_C\left(\bar{y}'\right)\right) = \left(\frac{1}{n} - \frac{1}{N}\right)S^2$$

$$V_C\left(\bar{y}'\right) = \left(\frac{n_2}{n}\right)^2\left(\frac{1}{r} - \frac{1}{n_2}\right)S_2^2$$

$$EV_C\left(\bar{y}'\right) = W_2\frac{1}{n}\left(\frac{n_2}{r} - 1\right)S_2^2 = \frac{W_2\left(K-1\right)}{n}S_2^2$$

writing $K = \frac{n_2}{r}$.

So, finally,

$$E_o\left(\bar{y}'\right) = \bar{Y} \quad \text{and} \quad V_o\left(\bar{y}'\right) = \left(\frac{1}{n} - \frac{1}{N}\right)S^2 + \frac{W_2\left(K-1\right)}{n}S_2^2.$$

Hansen and Hurwitz (1946) initiated this theory. They also utilized this variance function $V_o\left(\bar{y}'\right)$ in combination with the following cost function they introduced in order to answer the question as to how to fix the initial sample size n and the second sample-size r, rather the related number $k = \frac{n_2}{r}$. Their cost function is

$$C = \alpha n + \beta n_\cdot + \theta r.$$

Then,
$$C = n\left[\alpha + \beta\frac{n_1}{n} + \theta\frac{n_2}{n}\frac{1}{K}\right].$$

Since
$$E\left(\frac{n_1}{n}\right) = W_1 \quad \text{and} \quad E\left(\frac{n_2}{n}\right) = W_2, \quad \text{it follows that}$$

$$E\left(C\right) = n\left[\alpha + \beta W_1 + \theta W_2\frac{1}{K}\right] = nC_o, \quad \text{say.}$$

Let
$$V_o\left(\bar{y}'\right) + \frac{S^2}{N} = \frac{1}{n}\left[S^2 + W_2\left(K-1\right)S_2^2\right] = \frac{V_o}{n}, \quad \text{say.}$$

Then,
$$C_oV_o = \left[\alpha + \beta W_1 + \theta W_2\frac{1}{K}\right] \times \left[S^2 + KW_2S_2^2 - W_2S_2^2\right]$$

$$= \psi + \phi K + \frac{\lambda}{K}, \quad \text{say, observing}$$

$$\psi = \alpha\left(S^2 - W_2S_2^2\right) + \beta W_1\left(S^2 - W_W S_2^2\right) + \theta W_2 S_2^2$$

$$\phi = W_2 S_2^2\left(\alpha + \beta W_1\right), \quad \lambda = \theta W_2\left(S^2 - W_2 S_2^2\right).$$

Modeling

Now,
$$\phi K + \frac{\lambda}{K} = \left(\sqrt{\phi K}\right)^2 + \left(\sqrt{\frac{\lambda}{K}}\right)^2$$
$$= \left(\sqrt{\phi K} - \sqrt{\frac{\lambda}{K}}\right)^2 + 2\sqrt{\phi\lambda} \geq 2\sqrt{\phi\lambda}.$$

So, the value of $C_o V_o$ is minimized with respect to the choice of K if $\sqrt{\phi K} = \sqrt{\frac{\lambda}{K}}$ giving us the optimum choice of K as $K_{opt} = \sqrt{\frac{\lambda}{\phi}}$.

Only in case the values of λ and ϕ in terms of S^2, S_2^2, W_2, W_1 and α, β, θ are available, of course only then is an optimum choice of K possible.

Only if a good pilot survey may be executed yielding reliable values of these parameters, a rational choice of K is possible. Of the quantities V_o and C_o, if one is kept fixed, the other one is minimized for the choice of K as K_{opt}. Given this achieved minimal for V_o or C_o a suitably feasible value of n then follows with little difficulty.

This pioneering work by Hansen and Hurwitz (1949) has been studied and extended by various researchers in various directions. In this text we need not go into them. But one specific work by Politz and Simmons (1949) noted below in brief deserves a mention because of its intense novelty in a pioneering approach.

"Not-at-Homes": Politz and Simmons (1949)

If in a collected sample some persons are not found present to respond Politz and Simmons (1949) have given a device which is essentially an application of the post-stratification principle. Every respondent is asked to report the number of days previous to this day of the last week at this time he/she was present and hence in a position to answer similar queries. The possible answers, namely, 0,1,2,3,4,5,6, and 7 collected from all the respondents give estimates of the proportions W_h ($h = 0, 1, \cdots, 7$) of the people in the eigth distinguishable categories.

Simplistically Politz and Simmons (1949) used $\frac{h+1}{8}$, ($h = 0, 1, \cdots .7$) as the estimated probability that a respondent would be at home to respond given that he/she announced he/she was present on h days of the last week. Consequently, every sampled person's response was multiplied by the weight $\frac{8}{h+1}, h = 0, 1, \cdots, 7$ if the person said he/she was at home on the h of the seven days of the week just gone by.

Politz and Simmons (1949) coined the phrase "Not-at-Homes" to give the under-noted device to correct the bias in estimation of a finite population total vitiated by the occurrence of non-response because a possible respondent when sampled is not found at the spot to answer queries. It essentially runs as follows.

A sampled person if found to respond to the queries for the investigation, is asked in addition to report the number out of the previous days of the week he/she was present to answer similar queries if approached. Getting these numbers $o, 1, \cdots, 7$ from all the sampled and actual respondents the numbers $\frac{8}{1}, \frac{8}{2}, \cdots, \frac{8}{8}$ are taken as weights to be used for the values y_{hi} for a unit i announcing he/she was available on $h (= 0, 1, \cdots, 7)$, respectively, of the previous seven days because on the present, i.e., the eightth day he/she is actually giving out the value asked for. Then,

$$t_h = \sum_{i \in s} \frac{y_{hi}}{\pi_i}$$

is taken as an estimator for the total of those members of the population who are in the sample and report to fall in the hth $(h = 0, 1, \cdots, 7)$ class, namely those who were available on h days out of the previous seven days. Then,

$$\hat{Y} = \frac{\sum_{h=0}^{7} W_h t_h}{\sum_{0}^{7} W_h} \quad \text{is used to estimate} \quad Y = \sum_{1}^{N} y_i,$$

$V\left(\hat{Y}\right)$ and $\hat{V}\left(\hat{Y}\right)$ are then determined in usual ways. Here, clearly,

$$y_{hi} = y_i I_{hi}, \text{ writing } \quad I_{hi} = 1 \quad \text{if } i \text{ is in the } h^{th} \text{ class}$$
$$= 0 \quad \text{else,}$$

and W_h's are treated as fixed known numbers.

A reader may consult Cochran (1977) for other methods of correcting for incidences of non-responses. One of them is not scientific but widely practiced. This is called the Substitution method. Here the missing value for a unit is replaced by the value of a neighboring unit randomly chosen.

Let from $U = (1, \cdots i, \cdots, N)$ a sample s be chosen according to a design admitting positive inclusion-probabilities π_i, π_{ij} for single unit i and paired units (i, j) of U. Let $I_{ri} = 1/0$, if i responds/non-responds, $I_{si} = 1/0$ if $i \in s / i \notin s$. We have of course $E(I_{si}) = \pi_i > 0$. Let $E(I_{ri}) = q_i$ which is unknown but positive for every $i \in U$. Even if I_{ri} equals zero for certain units i in s we may still employ

$$\hat{t}_H = \sum_{i=1}^{N} y_i \frac{I_{si}}{\pi_i} \frac{I_{ri}}{q_i} \quad \text{as an unbiased estimator for} \quad Y = \sum_{1}^{N} y_i.$$

Since q_i's are unknown they may and should be estimated for those i's for which $0 < q_1 < 1$, in case we may find some auxiliary variate values z_i, known

for i in s with the variable z well-correlated with y. Let us have the log linear model
$$\log_e \left(\frac{q_i}{1 - q_i} \right) = a + b z_i, \text{ with } a, b \text{ known}.$$
For an estimator $\hat{q}_i$ for q_i one may postulate
$$\log_e \left(\frac{\hat{q}_i}{1 - \hat{q}_i} \right) = a + b z_i + e_i, i \in s$$
such that $0 < \hat{q}_i < 1$ and e_i's are distributed independently with 0 means and units variances, $i \in s$. Then, obtaining least square estimates $\hat{a}, \hat{b}$ for a, b we may take the modified estimates of q_i as $\tilde{q}_i = \hat{a} + \hat{b} z_i$, so that $0 < \tilde{q}_i < 1$ for $i \in s$.

Then, a model-based estimate of Y is taken as $\hat{Y} = \sum_{i \in s} \frac{y_i}{\pi_i} \frac{I_{ri}}{\tilde{q}_i}$.

Taking I_{ri}'s as independent variables with
$$E(I_{ri}) = q_i, V(I_{ri}) = q_i (1 - q_i), \text{ cov}(r_i, r_j) = 0, i \neq j,$$
approximate variance and variance-estimates for $\hat{Y}$ are not difficult to find, vide Chaudhuri (2010, pp 191–192).

5.9 WEIGHTING ADJUSTMENTS AND IMPUTATION

In practice we need to distinguish between two kinds of non-response in empirical survey sampling. Correspondingly, two ways of attacking these two kinds of problems have emerged. One of these we have substantially covered. This is called Unit Non-Response and its attempted control by Weighting Adjustments. We have considered so far estimators for Y of the general form

$$\hat{Y} = \sum_{i \in s} y_i b_{si} = \sum_{i=1}^{N} y_i b_{si} I_{si}$$

with b_{si}'s free of $\underline{Y} = (y_1, \cdots, y_i, \cdots, y_N)$.

In case of unit non-response, the "multiplier or weight" $b_{si} I_{si}$ of each y_i for i in the sample s in a general form of an estimator is

$$\tilde{Y} = \sum_{1}^{N} y_i b_{si} I_{si} \, C_{ri} I_{ri}$$

with a multiplier $C_{ri} I_{ri}$ reflecting the incidence of possible unit non-response.

The second kind of non-response is called Item Non-Response and a general way to tackle this issue is to employ what is generally called Imputation

Technique. In the fieldwork for a survey the sampled units may be addressed to elicit from them some responses. But out of all the relevant items of survey interest it may not be possible to gather information on some of the items of relevance while the same may be gathered on other relevant items of interest. We shall now discuss how to deal with such a situation of botheration but of practical importance.

Imputation means using a value for a missing value on an item concerning a unit surveyed in a sample chosen. This step is taken, as noted here, only in respect of sampled units for which values on various items are gathered but only some others are missing. Consequently, utilizing values on several variables gathered on the selected units they are classified forming groups of units, with common or similar and comparable values. They are called the Imputation Classes.

One imputation method called the Random Imputation method works as follows. Suppose in a sample s of size n the number of units in the hth imputation class formed is $n_h, h = 1, \cdots, H$ such that $n = \sum_{h=1}^{H} n_h$. Let for a specific item of interest out of the chosen n_h units above only for $r_h\,(< n_h)$ units values are available, but for the other $m_h = n_h - r_h$ units values on this item are missing. Writing

$$r = \sum_{1}^{H} r_h \quad \text{and} \quad m = \sum_{1}^{H} m_h$$

it is of course very undesirable to find $r < m$. But imputation-class-wise all the three possibilities

    i. $r_h > m_h$,
    ii. $r_h = m_h$,   and
    iii. $r_h < m_h$, $h = 1, \cdots, H$

may be admissible even if $r > m$. Then a random imputation procedure works as follows. For a particular h, let two integers k_h and t_h be chosen subject to $m_h = k_h r_h + t_h$ ($k_h, t_h \geq 0$). Then, an SRSWOR of t_h units is to be taken from the r_h values at hand, in case $0 <, t_h < r_h$. Then, these t_h values are each to be repeated $(k_h + 2)$ times and the remaining $(r_h - t_h)$ values are to be repeated $(k_h + 1)$ times each. Then, $t_h(k_h + 2) + (r_h - t_h)(k_h + 1)$ values are to be imputed values for the above m_h missing units in the hth imputation class. This solution is easy to see, taking, for example, $r_h = 7$ and $m_h = 4$, say.

Then to get in all $n_h = r_h + m_h = 11$ values, we should take $k_h = 0$ to find $2t_h + (7 - t_h) = 11$ or $t_h = 4$.

So, four of the seven responding units are to give or donate their values and the three other respondents need not give their values to fill together the values for the four missing units. The units that give their values are called the Donors and the units receiving them are called the Recipients. Again, let

$n_h = 13, r_h = 5$ and $m_h = 8$. So, we are to solve

$$8 = 5k_h + t_h \tag{1}$$
and
$$13 = t_h (k_h + 2) + (5 - t_h)(k_h + 1). \tag{2}$$

The solution $t_h = 3$ and $k_h = 1$ give the results.

Another method known as the Regression Method of imputation works as follows. Suppose for the variable y of interest there are r respondents and m missing for a sample of $n = r + m$ persons. But let there exist a correlated variable x for which values are available on all the n sampled units. Thus, let $(y_i, x_i), i = 1, \cdots, r$ values and $x_i = r+1, \cdots, n$ values be available. We need imputing the values $\hat{y}_i$, say, for $i = r+1, \cdots, n$. Now fitting a linear regression line $y = \alpha + \beta x$ by least squares to estimate α, β by

$$a = \bar{y}_r - b\bar{x}_r \quad \text{and} \quad b = \frac{\sum_{i=1}^{r}(y_i - \bar{y}_r)(x_i - \bar{x}_r)}{\sum_{i=1}^{r}(x_i - \bar{x}_r)^2},$$

writing
$$\bar{y}_r = \frac{1}{r}\sum_{i=1}^{r} y_i, \quad \bar{x}_r = \frac{1}{r}\sum_{1}^{r} x_i,$$

then,
$$\hat{y}_i = a + bx_i, \quad i = r+1, \cdots, n$$

give the imputed values.

Another popular method of imputation is known as Hot Deck device. Suppose in a given sample survey context for an hth imputation class a value at hand from a previous survey or on a current survey of an allied nature x_h be an available value called a Cold Deck value. But suppose for the current survey of relevance for a particular imputation class the units sampled are $i_1, i_2, i_3, i_4, i_5, i_6, i_7$ but only the y-values y_{i3}, y_{i5}, y_{i6} are gathered together while all the others are missing. Then, the hot-deck imputation produces the y-values as $\hat{y}_{i1}, \hat{y}_{i2}, \cdots \hat{y}_{i7}$ given by

$$\hat{y}_{i1} = \hat{y}_{i2} = x_h, \ \hat{y}_{i3} = y_{i3}, \ \hat{y}_{i4} = y_{i3}, \ \hat{y}_{i5} = y_{i5}, \ \hat{y}_{i6} = y_{i6} \text{ and } \hat{y}_{i7} = y_{i6}.$$

Subsequent analysis proceeds with these $\hat{y}_{i1}, \cdots, \hat{y}_i 7$ as the raw materials at hand.

Chaudhuri and Stenger (2005) is a reference where more materials and references may be gathered.

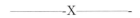

5.10 TIME SERIES APPROACH IN REPEATED SAMPLING

Blight and Scott (1973) were the first to apply a time series approach in estimating finite population totals over successive points of time on sampling the populations on repeated occasions, allowing populations to change over time in size and composition and not insisting on selection and survey of the same units more than once as required in panel surveys. They take a population total at time t as θ_t. Blight and Scott (1973) postulated a first-order autoregressive AR(1) process for θ_t and a moving average (MA) process for the associated process of the elementary estimates $\bar{y}_t$. Model variances for θ_t and design variances of $\bar{y}_t$ are utilized in deriving Bayes estimators and developing recursive formulae concerning them. Scott and Smith (1974) followed up this work to derive one-step advanced estimates for θ_{tH} along with variance formulae. Jones (1980) generalized these to derive Bayes' estimates for vectors of θ_t's over several values of t and variances of these estimates. These involve high-order matrices and their inverses which take unstable form when several points of time are involved. To simplify Jones introduced "state space" models and Kalman filtering so as to develop useful recursive relations with matrix orders reduced. Binder and Hidiroglou (1988) is an important reference covering these developments. As such a model-dependent approach has not been quite popular yet in the context of survey sampling and as the present author has no contribution to this subject we need not proceed further on it.

6 Stigmatizing Issues

Abstract. Introduction. Early growth of RR and the current status. Optional Randomized Response technique. Indirect questioning.

6.0 ABSTRACT

Special Survey Techniques are needed to capture trustworthy data relating to people bearing questionable and stigmatizing characteristics about which direct personal queries are embarrassing to try for and gather reliable data to be put to intelligent uses. The monograph by Chaudhuri and Christofides (2013) and an earlier one by Chaudhuri (2011) have given many detailed accounts of the problems along with their attempted solutions. Warner's (1965) first publication stimulated enthusiastic research in this field to be summarized in this volume.

6.1 INTRODUCTION

Addiction to drugs and alcoholism, habits of tax evasion, gambling, reckless and illegal driving of vehicles, testing positive in HIV, dope tests, AIDS, criminal antecedence, etc., are some of the features which are too sensitive and stigmatizing which their bearers like to hide from public exposure. So, a usual survey is unlikely to be suitably carried out asking direct questions expecting honest and truthful answers.

Warner (1965) in his debut to tackle such situations implemented a Randomized Response (RR) survey as opposed to a Direct Response (DR) survey. Under this scheme a sampled person is offered a box of a large number of identical cards differing only in the proportions p of them carrying the mark A and $(1-p)\left(0 < p \neq \frac{1}{2} < 1\right)$ of them bearing the mark A^C. Every respondent bearing either the stigmatizing characteristic A or its complement A^C on being requested by the investigator is to randomly draw one card from the box and respond as either "Yes" or "1" if his/her feature A or A^C matched the card mark or the response will be "No" or "0" if the feature "mis-matched" the card type. The respondent is not to disclose to the questioner the outcome of the draw from the box. When such RR data are collected from a mass of sampled persons, it will be easy to unbiasedly estimate the true proportion of the people bearing A in the community when the samples are suitably drawn as probability samples even though no single such RR may disclose the true characteristic of the person concerned. A theory developed from this simple beginning with a tremendous volume of current research output. Details are to follow.

6.2 EARLY GROWTH OF RR AND THE CURRENT STATUS

For the early growth of the subject Randomized Response (RR) Techniques (RRT) one may find it stimulating to go through Warner (1965); Horvitz, Shah, and Simmons (1967); Greenberg, Abul-Ela, Simmons, and Horvitz (1969), and the monographs by Fox and Tracy (1986) and by Chaudhuri and Mukerjee (1988). To form an up to date idea of the currently grown-up subject of Indirect Question Techniques including RRT's we believe a sharp glance through the following may be instructive: Chaudhuri (1987, 1992, 2001); Chaudhuri and Mukerjee (1985, 1987); Chaudhuri and Saha (2005); Chua and Tsui (2000); Droitcour, Larson, and Scheuren (2001); Eichorn and Hayre (1983); Eriksson (1973); Franklin (1989); Huang (2004); Kerkvliet (1994); Ljungqvist (1993); Mangat (1994); Mangat, Singh, Singh, Bellhouse; and Kashani (1995); O'Hagan (1987); Scheers (1992); Tracy and Mangat (1996); Van der Heiden, Gills, Bout, and Hox (2000); Barabesi, Franceschi, and Marcheselli (2012); and the monographs by Chaudhuri (2011) and by Chaudhuri and Christofides (2013).

Warner (1965) in order to apply his pioneering indirect gathering of responses to queries about sensitive issues restricted himself to Simple Random Sampling (SRS) With Replacement (SRSWR). Without specifying the population size whether finite or infinite left unprescribed, in order to unbiasedly estimate θ, the population proportion of people bearing the sensitive characteristic A, he gathered from an SRSWR in n draws the sample proportion $\frac{n_1}{n}$ "yes" responses to claim $\hat{\theta}_w = \frac{1}{2p-1}\left(\frac{n_1}{n} - 1 + p\right)$ as a good estimator for θ with a variance

$$V\left(\hat{\theta}_w\right) = \frac{\theta(1-\theta)}{n} + \frac{1}{n}\left[\frac{1}{16\left(p-\frac{1}{2}\right)^2} - \frac{1}{4}\right].$$

An unbiased estimator for $V\left(\hat{\theta}_w\right)$ is

$$v\left(\hat{\theta}_w\right) = \frac{\hat{\theta}_w\left(1-\hat{\theta}_w\right)}{(n-1)} + \frac{1}{(n-1)}\left[\frac{1}{16\left(p-\frac{1}{2}\right)^2} - \frac{1}{4}\right].$$

Though not made explicit by Warner (1965) himself the feature A is stigmatizing, not so is A^C. This recognition that A as well as A^C may be simultaneously sensitive like, say, being a supporter or opponent of a particular political party, Simmons and his colleagues (Greenberg et al., 1969, and Horvitz et al. 1967) introduced an alternative RRT known as Unrelated Question Model (URL) which briefly is as follows.

A sample person is offered a box of numerous identical cards marked either A or B in proportions $P_1 : (1 - P_1)$ in the first sample of n_1 persons and $P_2 : (1 - P_2)$ in the second similar box offered to each person in the second sample of n_2 persons ($0 < P_1 < 1, 0 < P_2 < 1, P_1 \neq P_2$). Here B represents an

Stigmatizing Issues

innocuous characteristic "unrelated" to the stigmatizing characteristic A. A sampled person is to respond "Yes" or "1" if his/her characteristic "matches" the card type A or B when drawn or "No" or "0" if it "mis-matches." The samples are both drawn by SRSWR in independent manners. Many other alternative devices have emerged since these two fundamental innovations each employing SRSWR's. Chaudhuri (2011) has emphasized that no RRT is needed to be tied to a specific way in which a sample is to be chosen in deriving an unbiased estimator for θ noted above and in deriving a formula for the variance of the estimated θ or for an unbiased estimator of this variance, only provided the design for the sample employed admits a positive probability of inclusion in a sample of every unit and every paired units. This idea and revised methods of sample selection were introduced by Chaudhuri (2001) covering various RRT's. Let us cite a few RR procedures.

6.2.1 WARNER (1965)

Let $U = (1, \cdots i, \cdots N)$ denote a finite survey population of N units bearing y-values $y_i, i \in U$, which is 1 if the ith person bears A or is 0 if the ith person bears A^C. We need to unbiasedly estimate $\theta = \frac{\sum y_i}{N} = \frac{Y}{N} = \bar{Y}$, say. A sample s of a number of persons is to be selected with a probability $p(s)$ and for this sampling design p, let $\pi_i = \sum_{s \ni i} p(s)$ and $\pi_{ij} = \sum_{s \ni i,j} p(s), i \in U$ and $i,j \in U, i \neq j$.

Let $I_i = 1$ if the ith person's feature A or A^C "matches" the card type when drawn, $= 0$ if it "mis-matches."

Thus, $\quad$ Prob $(I_i = 1) = py_i + (1-p)(1-y_i)$

or, $\quad$ Prob $(I_i = 1) = (1-p) + (2p-1) y_i$.

Let $r_i = \frac{I_i - (1-p)}{(2p-1)}$. Then, $E_R(I_i) = y_i$ writing E_R generically as the expectation operator with respect to an RR technique. Also,

$$V_R(r_i) = \frac{V_R(I_i)}{(2p-1)^2} = \frac{p(1-p)}{(2p-1)^2},$$

since $I_i^2 = I_i, y_i^2 = y_i$ writing V_R as the variance operator with respect to an RRT generically. RRT's are executed independently across the sample persons. For Y let $t = t(s, \underline{Y})$ be an unbiased estimator.

Let $\quad e = e(s, \underline{y})|_{\underline{Y} = \underline{R}}$,

writing $\quad \underline{R} = (r_1, \cdots, r_N) \quad$ and $\quad \underline{Y} = (y_1, \cdots, y_N)$.

Then, writing $E = E_p E_R = E_R E_p$, the overall expectation operator and $V = E_p V_R + V_p E_R = E_R V_p + V_R E_p$, the overall variance operator. For simplicity, let

$$t = t(s, \underline{Y}) = \sum_{i \in s} y_i b_{si}$$

with b_{si} subject to
$$\sum_{s \ni i} p(s) b_{si} = 1 \ \forall \ i.$$

Then, $E_p(t) = Y$ and $E(e) = E_R\left(\sum_1^N r_i\right) = Y$

and $E(e) = E_p(E_R(e)) = E_p(t) = Y$

obviously noting $e = e(s, \underline{R}) = \sum_{i \in s} r_i b_{si}.$

Then, $V(t) = \sum y_i^2 C_i + \sum_{i \neq s} y_i y_j C_{ij}$

writing $C_i = \sum_{s \ni i} p(s) b_{si}^2 - 1$ and $C_{ij} = \sum_{s \ni ij} p(s) b_{si} b_{sj} - 1.$

Then, since $y_i^2 = y_i$, it follows
$$V_p(t) = \sum y_i C_i + \sum\sum_{i \neq j} y_i y_j C_{ij}.$$

Then, an unbiased estimator for $V(t)$ turns out as
$$v_p(t) = \sum_{i \in s} r_i C_{si} + \sum\sum_{i \neq j} r_i r_j C_{sij}$$

writing C_{si} and $C_{sij}'s$ as real numbers free of $\underline{R}$ and $\underline{Y}$ satisfying
$$\sum_s p(s) C_{si} = C_i \quad \text{and} \quad \sum_s p(s) C_{sij} = C_{ij}.$$

Then, we note $V(e) = E_p V_R(e) + V_p E_R(e)$

$$= E_p \sum_{i \in s} V_R(i) b_{si}^2 + V_p(t)$$

$$= \frac{p(1-p)}{(2p-1)^2} \sum_1^N (1 + C_i) + \sum y_i C_i + \sum\sum_{i \neq j} y_i y_j C_{ij}.$$

So, an unbiased estimator of $V(e)$ is
$$v(e) = \frac{p(1-p)}{(2p-1)^2} \sum_{i \in s} \frac{1}{\pi_i}(1 + C_{si}) + \sum_{i \in s} r_i C_{si} + \sum\sum_{i \neq j \in s} r_i r_j C_{sij}.$$

6.2.2 UNRELATED QUESTION MODEL

Let $y_i, i \in U$ be as in 6.2.1 and let $x_i = 1$ if the ith person bears an innocuous characteristic B, say, preferring Football to Tennis and $x_i = 0$ if he/she bears the complementary characteristic B^C, this characteristic being presumably unrelated to the characteristic A.

Let an investigator approach a sampled person labeled i with two boxes containing, respectively, numerous cards marked A and B in proportions $p_1 : (1 - p_1)$ in the first box and $p_2 : (1 - p_2)$ in the second box, $(0 < p_1 < 1, 0 < p_2 < 1, p_1 \neq p_2)$.

Let
$$I_i = 1 \quad \text{if } i\text{th person gets a ``Match'' in the first box}$$
$$= 0 \quad \text{if a ``mis-match'' occurs}$$
$$J_i = 1 \quad \text{if the } i\text{th person gets a ``Match'' in the second box}$$
$$= 0 \quad \text{if a ``mis-match'' occurs.}$$

Then,
$$E_R(I_i) = p_1 y_i + (1 - p_1) x_i \quad \text{and}$$
$$E_R(J_i) = p_2 y_i + (1 - p_2) x_i.$$

Then,
$$E_R\left[(1 - p_2) I_i - (1 - p_1) J_i\right] = (p_1 - p_2) y_i.$$

Letting
$$r_i = \frac{(1 - p_2) I_i - (1 - p_1) J_i}{(p_1 - p_2)}$$

easily one gets $E_R(r_i) = y_i$. Also, $V_R(r_i) = V_i$, say, easily follows which we need not show. More importantly, since

$$V_R(r_i) = E_R(r_i^2) - E_R^2(r_i) = E_R(r_i^2) - y_i^2$$
$$= E_R(r_i^2) - y_i = E_R(r_i^2) - E_R(r_i) = E_R r_i (r_i - 1)$$

one gets an unbiased estimator for V_i as $v_i = r_i (r_i - 1)$.

Now $e = \sum\limits_{i \in s} r_i b_{si}$ has $E(e) = E_P E_R(e) = E_p\left(\sum y_i b_{si}\right) = Y$.

So, $\dfrac{e}{N}$ is an unbiased estimator for θ.

Next, $V(e) = E_R V_p(e) + V_R E_p(e)$

$$= E_R\left[\sum r_i^2 C_i + \sum\sum_{i \neq j} r_i r_j C_{ij}\right] + V_R\left(\sum r_i\right)$$

$$= \sum V_i C_i + \sum y_i C_i + \sum\sum_{i \neq j} y_i y_j C_{ij} + \sum V_i.$$

So, an unbiased estimator for this $V(e)$ is

$$v(e) = \sum_{i \in s} v_i c_{si} + \sum_{i \in s} r_i c_{si} + \sum_{i \neq j} \sum_s r_i r_j c_{sij} + \sum_{i \in s} \frac{v_i}{\pi_i}.$$

6.2.3 RRT WITH QUANTITATIVE VARIABLES

Suppose we feel like estimating the totals of bribes taken and paid or amounts of money gained or lost or amounts of underpayments of taxes or expenses on treatment of AIDS or fees paid on fighting legal battles against spouses, etc., generally may be deemed as stigmatizing. In such cases also RRT's are applicable.

Let y be a real variable with values y_i assumed on the respective persons labeled as in the finite survey population $U(1, \cdots, i, \cdots, N)$ and let these values be such that these may their bearers be generally expected not to reveal on queries. In order to estimate the total $Y = \sum_{i \in s} y_i$ let a sample s be selected with a probability $p(s)$ from U. A selected person labeled i is then approached by an investigator with two boxes, respectively, bearing a number of look-alike cards with numbers inscribed as $a_1, \cdots a_j \cdots a_T$ with

$$\mu_a = \frac{1}{t} \sum_{j=1}^{T} a_j \neq 0, \quad \sigma_a^2 = \frac{1}{T} \sum_{j=1}^{T} (a_j - \mu_a)^2$$

in one of the boxes and the numbers $b_1, \cdots, b_p, \cdots, b_L$ with

$$\mu_b = \frac{1}{L} \sum_{k=1}^{L} b_k \quad \text{and} \quad \sigma_b^2 = \frac{1}{L} \sum_{k=1}^{L} (b_K - \mu_b)^2 \quad \text{in the other.}$$

Taking randomly one of the cards from the first box and independently from the second box randomly one card from one other, the ith person on request is to respond as $z_i = a_j y_i + b_k, i \in s$.

Then, $\qquad E_R(z_i) = y_i \mu_a + \mu_b.$

Letting $\qquad r_i = (z_i - \mu_b)/\mu_a$

one gets $\qquad E_R(r_i) = y_i \quad \text{and} \quad V_R(r_i) = \dfrac{V_R(z_i)}{\mu_a^2}$

or $\qquad V_i = V_R(r_i) = \dfrac{1}{\mu_a^2}[y_i^2 \sigma_a^2 + \sigma_b^2].$

Then, as an unbiased estimator for Y one may take $e = \sum_{i \in s} r_i b_{si}$ with b_{si} as constants free of $\underline{Y} = (y_1, \cdots, y_i \cdots, y_n)$ subject to

$$\sum_{s \ni i} p(s) b_{si} = 1 \ \forall \ i \in U.$$

Then,
$$E(e) = E_p(E_R(e)) = E_p(t) = Y,$$
writing
$$t = \sum_{i \in s} y_i b_{si}.$$

Also,
$$V(e) = E_p V_R(e) + V_p E_R(e)$$
$$= E_p \left(\sum_{i \in s} V_i b_{si}^2 \right) + V_p(t)$$
$$= \sum_{i=1}^{N} V_i \sum_{s \ni i} p(s) b_{si}^2 + \sum y_i^2 C_i + \sum \sum_{i \neq j} y_i y_j C_{ij}$$

writing
$$C_i = \sum_{s \ni i} p(s) b_{si}^2 - 1 \quad \text{and} \quad C_{ij} = \sum_{s \ni i,j} p(s) b_{si} b_{sj} - 1.$$

An unbiased estimator for V_i above is
$$v_i = \left(r_i^2 \frac{\sigma_a^2}{\mu_a^2} + \frac{\sigma_b^2}{\mu_a^2} \right) \bigg/ \left[1 + \frac{\sigma_a^2}{\mu_a^2} \right]$$

because clearly, $E_p(v_i) = V_i$.

Let $C_{si}, Csij$'s free of $\underline{Y}$ be available satisfying
$$\sum_{s \ni i} p(s) C_{sij} = C_i \quad \text{and} \quad \sum_{s \ni ij} p(s) C_{sij} = C_{ij}.$$

Let
$$v(e) = \sum_{i \in s} r_i^2 c_{si} + \sum \sum_{i \neq j} r_i r_j c_{sij} + \sum_{i \in s} v_i \left(b_{si}^2 - c_{si} \right).$$

Then
$$E[v(e)] = E_p \left[\sum_{i \in s} y_i^2 c_{si} + \sum \sum_{i \neq j} y_i y_j c_{sij} + \sum_{i \in s} V_i \left(b_{si}^2 - C_{si} \right) \right] + \sum_{i \in s} V_i c_{si}$$
$$= \sum_i y_i^2 c_i + \sum \sum_{i \neq j} y_i y_j c_{ij} + \sum_i V_i \sum_{s \in i} p(s) b_{si}^2.$$

Alternatively, $V(e) = E_R V_p(e) + V_R E_p(e)$
$$= E_R \left[\sum_i r_i^2 C_i + \sum \sum_{i \neq j} r_i r_j C_{ij} \right] + V_R \left(\sum_1^N r_i \right)$$
$$= \sum y_i^2 C_i + \sum \sum_{i \neq j} y_i y_j C_{ij} + \sum_i V_i C_i + \sum V_i$$
$$= \sum y_i^2 C_i + \sum \sum_{i \neq j} y_i y_j C_{ij} + \sum_i V_i \left(\sum_{s \ni i} p(s) b_{si}^2 \right).$$

So,
$$v'(e) = \sum_{i \in s} r_i^2 C_{si} + \sum_{i \neq j} r_i r_j C_{sij} + \sum_{s \in s} v_i b_{si}$$

is an unbiased estimator of $V(e)$ because

$$E_p(v'(e)) = \sum_i r_i^2 C_i + \sum_{i \neq j} r_i r_j C_{ij} + \sum_1^N v_i$$

and so
$$Ev'(e) = E_R \left[E_p(v'(e)) \right]$$
$$= \sum y_i^2 C_i + \sum_{i \neq j} y_i y_j C_{ij} + \sum V_i + \sum_i V_i C_i = V(e).$$

6.3 OPTIONAL RANDOMIZED RESPONSE TECHNIQUES

An RRT is employed in the belief that an item of enquiry is sensitive from the perspective of a potential respondent. But when put into actual practice a respondent on being explained the background and motivation may assert that the issue does not appear at all sensitive to him/her and hence may volunteer to give out the correct answer directly to the query. Chaudhuri and Mukerjee (1985, 1988) gave their method to obtain an unbiased estimator for the proportion of the people bearing a supposedly stigmatizing characteristic A in a community on taking an SRSWR and giving every sampled person an option either to give out a direct response or an RR.

Corresponding variance and variance estimator formulae are given by Chaudhuri and Mukerjee (1985, 1988). But as Chaudhuri (2011) has emphasized that analysis of RR data need not depend on the sample-selection method we need not describe their method. Instead, Chaudhuri and Saha's (2005) procedure based on a general sampling scheme deserves a quote as follows. A sample s composed of a part s_1, giving DR's as y_is and the other part s_2 giving RR's yields for Y, an unbiased estimator

$$e*_b = \sum_{i \in s_1} y_i b_{si} + \sum_{i \in s_2} r_i b_{si}$$

with $y_i = 1/0$ for $i \in A/i \in A_C$, r_i subject to $E_R(r_i) = y_i$ and b_{si} subject to $1 = \sum_{s \in i} p(s) b_{si} \forall i \in U$.

This e_b^* based on optional RR's (i.e., $ORR's$) may be contrasted with the usual RR-based, i.e., totally RR-based unbiased estimator for Y, namely $e_b = \sum_{i \in s} r_i b_{si}$.

Now, $E(e_b | y_i, i \in s_1) = e_b^*$ and hence by Rao-Blackwell's theorem

$$V_R(e_b) = V_R(e_b^*) + E_R(e_b - e_b^*)^2$$

and also, given an unbiased estimator $v(e_b)$ for $V(e_b)$ an unbiased estimator for $V(e_b^*)$ follows as

$$v(e_b^*) = v(e_b) - (e_b - e_b^*)^2.$$

One is induced to consult Chaudhuri (2011) for further clarifications.

In order to unbiasedly estimate the total $Y = \sum_1^N y_i$ of real variable values supposedly stigmatizing also an optional RRT may be cleverly employed. Chaudhuri and Dihidar (2009) have given a procedure worthy of reporting here in brief.

Let a contemplated RRT be to request a sampled person to randomly draw from a box given to him/her a card out of several of them that are alike except that numbers $a_1, \cdots, a_M$ are inscribed on them and similarly but independently draw one card out of another box of cards marked by $b_1, \cdots, b_L$, respectively, with $\mu_a = \frac{1}{M}\sum_1^M a_i = 1$ and $\mu_b = \frac{1}{L}\sum_1^L b_i$. Similarly, the ith person is requested to repeat this exercise using a third box containing cards marked $b'_1, \cdots, b'_L$ with $\mu'_b = \frac{1}{L}\sum_1^L b'_i \neq \mu_b$. Every ith person in the sample is given an option either to give a DR about y_i with an unknown probability C_i $(0 < C_i < 1)$ or with probability $(1 - C_i)$ to give two independent RR's as

$$I_i = a_i\, y_i + b_i \quad \text{and} \quad I'_i = a_i\, y_i + b'_i.$$

Letting

$$z_i = y_i \quad \text{with probability} \quad C_i$$
$$ = I_i \quad \text{with probability} \quad (1 - C_i)$$

and

$$z'_i = y_i \quad \text{with probability} \quad C_i$$
$$ = I'_i \quad \text{with probability} \quad (1 - C_i).$$

Then observing,

$$E_R(z_i) = C_i y_i + (1 - C_i)(y_i + \mu_i),$$
$$E_R(z'_i) = C_i y_i + (1 - C_i)(y_i + \mu'_b) \text{ yielding}$$
$$r_{1i} = (\mu'_b z_i - \mu_b z'_i)/(\mu'_b - \mu_b) \text{ with } E_R(r_{1i}) = y_i.$$

Repeating such an exercise independently once again another quantity r_{2i} with $E_R(r_{2i}) = y_i$ may be derived independently of r_{1i} giving

$$r_i = \frac{1}{2}(r_{1i} + r_{2i}) \text{ with } E_R(r_i) = y_i \text{ and}$$
$$v_i = \frac{1}{4}(r_{1i} - r_{2i})^2 \text{ with } E_R(v_i) = V_R(r_i).$$

From this one may derive $e_b = \sum_{i \in s} r_i b_{si}$ satisfying $E(e_b) = Y$ with an unbiased estimator for $V(e_b)$.

Next, let us consider how to protect the privacy of a respondent.

First considering the case of a qualitative characteristic A which is stigmatizing, let L_i denote the prior probability that the ith person bears A, i.e., $y_i = 1$ and let $L_i(R)$ denote the posterior probability that the ith person bears A when the RR from him/her is R. By Bayes theorem then we get, e.g., for Warner's (1965) RR model

$$L_i(R) = \frac{L_i P(I_i = 1|y_i = 1)}{L_i P(I_i = 1|y_i = 1) + (1 - L_i) P(I_i = 1|y_i = 0)}$$

$$= \frac{pL_i}{pL_i + (1-p)(1-L_i)} = \frac{pL_i}{(1-p) + (2p-1)L_i}.$$

Then, the closer $L_i(R)$ is to L_i the better the RRT.

Then,
$$J_i(R) = \frac{L_i(R)/L_i}{[1 - L_i(R)]/(1 - L_i)}$$

is taken as a "jeopardy measure" for the RRT associated with the ith person in respect of the RR given out as Chaudhuri, Christofides, and Saha (2009) have recommended the use of

$$\bar{J}_i = \frac{\sum_R J_i(R)}{\# \text{ RR's prescribed by the RRT}},$$

$\sum_R$ denoting the sum over the set of RR's prescribed by the specific RRT. Chaudhuri et al. (2009) prescribe that the closer the $\bar{J}_i$ to unity the better the RRT. It is left "the ith respondent specific" because corresponding to each i one has to note the magnitude of $V_R(r_i) = V_i$, writing for the ith person the transformed RR as r_i for which $E_R(r_i) = y_i$ and in terms of r_i's for i in s the final estimator for Y is to be obtained and the variance of the estimator has its magnitude also influenced by V_i and the variance estimator by an unbiased estimator for $V_i, i \in s$.

Next we consider what to do about protection of privacy in case of dealing with quantitative variables that are anticipated to be stigmatizing.

Let a person i bear the real value y_i which may be a stigmatizing one or not, our intention being to unbiasedly estimate the total $Y = \sum_i y_i$. Let the investigator approach a sampled person with two boxes of cards, respectively, marked the values

$a_1, \cdots, a_j, \cdots, a_T$ with

$$\mu_a = \frac{1}{T} \sum_1^T a_j \neq 0 \quad \text{and} \quad \sigma_a^2 = \frac{1}{T} \sum_1^T (a_j - \mu_a)^2 > 0 \quad \text{and}$$

$b_1, \cdots, b_M$ with

$$\mu_b = \frac{1}{M} \sum_1^M b_k \quad \text{and} \quad \sigma_b^2 = \frac{1}{M} \sum_1^M (b_k - \mu_b)^2$$

with a request to independently draw just one card from each box. Then from a sampled person i the RR's are gathered

$$z_i = a_j y_i + b_k, j = 1, \cdots T, \cdots k = 1, \cdots, M$$

giving $E_R(z_i) = y_i \mu_a + \mu_b, V_R(z_i) = y_i^2 \sigma_a^2 + \sigma_b^2$

leading to $r_i = \dfrac{z_i - \mu_b}{\mu_a}, \quad E_R(r_i) = y_i, V_R(r_i) = \dfrac{y_i^2 \sigma_a^2 + \sigma_b^2}{\mu_a^2}$

and $v_R(r_i) = \dfrac{(r_i^2 \sigma_a^2 + \sigma_b^2)}{1 + \frac{\sigma_a^2}{\mu_a^2}} / \mu_a^2 \quad \text{with} \quad E_R(v_R(r_i)) = V_R(r_i).$

It should be noted that for any y_i which is fixed but unknown z_i assumes one of its possible TM values (for $j = 1, \cdots T$ and $k = 1, \cdots M$) each with a common probability $\frac{1}{TM}$.

Let $L(y_i) = L_i$ be the prior probability that the variable assumed discrete takes the value y_i for the ith person. Let $P(z_i/y_i)$ be the probability that z_i may take on a value as above given y_i or just the likelihood of y_i given $z_i, i \in s$. Then the posterior probability that y_i is revised given that the response z_i is at the hand of the investigator turning out as

$$L(y_i|z_i) = \dfrac{L_i P(z_i|y_i)}{P(z_i)}$$

writing $P(z_i)$ for the probability that z_i may assume one of the TM values as above. So, $L(y_i|z_i) = L_i \ \forall \ i \in U$.

Thus, a respondent's privacy is well protected for every T and M provided they are both at least moderately large. If, e.g., $T = 1$ and $M = 1$, then given z_i, the value of y_i is uniquely found as

$$y_i = \dfrac{z_i - b_1}{a_1} \forall \ i \in U.$$

In this case no protection of privacy is ensured. But even if $T = 2$ and $M = 2$, for example, privacy is hardly compromised. Chaudhuri and Dihidar (2009) have a second device to cover quantitative response. In this device an investigator carries a box of cards, a proportion C $(0 < C < 1)$ marked blank and the remaining numbers $x_1, \cdots, x_M$ such that their respective proportions q_j $(j = 1, \cdots, M)$ are such that $0 < q_j < 1$ but $\sum_1^M q_j = (1 - C)$. Then, on request, from the ith person the forthcoming response is

$$z_i = y_i \text{ if a blank is drawn}$$
$$= x_j \text{ if an } x_j\text{-marked card is drawn.}$$

Then,
$$E_R(z_i) = Cy_i + \sum_{j=1}^{M} q_j x_j.$$

Letting
$$r_i = \left(z_i - \sum_{1}^{M} q_j x_j\right)/C, i \in U,$$
$$E_R(r_i) = y_i \ \forall \ i \in U.$$

Also,
$$V_R(z_i) = Cy_i^2 + \sum_{1}^{M} q_j x_j^2 - \left(Cy_i + \sum_{1}^{M} q_j x_j\right)^2.$$

So,
$$V_i V_R(r_i) = \frac{1}{C^2} V_R(z_i) = \alpha y_i^2 + \beta y_i + \phi, \text{ say,}$$

with α, β, ϕ as known. Then,
$$v_i = \left(\alpha r_i^2 + \beta r_i + \phi\right)/(1+\alpha) \quad \text{has} \quad E_R(v_i) = V_i.$$

Now paralleling the situations as in the earlier device we may get

$$L(y_i|z_i) = \frac{L(y_i)C}{L(y_i)C + (1-C)(1-L(y_i))}$$
$$= \frac{CL_i}{CL_i + (1-C)(1-L_i)} = \frac{CL_i}{L_i(2C-1) + (1-C)}$$
$$= \frac{L_i}{L_i\left(2 - \frac{1}{C}\right) + \left(\frac{1}{C} - 1\right)}.$$

So, $L(y_i|z_i)$ matches L_i if $C = \frac{1}{2}$.

Taking $C = \frac{1}{2}$, the privacy is fully protected. So, C is to be appropriately fixed to keep $V_R(r_i)$ under control and $L(y_i|z_i)$ kept as close to L_i as practicable.

Next, let us discuss our method of Bayes estimation concerning $Y = \sum_{1}^{N} y_i$ as briefly reproduced from Chaudhuri and Christofides's (2013) monograph. With reference to Warner's (1965) RRT let

$$y_i = 1 \text{ if } i \text{ bears } A$$
$$= 0 \text{ if } i \text{ bears } A^C, i \in U.$$

Let $I_i = 1$ if the card type drawn by i matches i's characteristic A or A^C
$$= 0, \text{ if it mis-matches.}$$

Then,
$$E_R(I_i) = py_i + (1-p)(1-y_i) = (1-p) + (2p-1)y_i$$
and
$$V_R(I_i) = p(1-p).$$

Stigmatizing Issues

Letting
$$r_i = \left[I_i - (1-p)\right] \Big/ (2p-1), \; p \neq \frac{1}{2},$$
$$E_R(r_i) = y_i \text{ and}$$
$$V_i = V_R(r_i) = \frac{p(1-p)}{(2p-1)^2} \; \forall \, i \in U.$$

Let L_i be the prior unknown probability that y_i equals $1, i \in U$. Then, the posterior probability that y_i equals 1 when I_i turns up 1 takes the form

$$L_i(1) = \frac{pL_i}{pL_i + (1-p)(1-L_i)} = \frac{pL_i}{(1-p) + (2p-1)L_i}.$$

Whence
$$\frac{1}{L_i(1)} = \frac{1-p}{pL_i} + \frac{(2p-1)}{p} \quad \text{giving us}$$
$$L_i = \frac{1-p}{p}\left[\frac{1}{L_i(1)} - \frac{2p-1}{p}\right]^{-1}.$$

With a simplified Bayesian approach let r_i be taken as an estimator for $L_i(1)$.

Then,
$$\hat{L}_i = \frac{(1-p)}{p}\left[\frac{1}{r_i} - \frac{2p-1}{p}\right]^{-1}$$
$$= \alpha\left(\frac{1}{r_i} - \beta\right)^{-1} \quad \text{with}$$
$$\alpha = \frac{1-p}{p} \text{ and } \beta = \frac{2p-1}{p}$$

may be taken as an estimator for $L_i, i \in U$.

We may then take $\psi = \frac{1}{N}\sum_1^N L_i$ as our estimated parameter of interest. Then, taking π_i as a positive inclusion probability of the ith person for our chosen sampling design,

$$\hat{\psi} = \frac{1}{N}\sum_1^N \frac{\hat{L}_i}{\pi_i}$$

may be taken as an unbiased estimator for ψ. Writing π_{ij} as the positive inclusion-probability of the paired units (i,j), we may take

$$v(\hat{\psi}) = \frac{1}{N^2}\left[\sum\sum_{i<j \in s}\frac{(\pi_i\pi_j - \pi_{ij})}{\pi_{ij}}\left(\frac{\hat{L}_i}{\pi_i} - \frac{\hat{L}_j}{\pi_j}\right)^2 + \sum_{i \in s}V_i\left(\frac{\hat{L}_i}{\pi_i}\right)\right]$$

as an unbiased estimator for

$$V(\hat{\psi}) \simeq \frac{1}{N^2}\left[\sum_{i<j}^N\sum^N(\pi_i\pi_j - \pi_{ij})\left(\frac{L_i}{\pi_i} - \frac{L_j}{\pi_j}\right)^2 + \sum_{i=1}^N V_R(\hat{L}_i)\Big/\pi_i\right],$$

and here we may note

$$E_R\left(\hat{L}_i\right) \simeq \frac{1-p}{p}\left(\frac{1}{L_i(1)} - \frac{2p-1}{p}\right)^{-1} = L_i.$$

For other RRT's similar Bayesian methods may be tried.

6.4 INDIRECT QUESTIONING

In encountering stigmatizing issues, Randomized Response (RR) is of course a celebrated technique and is widely being cultivated ever since its inception in 1965 when Warner published his pioneering work. But because of certain practical criticism against its applicability, some other techniques of Indirect Questioning are also emerging as alternatives to RRT. Let us briefly recount a few of them. One important criticism against RRT is that it cannot be repeated yielding the same response to a same question from the same person on subsequent interviews following the first. Also a respondent in RR has to be well educated, well motivated, and also quite active to respond usefully. These traits are rather hard to find in practice. Often a respondent refuses to co-operate anticipating a trick being made in order to make a fool of a credulous respondent. So, RRT is not quite a popular device. We may briefly recount four techniques rivaling this.

One is Item Count Technique (ICT). Miller, Cisin, and Harrel (1986) introduced it. As usual they developed the theory demanding exclusively SRSWR. Chaudhuri and Christofides (2007) liberated it to open it out to the involvement of general sampling schemes admitting positive inclusion-probabilities of single and paired persons. Chaudhuri (2011) and Chaudhri and Christofides (2013) in their monographs have further dwelt upon this theme.

Following Miller et al. (1986) and Chaudhuri (2011) the ICT may be briefly described as one in which a respondent in one sample s_1 is given a list of $G(>1)$ items, each innocuous and in a second sample s_2 chosen independently of s_1, is given a list containing these G innocuous items plus one sensitive item. Each respondent is to report just the number of these listed items that apply to himself/herself.

Letting π_i, π_{ij} denote the inclusion-probabilities of i and j in a sample chosen according to the same design p, following which both s_1 and s_2 have been chosen and y_i equals the number reported by i in s_1 and x_j equals the number reported by j in s_2 chosen from the population U of N persons and θ_F, the known proportion of the people born in January to August, say, and θ the proportion in U bearing the sensitive attribute A, it follows that

$$\hat{\theta} = (t_1 - t_2) + (1 - \theta_F)$$

unbiasedly estimates θ.

Here $t_1 = t_1(s_1)$ such that

$$Nt_1 = \sum_{i \in s_1} \frac{y_i}{\pi_i}$$

and $t_2 = t_2(s_2)$ such that

$$Nt_2 = \sum_{j \in s_2} \frac{x_j}{\pi_j}.$$

This is because
$E_p(t_1)$
= Proportion bearing A + Proportion bearing F − Proportion bearing $A \cap F$,

$E_p(t_2) = 1 -$ Proportion bearing $A \cap F$ using DeMorgan's Law giving

$$A^C \cup F^C = (A \cap F)^C.$$

So, $E_p(t_1) - E_p(t_2) = \theta + \theta_F - 1$. Hence, $\hat{\theta}$ is unbiased for θ.

Letting
$$\beta_k = 1 + \frac{1}{\pi_K} \sum_{\substack{l=1, \\ l \neq k}}^{N} \pi_{kl} - \sum_{1}^{N} \pi_k, \quad \text{easily,}$$

$$V_p(\hat{\theta}) = \frac{1}{N^2} \left[\sum_{k=1}^{N} \sum_{k+1}^{N} (\pi_k \pi_l - \pi_{kl}) \left\{ \left(\frac{y_k}{\pi_k} - \frac{y_l}{\pi_l} \right)^2 + \left(\frac{x_k}{\pi_k} - \frac{x_l}{\pi_l} \right)^2 \right\} \right.$$

$$\left. + \sum_{1}^{N} \frac{\beta_K}{\pi_k} \left(y_k^2 + x_k^2 \right) \right], \quad \text{vide Chaudhuri and Pal (2002)}$$

and
$$v_p(\hat{\theta}) = \frac{1}{N^2} \left[\sum_{\substack{k \in s_1 \\ l > k}} \sum_{l \in s_1} \left(\frac{\pi_k \pi_l - \pi_{kl}}{\pi_{kl}} \right) \left(\frac{y_k}{\pi_k} - \frac{y_l}{\pi_l} \right)^2 \right.$$

$$+ \sum_{\substack{k \in s_2, \ l \in s_2 \\ l > k}} \left(\frac{\pi_k \pi_l - \pi_{kl}}{\pi_{kl}} \right) \left(\frac{x_k}{\pi_k} - \frac{x_l}{\pi_l} \right)^2$$

$$\left. + \sum_{k \in s_1} \frac{y_k^2}{\pi_k^2} \beta_K + \sum_{k \in s_2} \frac{x_k^2}{\pi_k^2} \beta_k \right]$$

is an unbiased estimator for $V_p(\hat{\theta})$.

Another Indirect Questioning Technique as an alternative to RRT is known as Nominative Technique introduced by Miller (1985). In a forthcoming monograph Chaudhuri (2014) has given details of Network sampling introduced by Thompson (1992) further elaborated by Chaudhuri (2000). The Nominative Technique (NT) is essentially an extended exercise of Network sampling in our view. This NT is discussed as follows in brief. Here a sampled person is required to give the whereabouts of his/her acquaintances who are known bearers of the stigmatizing attribute A in question but that person is not asked to reveal any truth about himself/herself. In a finite universe $U = (1, \ldots, i, \ldots, N)$

let $\quad r_{ij} = 1 \quad$ if jth person reports that an ith person bears A
$\qquad\quad\; = 0, \quad$ else; $\quad i \neq j \in U$.

Then, $\displaystyle\sum_{\substack{i=1 \\ j \neq i}}^{N} \sum_{j=1}^{N} r_{ij} = $ Total number in the community U of N people

are reported as bearing A

Then, $\displaystyle T = \sum_{i=1}^{N} \sum_{\substack{j=1 \\ j \neq i}}^{N} \left(\frac{r_{ij}}{\sum_{k=1}^{N} r_{iK}} \right) = N_A = $ Total number bearing A

By convention, take $\displaystyle \frac{r_{ij}}{\sum_{k=1}^{N} r_{iK}} = 0$, if $\displaystyle\sum_{1}^{N} r_{iK} = 0$.

Let $A_j = $ Number bearing A as reported by jth person and $B_j = $ Number nominated by jth person who know that jth person also bears A. Let $x_j = \frac{A_j}{1+B_j}$. For an SRSWR in n draws let

$$\hat{\theta} = \frac{1}{n} \sum_{j=1}^{n} x_j \quad \text{and} \quad t = N\hat{\theta}.$$

Then $\hat{\theta}$ unbiasedly estimates $\theta = \frac{N_A}{N}$ and t unbiasedly estimates $N_A \equiv$ the number bearing A among people in U.

Chaudhuri and Christofides (2008) modify this NT as follows.

A selected person j reports (1) that one i in U bears A and also (2) a number of close friends of i who know that i bears A. Here j is a selection unit (SU) and i is an observation unit (OU) and is "nominee" of jth person.

Let $M_j = $ the set of j's nominees, and $m_i = $ Total number of i's friends knowing his features and ready to report so.

Let $\quad I_i(A) = 1 \quad $ if i bears A
$\qquad\quad\;\; = 0, \quad $ else.

Stigmatizing Issues

Let
$$y_j = \sum_{i \in M_j} \frac{I_i(A)}{m_i}.$$

Then, $\sum_1^N y_j = N_A$ and $t = \sum_{j \in s} \frac{y_j}{\pi_j}$ unbiasedly estimates N_A.

$$v_p(t) = \sum\sum_{\substack{k,l \in s \\ l > K}} \left(\frac{\pi_k \pi_l - \pi_{kl}}{\pi_{kl}}\right)\left(\frac{y_k}{\pi_k} - \frac{y_l}{\pi_l}\right)^2 + \sum_{k \in s} \frac{y_k^2}{\pi_k^2} \beta_k, \quad \text{if} \quad \pi_{kl} > 0,$$

is an unbiased variance estimator for t,

$$\beta_K = 1 + \frac{1}{\pi_K}\sum_{l \neq K} \pi_{Kl} - \sum_1^N \pi_k.$$

Another Indirect Question Technique called Three Sample Item Count Technique (TSICT) is described briefly as follows by Chaudhuri and Christofides (2013).

Three independent SRSWR's of sizes n_1, n_2, n_3 are drawn as s_1, s_2, s_3. Each sampled person is requested to truthfully report just the number k of items applicable to him/her out of G items in the list for s_3 but those G items, plus a $G+1_{st}$ item, which says

(i) I bear at least one of A and F
and (ii) I bear at least one of A and F^C,

respectively, in lists given to one in s_1 and one in s_2. Here as in case of Item Count Technique (ICT) A is the sensitive item with unknown proportion θ for the population $U = (1, \ldots, i, \ldots, N)$ and F with a known proportion θ_F for the people in U.

Let $n_K(1), n_K(2), n_K(3)$ be the observed numbers in respective samples s_1, s_2, s_3 reporting that each of them bears exactly k items out of the list given; $k = 0, 1, \cdots, G, G+1$ for s_1, s_2 and $k = 0, 1, \cdots, G$ for people in s_3. Let p_k be the probability that a person bears k of the innocuous items $1, 2, \cdots, G$, $q_k(1)$ be the probability that a person in s_1 bears k items and $q_k(2)$ be the corresponding probability for one in s_2. Then since the samples are independent and characteristics A and F are also independent using the properties of Multinomial distribution Chaudhuri and Christofides find

$$\hat{\theta} = \frac{1}{n_1}\sum_1^{G+1} k n_k(1) + \frac{1}{n_2}\sum_1^{G+1} k n_k(2) - \frac{2}{n_3}\sum_1^{G} k n_k(3) - 1$$

as an unbiased estimator for θ and

$$v\left(\hat{\theta}\right) = \frac{s_1^2}{n_1} + \frac{s_2^2}{n_2} + \frac{s_3^2}{n_3}$$

as an unbiased estimator for $V\left(\hat{\theta}\right)$, writing $s_i^2, i = 1, 2, 3$ as the sample variances for the reported numbers by the persons in the respective samples.

The next Indirect Method we briefly present is known as the Three Card Method. This method also needs three independent SRSWR's as in the previous device. Droitcour, Larson, and Scheuren (2001) introduced this. The method uses three boxes presented to each person in each sample. One sensitive characteristic A is considered along with three others B, C, D which are innocuous.

A sampled person is to report only the Box number truthfully announcing the characteristic borne by him/her. A person in sample s_1 is presented three boxes as Box 1 : I bear A, Box 2 : I bear C or D or A, Box 3 : I bear nothing noted in Box 1 or Box 2. A person in sample s_2 is presented Box 1 : I bear C, Box 2 : I bear B or D or A, Box 3 : I bear nothing in Boxes 1 and 2. A person in sample s_3 is presented Box 1 : I bear D, Box 2 : I bear B or C or A, and Box 3 : I bear nothing as in Boxes 1 and 2. Obtaining the truthful responses from all the sampled persons about the Box that reflect his/her feature the proportion in the population is easily estimated unbiasedly, as discussed by Chaudhuri and Christofides (2013).

One more Indirect Response Procedure of interest is known as Non-Randomized Techniques. Tian, Yu, Tang, and Gang (2007); Yu, Tian, and Tang (2008); Tan, Tian, and Tang (2009); Christofides (2009); Chaudhuri (2012); and Chaudhuri and Christofides (2013); among others are some of the important references where one may find details of Non-Randomized Methods useful in gathering and analytically tracking down data on sensitive issues by methods other than RRTs.

Let y denote a sensitive characteristic, A take the value 1 for a person bearing A, and the value 0 for one bearing A^C. Let x be another characteristic B which is innocuous such that when a person bears B his/her x-value is 1 and when he/she bears B^C, his/her x-value is 0, independent of what value y takes for a person with $x = 1$ or with $x = 0$. We are interested in the situations $(y = 0, x = 0), (y = 1, x = 0)$ and $(y = 1, x = 1)$ in order to estimate the

$$\text{Prob}\,(y = 1) = \theta = \frac{1}{N} \sum_{i=1}^{N} y_i \quad \text{for} \quad i \in U = (1, \cdots, i, \cdots, N).$$

Let
$$\text{Prob}\,(x = 1) = \frac{1}{N} \sum_{1}^{N} x_i = p$$

which is supposed to be known. Because of independence assumption one has

$$P\,(y = 1, x = 1) = \theta p, \quad P\,(y = 1, x = 0) = \theta\,(1 - p),$$
$$P\,(y = 0, x = 1) = (1 - \theta)\,p, \quad P\,(y = 0, x = 0) = (1 - \theta)\,(1 - p).$$

Stigmatizing Issues

For a person labeled i let

$$C_{00i} = (y_i = 0, x_i = 0), \quad C_{01i} = (y_i = 0, x_i = 1),$$
$$C_{10i} = (y_i = 1, x_i = 0), \quad C_{11i} = (y_i = 1, x_i = 1) \quad \text{and let}$$
$$d_i = C_{01i} U C_{10i} U C_{11i}.$$

Consequently, $\theta = \dfrac{1}{N}\sum_1^N y_i = \dfrac{1}{N}\left(\sum_1^N C_{10i} + \sum_1^N C_{11i}\right).$

A sampled person i is requested to report truly confounded features about himself/herself like $(y = 1, x = 0)$ or $(y = 1, x = 1)$ or $(y = 0, x = 1)$ at a time; if the answer is "Yes" we record the response as 1; if "No" we record it as 0. This is known as the triangular model because graphically taking $x = 0, x = 1$ horizontally and $y = 0, y = 1$ vertically, the above Reporting model takes a "triangular" shape. An alternative model called the Cross-wise model asks for a "Yes" or "1" response about either $(y = 0, x = 1)$ or $(y = 1, x = 1)$ and "No" or "0" in the contrary case. With the Triangular Model one estimates θ by

$$\hat{\theta} = \frac{1}{N}\left[\sum_{i \in s}\frac{d_i}{\pi_i} - \frac{p}{1-p}\sum_{i \in s}\frac{C_{00i}}{\pi_i}\right] = \frac{1}{N}\sum_{i \in s}\frac{e_i}{\pi_i}, \quad \text{say,}$$

with $\quad e_i = d_i - \left(\dfrac{p}{1-p}\right)C_{00i}.$

Then, it is easy to see that

$$\hat{V}\left(\hat{\theta}\right) = \frac{1}{N^2}\left[\sum\sum_{i<j}\left(\frac{\pi_i\pi_j - \pi_{ij}}{\pi_{ij}}\right)\left(\frac{e_i}{\pi_i} - \frac{e_j}{\pi_j}\right)^2\right], \quad \text{if } \pi_{ij} > 0.$$

For the Cross-wise Model, from i Response will be either

$$b_i = (y_i = 0, x_i = 0) U (y_i = 1, x_i = 1)$$
$$\text{or } g_i = (y_i = 0, x_i = 1) U (y_i = 1, x_i = 0),$$

the response being formally 1 or 0 as usual.

Chaudhuri and Christofides (2013) have given

$$\theta^* = \frac{1}{N}\sum_{i \in s}\frac{f_i}{\pi_i}$$

as an unbiased estimator for θ taking

$$f_i = \frac{pb_i - (1-p)g_i}{(2p-1)}.$$

Then, $\quad \hat{V}\left(\theta^*\right) = \dfrac{1}{N^2}\sum\sum_{i<j}\left(\dfrac{\pi_i\pi_j - \pi_{ij}}{\pi_{ij}}\right)\left(\dfrac{f_i}{\pi_i} - \dfrac{f_j}{\pi_j}\right)^2,$

taking throughout a Fixed sample-size, sampling design admitting all units are distinct.

7 Developing Small Domain Statistics

Abstract. Introduction. Some details.

7.0 ABSTRACT

The Model Assisted approach is here considered in estimating suitably the total or mean of a variable related to a "specific section" of a "survey population" on surveying a probability sample drawn from the latter. A classical design-based estimator does not suit this situation because the sample-size coming as a chunk from this specific domain may be inadequate, yielding too high a measure of error and implying a poor efficiency level. The current research deals with this situation only through model-based and more specifically the Bayesian approach.

7.1 INTRODUCTION

Rao (2003) and Mukhopadhyay (1998) are, to our knowledge, the only authors of books on this topic. The present author also, however, published an e-Book entitled "Developing Small Domain Statistics: Modelling in Survey Sampling" in 2012, *vide* Lambert Academic Publishing (LAP) (ISBN: 978-3-659-13676-4) GMb H & Co. KG Heinrich-Bocking-Str 6-8, 66121 Saarbrucken, Germany, email: info@ Lap-publishing.com. It is a common practice nowadays to have a brief section on this topic in Sample Survey Texts.

The problem of Small Area Estimation arises as follows. Suppose according to necessity a sample has been duly chosen for survey from a population in a given context. But in addition it is desired to obtain serviceable estimators not only for the population total or mean but also for the totals and means of variables of interest valued on one or more mutually exclusive sections of the same population. In such circumstances often it so happens that though the population parameter is estimated well, because too small samples may happen to come from certain population sections of interest, the corresponding estimates of the totals/means for some of the sections may turn out too poor in terms of their estimated measures of error. This is recognized as a Small Area Estimation problem.

7.2 SOME DETAILS

We shall deal briefly with the following: Small Area Estimation, Generalized Regression, Empirical Bayes Methods, Fay-Herriot Methods, and Kalman

Filtering Estimation.

Let a finite population $U = (1, \cdots, i, \cdots N)$ of N units be split up into a large number D of constituent disjoint parts $U_d, d = 1, \cdots, D$, to be called the domains of respective sizes

$$N_d, \quad \sum_{d=1}^{D} N_D = N.$$

Let a sample s of size n be drawn from U with a selection-probability $p(s)$. Let one may determine the units of s being members of the respective U_d for $d = 1, \cdots, D$ and hence numbers n_d of persons of $s_d = s \cap U_d, d = 1, \cdots, D$ be ascertained. Clearly n_d and s_d are random variables and

$$n = \sum_{d=1}^{D} n_d, \quad n_d \geq 0 \ \forall \ d = 1, \cdots, D.$$

Let, for a real variable y, the population total and the domain totals

$$Y = \sum_{1}^{N} y_i \text{ and } Y_d = \sum_{i \in U_d} y_i, d = 1, \cdots, D$$

be both required to be suitably estimated. Even in case n is suitably large so that a reasonably efficient estimator $t = t(s, \underline{Y}), \underline{Y}(y_1, \cdots, y_i, \cdots, y_N)$ may be found the corresponding estimator $t_d = t(s_d, \underline{Y})$ for Y_d may not be good enough because n_d may be too low relative to n. For example, the level of efficiency may be ascertained in terms of magnitude of the estimated coefficient of variation (CV) which is

$$CV = 100 \times \frac{\text{Estimated standard deviation of } t_d}{|t_d|}.$$

Conventionally, t_d as an estimator of Y_d is

i. Excellent if $CV \leq 10\%$,
ii. Good enough if $10\% < CV \leq 20\%$,
iii. Tolerable if $20\% < CV \leq 30\%$, and
iv. Unacceptable if $CV > 30\%$.

This is recognized as a problem of Small Area Estimation (SAE) or more generally one of Developing a Small Domain Statistic.

Let us consider one example. The Horvitz-Thompson (1952) estimator for Y is

$$t_H = \sum_{i \in s} \frac{y_i}{\pi_i} \text{ with a variance}$$

$$V_p(t_H) = \sum_{i<j}^{N} \sum^{N} (\pi_i \pi_j - \pi_{ij}) \left(\frac{y_i}{\pi_i} - \frac{y_j}{\pi_j} \right)^2 + \sum_{1}^{N} \frac{y_i^2}{\pi_i} \beta_i$$

with $\qquad \beta_i = 1 + \frac{1}{\pi_i} \sum_{j \neq i}^{N} \pi_{ij} - \sum_{1}^{N} \pi_i$

admitting an unbiased estimator

$$v(t_H) = \sum_{i<j}^{N}\sum^{N} (\pi_i \pi_j - \pi_{ij}) \left(\frac{y_i}{\pi_i} - \frac{y_j}{\pi_j}\right)^2 \frac{I_{sij}}{\pi_{ij}} + \sum_1^N \frac{y_i^2}{\pi_i}\beta_i \frac{I_{si}}{\pi_i}$$

assuming $\pi_i > 0, \pi_{ij} > 0 \ \forall \ i,j \ (\neq i)$.

Let
$$I_{di} = 1 \quad \text{if} \quad i \in U_d.$$
$$= 0, \quad \text{else}.$$

Let
$$y_{di} = y_i I_{di}.$$

Then, for Y_d the Horvitz-Thompson Estimator (HTE) for Y_d with its variance and variance estimator are

$$t_{Hd} = \sum_{i \in s} \frac{y_i}{\pi_i} I_{di} = \sum_1^N y_{di} \frac{I_{si}}{\pi_i},$$

$$V_p(t_{Hd}) = V_p(t_H)|_{y_i=y_{di}}, \quad v(t_{Hd}) = v(t_H)|_{y_i=y_{di}}.$$

This is too simple but does not reveal any problem before us. So, let us suppose that from U an SRSWOR of n units has been taken. Then, for the population mean $\bar{Y} = \frac{Y}{N}$ an unbiased estimator is

$$\bar{y} = \frac{1}{n}\sum_{i \in s} y_i = \frac{1}{N}\sum_{i \in s} \frac{t_i}{\pi_i} \quad \text{because here} \quad \pi_i = \frac{n}{N}.$$

But by the preceding arguments for

$$\bar{Y}_d = \frac{1}{N_d}\sum_1^N y_{di} = \frac{1}{N_d}\sum_{i \in U_d} y_i$$

a suitable estimator should be

$$\hat{\bar{Y}}_d = \frac{1}{N} \cdot \frac{N}{n}\sum_{i \in s} y_{di} = \frac{1}{n}\sum_{i \in s} y_i I_{di} = \frac{1}{n}\sum_{i \in s_d} y_i$$

and not the domain sample mean, namely,

$$\bar{y}_d = \frac{1}{n_d}\sum_{i \in s_d} y_i.$$

If $n_d = 0$, then $\hat{Y}_d$ will be defined but valued zero while $\bar{y}_d$ cannot be defined.

Also, $\quad V\left(\hat{\bar{Y}}_d\right) = \frac{1}{N^2} v\left(t_{Hd}\right) \Big|_{\substack{\pi_i = \frac{n}{N} \\ \pi_{ij} = \frac{n(n-1)}{N(N-1)}}}$

and $\quad v\left(\hat{\bar{Y}}_d\right) = \frac{1}{N^2} v\left(t_{Hd}\right) \Big|_{\substack{\pi_i = \frac{n}{N} \\ \pi_{ij} = \frac{n(n-1)}{N(N-1)}}}.$

But $\quad V\left(\bar{y}_d\right) = \frac{\left(\frac{1}{n_d} - \frac{1}{N_d}\right)}{(N_d - 1)} \sum_{i=1}^{N_d} \left(y_{di} - \bar{Y}_d\right)^2, \quad \text{if } n_d > 0,$

and $\quad \hat{V}\left(\bar{y}_d\right) = \left(\frac{1}{n_d} - \frac{1}{N_d}\right) \frac{1}{(n_d - 1)} \sum_{i \in s_d} \left(y_{di} - \bar{y}_d\right)^2, \quad \text{if } n_d \geq 2.$

Thus, $\hat{\bar{Y}}_d$ is quite different from $\bar{y}_d$ in their characteristics.

If a statistic $t_d = t_d(s_d, \underline{Y})$ produces an unacceptable CV it should be revised by a suitable alternative. How to find such an alternative is now our subject of discussion.

Without drawing a supplementary sample, we intend to derive a revised estimator with an effectively enhanced sample-size by postulating appropriate models which may enable us to borrow strength from the sample at hand but not confine us to the intersection of the sample with the domain of relevance, the technique is called "Borrowing Strength" from domains supposed "Alike" in certain characteristics.

Let in addition to the variable of interest y a real variable x be available well-correlated with y and with known totals for the relevant domains, namely, $X_d = \sum_{i \in U_d} x_i, d = 1, \cdots, D$.

Let us then postulate the model:

$$y_i = \beta_d x_i + \epsilon_i \quad \text{for} \quad i \in U_d$$

with β_d as U_d-specific unknown constants and ϵ_d's are independently distributed with unknown variances $\sigma_d^2, d = 1, \cdots, D$. Then Cassel, Sarndal, and Wretman's (1976) estimator, rather predictor because Y_d's are now random variables and not constants, called the Generalized Regression (Greg) estimator is

$$t_g = \sum_{i \in s} \frac{y_i}{\pi_i} + b_{Qd} \left(X_d - \sum_{i \in s_d} \frac{x_i}{\pi_i} \right)$$

with $\quad b_{Qd} = \frac{\sum_{i \in s_d} y_i x_i Q_i}{\sum_{i \in s_d} x_i^2 Q_i}, Q_i's \quad$ are positive constants.

Developing Small Domain Statistics

From Sarndal (1982) we know that the generalized Regression (Greg) predictor for a finite population total $Y = \sum_1^N y_i$, namely,

$$t_g = \sum_{i \in s} \frac{y_i}{\pi_i} + b_Q \left(X - \sum_{i \in s} \frac{x_i}{\pi_i} \right)$$

assisted by the model for which $y_i = \beta x_i + \in_i, i \in U = (1, \cdots i, \cdots, N)$ is model unbiased for Y because, on taking

$$b_Q = \frac{\sum_{i \in s} y_i x_i Q_i}{\sum_{i \in s} x_i^2 Q_i}, \quad \text{with } Q_i > 0,$$

$$E_m(t_g) = \beta \sum_{i \in s} \frac{x_i}{\pi_i} + \beta \left(X - \sum_{i \in s} \frac{x_i}{\pi_i} \right) = \beta X,$$

and $\qquad E_m(Y) = \beta X.$

Also, approximately the design-based mean square error (MSE) of t_g about Y is

$$E_p(t_g - Y)^2 \simeq V_p \left(\sum_{i \in s} \frac{E_i}{\pi_i} \right),$$

$$E_i = y_i - B_Q x_i, \quad \text{writing} \quad B_Q = \frac{\sum_1^N y_i x_i Q_i \pi_i}{\sum_1^N x_i^2 Q_i \pi}$$

or

$$M_p(t_g) = \sum_{i<j}^N \sum^N (\pi_i \pi_j - \pi_{ij}) \left(\frac{E_i}{\pi_i} - \frac{E_j}{\pi_j} \right)^2$$

$$\beta_i = 1 + \frac{1}{\pi_i} \sum_{i \neq i}^N \pi_{ij} - \sum \pi_i.$$

Next, writing $e_i = y_1 - b_Q x_i$, a useful estimator for $M_p(t_g)$ is

$$m(t_g) = \sum_{i<j} \sum (\pi_i \pi_j - \pi_{ij}) \frac{I_{sij}}{\pi_{ij}} \left(\frac{e_i}{\pi_i} - \frac{e_j}{\pi_j} \right)^2 + \sum \beta \frac{e_i^2}{\pi_i} \frac{I_{si}}{\pi_i}.$$

Drawing upon these developments in the literature, for the domain total Y_D, the Greg predictor, namely,

$$t_{gd} = \sum_{i \in s} \frac{y_{di}}{\pi_i} + b_{Qd} \left(X_d - \sum_{i \in s_d} \frac{x_i}{\pi_i} \right)$$

assisted by the model $y_i = \beta_d x_i + \in_i, i \in U_d$, has an approximate design MSE about Y_d as

$$M_p(t_{gd}) \simeq \sum_{i<j}^{N_d} \sum^{N_d} (\pi_i \pi_j - \pi_{ij}) \left(\frac{E_i I_{di}}{\pi_i} - \frac{E_j I_{dj}}{\pi_j} \right)^2 + \sum \frac{E_i^2 I_{di}}{\pi_i} \beta_i$$

and an estimator for it is

$$m(t_{gd}) = \sum\sum_{i<j\in s_d}(\pi_i\pi_j - \pi ij)\frac{I_{sij}}{\pi_{ij}}\left(\frac{e_i I_{di}}{\pi_i} - \frac{e_j I_{dj}}{\pi_j}\right)^2 + \sum_{i\in s_d}\frac{e_i^2}{\pi_i}I_{di}\frac{\beta_i}{\pi_i}.$$

In case s_d and its size n_d are small, this $m(t_{gd})$ often turns out too large giving the $CV = 100\frac{\sqrt{m(t_{gd})}}{|t_{gd}|}$ unacceptably too high.

If we are too courageous to suppose that the domains are too similar to each other so that the y's may be modeled as $y_i = \beta x_i + \in_i \ \forall \ i \in U$, then one may employ for Y_d the "Synthetic Greg" predictor, namely,

$$t_{sgd} = \sum_{i\in s_d}\frac{y_i}{\pi_i} + b_Q\left(X_d - \sum_{i\in s_d}\frac{x_i}{\pi_i}\right), \quad \text{with} \quad b_Q = \frac{\sum_{i\in s}y_i x_i Q_i}{\sum_{i\in s}x_i^2 Q_i}, Q_i > 0$$

as it is assisted by the model $y_i = \beta x_i + \in_i \ \forall \ i \in U$. This predictor is based on the effective sample size n and not n_d. For simplicity, let

$$t_{sgd} = \sum_{i\in s}\frac{y_i}{\pi_i}G_{sdi}, \quad \text{writing} \quad G_{sdi} = I_{di} + \left(X_d - \sum_{i\in s}\frac{x_i}{\pi_i}I_{di}\right)\frac{x_1 Q_i \pi_i}{\sum_{i\in s}x_i^2 Q_i}.$$

We have, taking $k = 1, 2$; $b_{1i} = 1, b_{2i} = G_{sdi}$, two estimated MSE's of t_{sgd} as

$$m_{skd} = \sum\sum_{i<j\in s}\left(\frac{\pi_i\pi_j - \pi_{ij}}{\pi_{ij}}\right)\left(\frac{b_{ki}e_{sdi}}{\pi_i} - \frac{b_{kj}e_{sdj}}{\pi_j}\right)^2 + \sum_{i\in s_d}\beta_i\frac{e^2 s_{di}}{\pi_i}\frac{1}{\pi_i},$$

with $e_{sdi} = (y_i - b_Q x_i), i \in s$.

The greg predictor t_{sgd} is called the synthetic generalized regression predictor in contrast with the greg predictor t_{gd} both for Y_d. This is because compared to b_{Qd} which uses y_i-values only for $i \in s_d$ and t_{gd} is motivated by the model involving β_d the component b_Q in t_{sgd} assisted by the model involving β, utilizes b_Q which uses y_i for i in $s = U_{d=1}^s s_d$. Thus, essentially s_{gd} is used on postulating a common regression coefficient of y on x for all the domains, and thus the model assisted synthetic greg predictor amalgamates all the y-values for i in the union of s_d over all the $U_d's$, $d = 1, \cdots, D$. If instead a common regression seems more appropriate for only a group of U_d's rather than all the U_d's, then instead of b_{Qd} an alternative estimate b_{QG}, say, denoting G by a group of U_d's for which a common regression coefficient seems plausible, then a revised synthetic greg predictor may be employed as more appropriate than b_{Qd} and b_Q leading to a revised synthetic greg predictor, say, e_{sGd}. Thus, by a synthetic estimator we mean an estimator which instead of confining to the use of y_i-values for i in a specific s_d utilizes y_i-values in a union of s_d's, on postulating the domain U_d's in the union to be "alike" in certain specified senses.

A synthetic estimator then by its nature is dependent on a larger sample-size than the domain-specific sample-size n_d alone. If the amalgamation is plausibly executed the synthetic estimator is likely to outperform the estimator utilizing the y-values exclusively for i in s_d alone.

A strictly model-based approach is more effective in Small Area Estimation. Let us proceed to demonstrate some aspects of it.

Suppose our initial design-based estimator for a domain total Y_d is t_d with a variance V_d, say, $d = 1, \cdots, D$. Let us postulate the model for which we may write

$$t_d = Y_d + e_d, \; d = 1, \cdots, D \qquad (i)$$
and
$$Y_d = \beta X_d + \in_d, \; d = 1, \cdots, D \qquad (ii)$$

Here, in (i) Y_d is a constant and e_d's are independently distributed with variances $V(e_d) = V(t_d) = V_d$. But in (ii) β is an unknown constant, X_d's are all known values on a real variable x well-correlated with y and t_d's are independently distributed random variables with zero means but a common unknown constant variance A. Thus, (ii) $\Rightarrow Y_d$'s are random variables with model expectations βX_d and a common unknown constant variance A. Supposing t_d as a linear function of y_i-values for i in s_d this model called Fay and Herriot's (1979) model under which a Best Linear unbiased predictor for Y_d is easily derived by them. This BLUP for Y_d involves V_d, β, and A. For simplicity V_d is supposed to be known though it is supposed to be taken as a suitably chosen design-based estimator v_d, say, for it. But β and A are estimated suitably from the sample survey data. On substituting these estimated values for β and A in the BLUP for Y_d what is derived is called an Empirical Best Linear Unbiased Predictor (EBLUP) for Y_d. A formally equivalent result but with a drastically different approach and interpretation is obtained as follows, which we intend to show in some detail.

Let us modify the Fay-Herriot model in the following way, facilitating the way to derive Bayes estimators and their modifications as Empirical Bayes Estimators (EBE).

Let
$$t_d | Y_d \overset{ind}{\sim} N(Y_d, V_d) \qquad (i)$$
and
$$Y_d | \beta, A \overset{ind}{\sim} N(\beta X_d, A) \qquad (ii)$$

as modified from (i) and (ii), respectively.

Then,
$$E(t_d | Y_d) = Y_d, Y_d \sim N(\beta X_d, A).$$

So
$$E(t_d) = E(Y_d) = \beta X_d,$$
$$V(t_d) = E[V(t_d | Y_d)] + V[E(t_d | Y_d)]$$
$$= A + V_d$$

$$E[(t_d - E(t_d))(Y_d - E(Y_d))] = E[((Y_d + e_d) - \beta X_d)(\beta X_d + \in_d - \beta X_d)]$$
$$= E[(\beta X_d + \in_d + e_d - \beta X_d) \in_d] = V(\in_d)$$
$$= A.$$

So,
$$\begin{pmatrix} t_d \\ Y_d \end{pmatrix} \sim N_2 \left(\begin{pmatrix} \beta X_d \\ \beta X_d \end{pmatrix}, \begin{pmatrix} A + V_d & A \\ A & A \end{pmatrix} \right)$$

Then,
$$Y_d | t_d \sim N \left(\beta X_d + \frac{A}{A + V_d} (t_d - \beta X_d), \left(A - \frac{A^2}{A + V_d} \right) \right)$$

by the properties of the multi-variate normal distribution, vide Anderson (1958).

So,
$$\left(\frac{A}{A + V_d} \right) t_d + \left(\frac{V_d}{A + V_d} \right) \beta X_d$$

is the Bayes Estimator of Y_d on taking a squared error loss function $(f_d - Y_d)^2$ if we contemplate employing an estimator f_d for Y_d.

This BE of Y_d is of course a convex combination of the initial estimator t_d for Y_d and βX_d which is the model-based mean of Y_d. Looking at their respective weights

$$\frac{A}{A + V_d} \quad \text{and} \quad \frac{V_d}{A + V_d},$$

it is clear that if the initial estimator t_d is good having a relatively small measure of error, namely V_d, then it will have a higher weight, i.e., A is greater relative to V_d. Conversely if one starts with a weak estimator t_d having a large variance, then it should receive a relatively less weight, its companion βx_D should receive a relatively higher weight, namely

$$\frac{V_d}{A + V_d}.$$

However, this Bayes estimator cannot be put to use in practice. So, it should be revised as follows. Let us write

$$\tilde{\beta} = \frac{\sum t_d X_d / (A + V_d)}{\sum X_d^2 / (A + V_d)}$$

and observe that

$$\sum_{d=1}^{D} \left(t_d - \tilde{\beta} X_d \right)^2 / (A + V_d)$$

is a Chi-square statistic with $(D - 1)$ degrees of freedom. So, solving the equation

$$\sum_{d=1}^{D} \left(t_d - \tilde{\beta} X_d \right)^2 / (A + V_d) = (D - 1)$$

in respect of A by iteration using the $\tilde{\beta}$ above starting with initial A as zero as recommended by Fay and Herriot (1979) using Newton and Raphson's method

Developing Small Domain Statistics

of solving numerical equations estimates $\hat{\beta}$ and $\hat{A}$ may be derived to produce for Y_d the empirical Bayes estimator, namely,

$$\hat{Y}_{EB_d} = \left(\frac{\hat{A}}{\hat{A}+V_d}\right) t_d + \left(\frac{V_d}{\hat{A}+V_d}\right) \hat{\beta} X_d.$$

Gracefully this is also a combination of the initial estimator t_d and a model-unbiased estimator $\hat{\beta} X_d$ of Y_d because

$$E_m\left(\hat{\beta} X_d\right) = E_m(Y_d) \quad \text{since} \quad E_m\left(\hat{\beta}\right) = \beta.$$

The interpretation of their associated weights also matches that of the weights of the Bayes estimator

$$\hat{Y}_{Bd} = \left(\frac{A}{A+V_d}\right) t_d + \left(\frac{V_d}{A+V_d}\right) \beta X_d.$$

Prasad and Rao (1990) have given us the following expressions for the MSE of $\hat{Y}_{EBd}$ about Y_d and of an estimator thereof, on writing

$$v_d = \frac{A}{A+V_d}$$

$$M_d = E_m\left(\hat{Y}_{EBd} - Y_d\right) = M_{1d} + M_{2d} + M_{3d} \quad \text{with}$$

$$M_{1d} = E_m\left(\hat{Y}_{Bd} - Y_d\right)^2 = \gamma_d V_d,$$

$$M_{2d} = E_m\left(\tilde{Y}_{Bd} - \hat{Y}_{Bd}\right)^2 = (1-\gamma_d)^2 \, X_d^2 \sum_d \left(\frac{X_d^2}{A+V_d}\right)$$

writing $\quad \tilde{Y}_{Bd} = \hat{Y}_{Bd}|_{\beta=\tilde{\beta}}$

and $\quad M_{3d} = E_m\left(\hat{Y}_{Bd} - \tilde{Y}_{Bd}\right)^2 = \frac{V_d^2}{(A+V_d)^3} V\left(\tilde{A}\right),$

writing $\quad V\left(\tilde{A}\right) = \frac{2}{D^2} \sum_d (A+V_d)^2.$

Prasad and Rao's (1990) estimator for M_d is $m_d = m_{1d} + m_{2d} + 2m_{3d}$, writing $m_{jd} = M_{jd}|_{A=\hat{A}}$ for $j = 1, 2, 3$.

Let us consider James-Stein estimators for Y_d which are non-Bayesian estimators but are somewhat alike of empirical Bayes estimators.

Let us consider the model for which $t_d \overset{ind}{\sim} N\left(Y_d, \sigma^2\right).$

Here to start with, σ^2 is supposed to be known.

Let
$$S^2 = \sum_{d=1}^{D} t_d^2.$$

Then,
$$\delta_d = \left(1 - \frac{D-2}{S}\sigma^2\right) t_d, \quad d = 1, \cdots, D$$

is called a James-Stein (1961) estimator for Y_d. Then, compared to $\underset{\sim}{t} = (t_1, \cdots, t_D)$, $\underset{\sim}{\delta} = (\delta_1, \cdots, \delta_D)$ is a better estimator for $\underset{\sim}{Y_d} = (Y_1, \cdots, Y_d, \cdots Y_D)$ in the sense that

$$D\sigma^2 = \sum_{d=1}^{D} E_m (t_d - Y_d)^2 \geq \sum_{d=1}^{M} E_m (\delta_d - Y_d)^2.$$

Thus, this δ_d is a shrinkage estimator provided σ^2 is known. To see this James-Stein (JS, say) estimator's connection with empirical Bayes estimate (EBE), let $Y_d \overset{ind}{\sim} N(\beta X_d, A)$. Thus, $\delta_d^* = \hat{\beta} t_d$ is the regression estimator for Y_d and the Bayes estimator with a square error loss for Y_d is

$$Y_{Bd}^* = \delta_d^* + \left(1 - \frac{\sigma^2}{\sigma^2 + A}\right)(\delta_d - \delta_d^*)$$
$$= \frac{\sigma^2}{\sigma^2 + A}\delta_d^* + \frac{A}{\sigma_2 + A}\delta_d.$$

Let
$$S^* = \sum_{d=1}^{D} (\delta_d - \delta_d^*)^2.$$

Then,
$$E\left[\frac{D-3}{S^*}\sigma^2\right] = \frac{\sigma^2}{\sigma^2 + A}.$$

Then,
$$\tilde{\delta}_d = \delta_d^* + \left[1 - \frac{D-3}{S^*}\sigma^2\right](\delta_d - \delta_d^*)$$
$$= \left(\frac{D-3}{S^*}\right)\sigma^2 \delta_d^* + \left(1 - \frac{D-3}{S^*}\sigma^2\right)\delta_d$$

is of course an Empirical Bayes estimator as well. If, particular, $X_d = 1$ then

$$\delta_d^* = \frac{1}{D}\sum_{1}^{D} \delta_d = \delta, \quad \text{say,}$$

$$S^* = \sum (\delta_d - \delta)^2 \quad \text{and}$$

$$\delta_d^* = \frac{D-3}{S^*}\sigma^2 \delta + \left(1 - \frac{D-3}{S^*}\sigma^2\right)\delta_d.$$

The deficiency of JSE versus EBE is the need to have a common σ^2 in the former while domain-specific variability in σ_d^2 is permitted with EBE on the contrary.

Borrowing across time rather than domains alone and Kalman Filtering

Let y_α be a variable of interest at a point of time $\alpha = 0, 1, \cdots, t-1, t$ and $Y_{d\alpha}$ = The total for the dth domain U_d at time α. Let $X_{dj\alpha}$ $(j = 1, \cdots, K)$ be the total of known values $x_{j\alpha}$ for all the units in U_d. Let $m_{d\alpha}$ be an initial estimator for $Y_{d\alpha}$ at time α. We intend to improve upon it on using the past values.

Let us postulate the model to write

$$(i) \quad m_{d\alpha} = Y_{d\alpha} + \epsilon_{d\alpha}$$

$$(ii) \quad Y_{d\alpha} = \sum_{j=1}^{K} X_{dj\alpha} \beta_{j\alpha} + u_{d\alpha}.$$

Here, $\beta_{j\alpha}$ are unknown regression coefficients common for every domain enabling us to borrow strength across the domains. Here $\epsilon_{d\alpha}$ are independently distributed with zero means, so are the $u_{d\alpha}$'s and moreover $\epsilon_{d\alpha}$'s and $u_{d\alpha}$'s are independent of each other.

Let

(iii) $Y_{d\alpha} = \underline{X}_{d\alpha} \underline{\beta}_\alpha + u_{d\alpha} = \underline{F}_{d\alpha} \underline{\theta}_{d\alpha}$

writing

$\underline{F}_{d\alpha} = (X_{d1\alpha}, \cdots, X_{dK\alpha}, 1)_{1 \times (K+1)}$

$\underline{\theta}'_{d\alpha} = (\beta_{i\alpha}, \cdots, \beta_{K\alpha} u_{d\alpha})_{1 \times (K+1)}$.

Then, $m_{d\alpha} = \underline{F}_{d\alpha} \underline{\theta}_{d\alpha} + \epsilon_{d\alpha}, \alpha = 1, \cdots, t$.

For simplicity in notation let us omit d now onwards.

Thus, let

1. $m_\alpha = \underline{F}_\alpha \underline{\theta}_\alpha + \epsilon_\alpha, \alpha = 1, \cdots, t$. Let $\underline{\theta}_\alpha$ be subject to ARMA model, vide Box and Jenkins (1970) and ϵ_α be the white noise.
So, let
2. $\underline{\theta}_\alpha = \underline{G}_\alpha \underline{\theta}_{\alpha-1} + \underline{u}_\alpha, \alpha = 1, \cdots, t$. $\underline{G}_\alpha$ is a known matrix (cf. Pfeffermann 1989). Let $\epsilon_\alpha \stackrel{ind}{\sim} N(\underline{0}, V_\alpha), \underline{u}_\alpha \stackrel{ind}{\sim} N(\underline{0}, \phi_\alpha), \underline{\epsilon} \perp \underline{u}_\alpha$.

Now further results follow from applying the Bayesian approach of Kalman Filtering which is a recursive method described by Meinhold and Singpurwalla (1983).

Let
$\underline{\theta}_t = (\theta_1, \cdots, \theta_t)$
$\underline{m}(t-1) = (m_{d1}, \cdots, m_{d\alpha}, \cdots, m_d(t-1))$
$P(\underline{\theta}_t \mid \underline{m}(t-1))$ = Prior of $\underline{\theta}_t$ given $\underline{m}(t-1)$

$$L(\underline{\theta}_t m_t, \underline{m}(t-1)) = P(m_t \underline{\theta}_t, \underline{m}(t-1))$$
$$= \text{Likelihood of } \underline{\theta}_t \text{ given } \underline{m}(t-1),$$

formally the probability distribution of m_t given $\underline{m}(t-1)$. Then the posterior of θ_t given $\underline{m}(t-1)$ is

$$P(\theta_t | m_t) \propto P(m_t \theta_t, \underline{m}(t-1)) P(\theta_t \underline{m}(t-1)).$$

The problem is first to derive a neat formula for $P(\theta_t|m_t)$ and then to estimate the parameters involved in that.

Let
$$\hat{\theta}(t-1) = E_m[\theta(t-1) m(t-1)] \quad \text{and}$$
$$\sum(t-1) = V_m[\theta(t-1) m(t-1)].$$

Now, $\hat{\theta}(0)$ and $\sum(0)$ are at the investigator's hands. Since

$$\theta(t-1) m(t-1) \sim N\left(\hat{\theta}(t-1), \sum(t-1)\right)$$

it follows that

$$\theta(t) m(t-1) \sim N\left(G(t)\hat{\theta}(t-1), G(t)\sum(t-1)G'(t) + \phi_t\right).$$

Let $\underline{\theta}(t)$ be estimated using $m(t-1)$ by $F_t G_t \hat{\theta}(t-1)$ and let

$$e_t = m_t - F_t G_t \hat{\theta}(t-1).$$
$$= F_t\left(\theta(t) - G_t \hat{\theta}(t-1)\right) + \epsilon_t.$$

So,
$$e_t(\theta_t, \underline{m}(t-1)) \sim N\left(F_t\left(\theta_t - G_t \hat{\theta}(t-1), V_t\right)\right)$$
$$\underline{R}_t = \underline{G}_t \sum(t-1)\underline{G}_t + \phi_t.$$

Then,
$$\theta_t m_t \sim N\left[G_t \hat{\theta}(t-1) + R_t\left(F_t R_t^{-1} F_t + V_t\right)^{-1} e_t,\right.$$
$$\left. R_t - R_t F_t'\left(F_t R_t^{-1} F_t' + V_t\right)^{-1} F_t\right].$$

So, an improved estimate of θ_t using m_t in addition to $\underline{m}(t-1)$ is given by the Bayes estimator

$$G_t \hat{\theta}(t-1) + R_t\left(F_t R_t^{-1} F_t + V_t\right)^{-1} e_t.$$

Since, $\hat{\theta}_0, \sum_0$ are known as also $\underline{F}\alpha$ an $\underline{e}_t$ may be worked out.

More references for Kalman Filtering include Binder and Hidiroglou (1988), Binder and Dick (1989), Choudhry and Rao (1989), and Chaudhuri (2010), among others.

Developing Small Domain Statistics

Let us finally present one simple illustration.

Let
(i) $g_{dt} = \beta_t X_{dt} + e_{dt}$, $e_{dt} \sim N(0, v_{dt})$
(ii) $\beta_t = \beta_{t-1} + w_t$, $t = 1, \cdots, T$.
β_t's are random variables and
$w_t \sim N(0, W_t)$
(iii) $e_{dt} \perp w_t$,
(iv) $\beta_0 \sim N(\phi_{d0}, \Sigma_{d0})$, $\phi_{do} = \dfrac{g_{d0}}{X_{d0}}$, $\Sigma_{d0} = \dfrac{v_{d0}}{X_{d0}^2}$.

Let
$$g^*_{d1} = X_{d1}\phi_{d0}, \quad \Delta_{d1} = g_{d1} - g^*_{d1},$$
$$R_{d1} = \Sigma_{d0} + W_1.$$

Then follow
(a) $\Delta_{d1}g_{d0} \sim N(0, X_{d1}^2 R_{d1} + v_{d1})$
(b) $\beta_1 g_{d0} \sim N(\phi_{do}, \Sigma_{d0} + W_1) = N(\phi_{d0}, R_{d1})$
(c) $[C_m(\Delta_{d1}, \beta_1)g_{d0}] = X_{d1}R_{d1}$
(d) $\left[\begin{pmatrix}\Delta_{d1}\\ \beta_1\end{pmatrix}g_{d0}\right] \sim N_2\left(\begin{pmatrix}0\\ \phi_0\end{pmatrix}\begin{pmatrix}X_{d1^2} & R_{d1}+v_{d1} & X_{d1}R_{d1}\\ X_{d1} & R_{d1} & Rd1\end{pmatrix}\right)$
(e) $\beta_1\Delta_{d1} \sim N\left(\phi_{d0} + \dfrac{X_{d1}R_{d1}}{X_{d1}^2 R_{d1} + v_{d1}}\Delta_{d1}, R_{d1} - \dfrac{(X_{d1}R_{d1})^2}{X_{d1}^2 R_{d1} + v_{d1}}\right)$,

i.e., $\beta_1(g_{d0}, g_{d1}) \sim N\left(\left(\dfrac{X_{d1}^2 R_{d1}}{X_{d1}^2 R_{d1} + v_{d1}}\right)\begin{pmatrix}g_{d1}\\ X_{d1}\end{pmatrix}\right.$
$\left. + \left(\dfrac{v_{d1}}{X_{d1}^2 R_{d1} + v_{d1}}\right)\begin{pmatrix}g_{d0}\\ X_{d0}\end{pmatrix}, \dfrac{R_{d1}v_{d1}}{X_{d1}^2 R_{d1} + v_{d1}}\right).$

So, an updated predictor for Y_{d1} is
$$\hat{Y}_{d1} = X_{d1}\hat{\beta}_1 \quad \text{where}$$
$$\hat{\beta}_1 = E_m(\beta_1 g_{d0}, g_{d1}) \text{ given above.}$$

So, $\hat{Y}_{d1} = X_{d1}\hat{\beta}_1 = \left(\dfrac{X_{d1}^2 R_{d1}}{X_{d1}^2 R_{d1} + v_{d1}}\right)g_{d1} + \left(\dfrac{v_{d1}}{X_{d1}^2 R_{d1} + v_{d1}}\right)\begin{pmatrix}X_{d1}\\ X_{d0}\end{pmatrix}$

is the Kalman Filter estimator for Y_{d1}.

Its measure of error is
$$M_{d1} = \dfrac{X_{d1}^2 R_{d1} v_{d1}}{X_{d1}^2 R_{d1} + v_{d1}}.$$

This procedure is to be implemented for the successive time points $t = 1, 2, ..., T$. For Y_{dt} the Kalman Filter estimator is

$$\hat{Y}_{dt} = \hat{\beta}_t X_{dt} \quad \text{with} \quad M_{dt} = X_{dt}^2 \Sigma_{dt} \quad \text{as its measure of error.}$$

It is of interest now to check the relative performances of

$$(g_{dt}, v_{dt}) \quad \text{versus} \quad \left(X_{dt}\hat{\beta}_t, X_{dt}^2 \Sigma_{dt} \right).$$

---

8 Network and Adaptive Procedures

Abstract. Introduction. Estimation by network sampling and estimation by adaptive sampling. Constraining network sampling and constraining adaptive sampling.

8.0 ABSTRACT

Network Sampling involves two kinds of units namely (1) Selection Units (SU) and (2) Observation Units (OU). The SU's with their numbers and identities known and for them frames may be constructed and utilized in choosing elements of SU's. The total number of OU's is not known and they are not identified except through the selected SU's to which they are "linked." The sets of OU's mutually linked with the SU's constitute what are known as "Networks." From these networks samples of OU's are chosen and surveyed for estimation of parameters related to the OU's. This is Network Sampling.

In Adaptive sampling fearing inadequate representation in a sample for the units with requisite properties of relevance, a provision is built up before the survey. This is an objective procedure to extend the initial sample by defining their neighbors and concepts of networks composed of initial units with the requisite property along with sequences of neighboring units with the same property as reached in the course of the survey is extended over the neighbors in succession.

For both the sampling procedures unbiased estimation methods have been developed. In both, the sample-sizes turn out random and their sizes may go beyond control. So, some requisite procedures have been developed to apply appropriate breaks on the advancing sample-sizes. Thus, constrained Network and constrained Adaptive sampling techniques are worked out.

8.1 INTRODUCTION

Suppose one intends to estimate by a sample survey the household expenses in a given time period on the hospital treatment of its inmates who suffered from some diseases like cancer that need protracted care and attention in an expansive region of one's interest. Here usual household surveys by dint of stratified multi-stage sampling are inadequate. This is because in many randomly selected houses there may not have been any such patients at all, so that information content in such samples may not be substantial enough vis-à-vis the expenses incurred and hassles faced in such surveys.

It may be advisable to take a sample of hospitals, nursing homes, etc., wherefrom for example we may gather the addresses within the region of interest of the patients who were treated therein for the specific disease, say, cancer during the specified period of our interest. In the addresses thus gathered, we may encounter the household respondents who may deliver the messages needed. It should be noted that the patients are not of our interest, rather the households of the persons who happened to have been treated in the hospitals giving us the information are our real interest because we need to procure the data on household expenses on the treatments of their inmates who are or were the hospital in-patients. We may employ a Network Sampling in this context. The hospitals, nursing homes, and clinics that admit cancer in-patients are the selection units (SU) out of which samples may be conveniently gathered. The households whose inmates as cancer in-patients were treated in the specified period of our concern are the Observation Units (OU). Those OU's may on enquiry give us additional messages about other SU's in the area of interest where the same or other household members were also the cancer in-patients receiving care and treatment in the relevant time period. Thus, a comprehensive network connecting the OU's contacted through the sampled SU's and also more SU's contacted through the OU's actually surveyed may be transgressed exploiting the links connecting numerous SU's and OU's. Surveying the OU's by this way of establishing links among OU's and SU's is called Network Sampling following the primary works by Thompson (1990, 1992) and by Thompson and Seber (1996) and further works by Chaudhuri (2000).

Another sampling procedure of a similar nature to enhance the information content of an initial sample by further sampling of special units is known as Adaptive sampling to be discussed now. For every unit in the population a neighboring unit is uniquely defined. If a unit happens to be selected bearing the property of interest each one of his/her neighbors is tested about bearing the trait in question. If any of these neighbors bears it, then all the neighbors of each of these neighbors bearing the trait is again tested about bearing this trait in its turn, and thus the process continues. A unit bearing the trait along with all the units bearing it as reached on examining these successive neighbors constitutes what is called a Network. Any unit without the trait of specification is called an edge unit. A unit bearing the trait with none of his/her neighbors bear it is a Singleton Network. By courtesy each edge unit is also treated as a Network. With this convention it follows that each Network is disjoint with every other. Also the union of all the networks coincides with the entire population of units. Then, starting with an initial sample s the set of all the units in the networks of all the units in s constitutes the Adaptive sample $A(S)$. The advantage of an adaptive sample is that the content of the information is enormously enhanced. This exercise in reaching the Adaptive sample $A(s)$ starting with the initial sample s is called the technique of adaptive sampling. We shall see in Section 8.2 how to unbiasedly estimate the finite population total and derive a formula for its variance and also get an unbiased estimator for this variance. Another vital problem is that given the size of the initial sample s that of the corresponding adaptive sample $A(s)$ may turn out exorbitantly larger. In such a case we need an in-built procedure to keep a check in the enormous growth in the size of $A(s)$. A procedure has been provided in Section 8.3, giving a method of constrained Adaptive Sampling. Finally we may add a note in clarification. A neighborhood of a unit consists of the unit itself and all the units in the uniquely defined neighborhood of it. Also a network consists of the initial unit itself and all the units in its network. A recent book *Adaptive Sampling Designs* by Seber and Salehi (2012) and the forthcoming one by Chaudhuri (2014) may be consulted for reference in the present context.

8.2 ESTIMATION BY NETWORK SAMPLING AND ESTIMATION BY ADAPTIVE SAMPLING

8.2.1 NETWORK SAMPLING AND ESTIMATION

Let M be the known number of SU's in a given context when the SU's are identified by the labels $j = 1, \cdots, M$. Let N be an unknown number of OU's linked to these M SU's in a well-defined manner and assigned the labels $i = 1, \cdots, N$ not identified to start with before studying how they are actually linked to these M SU's.

Let m_i be the number of SU's to which an ith OU is linked. Let $A(j)$ be the "Set" of OU's linked to a jth SU ($i = 1, \cdots, N; j = 1, \cdots, M$). We need to suitably estimate $Y = \sum_1^N y_i$ on writing y_i to denote the value of a real variable y taken on the ith OU. Let us then define

$$w_j = \sum_{i \in A(j)} \frac{y_i}{m_i}, i \in U = (1, \cdots, j, \cdots, M).$$

Then, it follows that

$$\sum_{j=1}^M w_j = \sum_{j=1}^M \left(\sum_{i \in A(j)} \left(\frac{y_i}{m_i} \right) \right)$$
$$= \sum_{j=1}^N y_i \left(\frac{1}{m_i} \sum_{j \mid A(j) \ni i} 1 \right) = \sum_{i=}^N y_i = Y$$

because $\sum_{j \mid A(j) \ni i} 1 = m_i$ by definition.

Since W equals Y and our intention is to suitably estimate Y, we may transfer our problem to the estimation of W in an equivalent way utilizing the values of w_j for $j \in s$. The determination of these w_j's, $j \in s$ of course needs the ascertainment of all the y_i-values for i in $\bigcup_{j \in s} A(j)$ which is the network sample needed to be surveyed. Thus, we may estimate Y or equivalently W by

$$e_b = \sum_{j \in s} b_{sj} w_j$$

with the coefficients b_{sj} free of $\underline{W} = (w_1, \cdots, w_M)$ but subject to

$$\sum_{s \ni j} p(s) b_{sj} = 1 \ \forall \ j \in U = (1, \cdots, M).$$

Here s is the sample of SU's chosen according to a sampling design p with probability $p(s)$.

Then, $\quad E_p(e_b) = \sum_s p(s) e_b = W.$

Then, $\quad V_p(e_b) = -\sum_{j<j'}^{M}\sum^{M} a_j a_{j'} \left(\frac{w_j}{a_j} - \frac{w_{j'}}{a_{j'}}\right)^2 + \sum_1^M \frac{w_j^2}{a_j}\alpha_j.$

Here
$$d_{jj'} = E_p(b_{sj}I_{sj} - 1)(b_{sj'}I_{sj'} - 1),$$
$$I_{sj} = 1 \text{ if } j \in s, \ I_{sj} = 0 \text{ if } j \notin s$$
$$I_{sjj'} = 1 \text{ if } j, j' \in s, j \neq j', a_j (\neq 0), \text{ constants,}$$
$$= 0, \text{ else}$$

$$\alpha_j = \sum_{j'=1}^{M} d_{jj'} a_j.$$

This is due to Rao (1979) and Chaudhuri and Pal (2002). One may find some details in Chaudhuri and Stenger (2005) and Chaudhuri (2010).

Also, writing $\quad C_j = \sum_s p(s) b_{sj}^2 - 1, \ C_{jj'} = \sum_s p(s) b_{sj} b_{sj'} - 1, \ C_{sj}$ and $C_{sjj'}$

such that $\quad \sum_s p(s) C_{sj} = C_j$ and $\sum_s p(s) C_{sjj'} = Cjj',$

we have an alternative formula for variance of t_b as

$$V_p'(P_b) = \sum_1^M C_j w_j^2 + \sum\sum_{j \neq j'} C_{jj'} w_j w_{j'}.$$

For $V_p(P_b)$ and $V_p'(e_b)$, respective unbiased estimators are

$$v_1(e_b) = -\sum\sum_{j<j'} a_j a_j' \left(\frac{w_j}{a_j} - \frac{w_j'}{a_j'}\right)^2 d_{sjj'} \frac{I_{sjj'}}{\pi_{jj'}} + \sum_j^M \frac{w_j^2}{a_j}\alpha_j \frac{I_{sj}}{\pi_j} \quad \text{and}$$

$$v_2(e_b) = \sum C_{sj} w_j^2 \frac{I_{sj}}{\pi_j} + \sum\sum_{j \neq j'} C_{sjj'} w_j w_{j'} \frac{I_{sjj'}}{\pi_{jj'}}$$

provided $\pi_{jj'} \neq 0$ and $\pi_{j \neq 0}, j \neq j'$. Here $d_{sjj'}, C_{sj}C_{sjj'}$ are supposed to be so chosen that $E_p(d_{sjj'}I_{sjj'}) = d_{jj'}, E_p(C_{sj}I_{sj}) = C_j$ and $E_p(C_{sjj'}I_{sjj'}) = C_{jj'}$.

As special cases of e_b the Horvitz and Thompson's (1952) scheme based on Hartley and Rao's sampling scheme and Rao, Hartley, and Cochran's (1962) strategy among others may be tried.

8.2.2 ADAPTIVE SAMPLING AND ESTIMATION

Let $U = (1, \cdots, i, \cdots, N)$ denote a finite population of units bearing y-values y_i for $i \in U$. Let $A(i)$ denote the network of i in U and C_i be the cardinality of A(i), that is the number of units this A(i) contains.

Let us define
$$t_i = \frac{1}{C_i} \sum_{j \in A(i)} y_j \text{ for } i \in U.$$

Then
$$T = \sum_{1}^{N} t_i \text{ equals } \sum_{j=1}^{N} y_j \left(\frac{1}{C_i} \sum_{j|A(j) \ni i} \right)$$
$$= \sum_{j=1}^{N} y_j = Y \quad \text{by definition of } C_i.$$

Our original problem of estimating Y is thus clearly one equivalently of estimating T. Following Rao (1979), Chaudhuri and Pal (2002), Chaudhuri and Stenger (2005), Chaudhuri (2010), among others, we get the following.

With $I_{si} = 1/0$ if $i \in s | i \notin s$ and b_{si}'s as free of $\underline{T} = (t_1, \cdots, t_i, \cdots, t_N)$ it follows that

$$t = \sum t_i b_{si} I_{si} \quad \text{with} \quad \sum_s p(s) b_{si} I_{si} = 1 \, \forall \, i \in U \quad \text{has}$$

$E_p(t) = T = Y$,

$$M_p(t) = e_p(t - T)^2 = -\sum\sum_{i<j} d_{ij} a_i a_j \left(\frac{t_i}{a_i} - \frac{t_j}{a_j} \right)^2 + \sum \frac{t_i^2}{a_i} \alpha_i.$$

Here, $s, p(s), d_{ij}, \alpha_i$ and $a_i \, (\neq 0)$ are analogues to those considered in Section 8.2.1. Also, an unbiased estimator for $M_p(t)$ is

$$m(t) = -\sum\sum_{i<j} d_{sij} I_{sij} a_i a_j \left(\frac{t_i}{a_i} - \frac{t_j}{a_j} \right)^2 + \sum C_{si} I_{si} \frac{t_i^2}{a_i} \alpha_i$$

with d_{sij}, C_{si} as in Section 8.2.1.

8.3 CONSTRAINING NETWORK SAMPLING AND CONSTRAINING ADAPTIVE SAMPLING

8.3.1 NETWORK SAMPLING

We noted that
$$w_j = \sum_{i \in A(j)} \frac{y_i}{m_i}, j = 1, \cdots, M.$$

For a particular j the set $A(j)$ of OU's in it may be exorbitantly extensive and thus too difficult to be surveyed and likewise for many other j's in the initial samples. So, to completely survey all the OU's in the Network sample, $N(s) = \sum_{j \in s} A(j)$ may demand an enormous amount of effort, time, and resources. Something needs to be done to mitigate such a burden.

Let us present here an exercise toward that end, essentially copying from our forthcoming text, namely Chaudhuri (2014).

Let $C_j =$ The total number of OU's in $A(j)$ for a j in $U = (1, \cdots j, \cdots, M)$. Let $C(s) = \sum_{i \in s} C_j$, the total number of OU's in the network sample $N(s)$, and let this be unbearably gigantic in size. So, let us decide to choose independently from respective $A(j)$ for j in s sub-samples $B(j)$ of sizes d_j for j in s such that $2 \leq d_j \leq C_j, j \in s$ such that $\sum_{j \in s} d_j = d(s)$ which is small enough compared to $C(s)$ so that the y_i-values for $i \in U_{j \in s} B(j)$ may be determined with a manageable effort.

Let for the design used to choose a sub-sample $B(j)$ from $A(j)$ of size d_j, the inclusion-probability of the ith OU be

$$\pi_i \left(0 < \pi_i < 1, \sum_{i \in A(j)} \pi_i = d_j \text{ for } j = 1, \cdots, M \right).$$

Suppose, for example, $B(j)$ is an SRSWOR from $A(j)$.

Then, $$\pi_i = \frac{d_j}{C_j} \quad \text{for } i \text{ in } A(j) \quad \text{and}$$

$$\pi_{ii'} = \frac{d_j(d_j - 1)}{C_j(C_j - 1)} \quad \text{for } i, i' \, (i \neq i') \text{ in } A(j)$$

$$\text{for } j = 1, \cdots, M.$$

Suppose the original sample s from $U = (1, \cdots, j, \cdots, M)$ is taken in m draws ($2 < m < M$), using available normed size-measures

$$p_j \left(0 < p_j < 1, \sum_{1}^{M} p_j = 1 \right)$$

so that, writing Q_i as the sum of the p-values falling in the ith group while drawing the Rao, Hartley, Cochran (1962) sample giving

$$t_{RHC} = \sum_m w_i \frac{Q_i}{p_i}$$

as an unbiased estimator of

$$W = \sum_{j=1}^{N} w_j, \quad w_j = \sum_{i \in A(j)} \frac{y_i}{m_i} \quad \text{and} \quad \sum_m$$

denoting the sum over the m groups formed in taking a sample of m SU's form U. The variance of t_{RHC} is then

$$V(t_{RHC}) = A\left[\sum_m \frac{w_j^2}{p_j} - W^2\right], \quad A = \frac{\sum_m M^2 - M}{M(M-1)},$$

and an unbiased estimator for it is

$$\hat{V}(t_{RHC}) = B\left[\sum_m Q_i \frac{w_i^2}{p_i^2} - t_{RHC}^2\right]$$

$$= B\left[\sum_m \sum_m Q_i Q_{i'} \left(\frac{w_i}{p_i} - \frac{w_{i'}}{p_{i'}}\right)^2\right],$$

$$B = \frac{\sum_m M_i^2 - M}{M^2 - \sum_m M_i^2};$$

here M_i is the number of SU's falling in the ith group while choosing the RHC sample of m SU's from U, $\sum_m M_i = M$.

Let E_1, E_2, be the expectation operators in respect of RHC sampling and SRSWOR, respectively; V_1, V_2 the corresponding variance operators; and E, V the overall expectation, variance operators.

Let $\quad u_j = \sum_{i\in B(j)} \frac{a_i}{\pi_i}, a_i = \frac{y_i}{m_i},\quad$ generically.

Then, $\quad E_2(u_j) = \sum_{i\in A(j)} a_i = w_j.$

Also, $\quad V_2(u_j) = \sum\sum_{i<i'\in A(j)} (\pi_i \pi_{i'} - \pi_{ii'})\left(\frac{a_i}{\pi_i} - \frac{a_{i'}}{\pi_{i'}}\right)^2 + \sum_{i\in A(j)} \frac{a_i^2}{\pi_i}\alpha_i$

$$\alpha_i = 1 + \frac{1}{\pi_i}\sum_{\substack{(i=1)\\(\neq i')}}^{C_j} \pi_{ii'} - \sum_1^{C_j} \pi_i.$$

Let $\quad e_{RHC} = \sum_m u_i \frac{Q_i}{p_i}.$

Then, $\quad E_2(e_{RHC}) = t_{RHC}$
and $\quad E(e_{RHC}) = E_1(t_{RHC}) = W = Y.$

Also,
$$V_2(e_{RHC}) = \sum_m V_2(u_i)\left(\frac{Q_i}{P_i}\right)^2$$
$$V(e_{RHC}) = E_1[V_2(e_{RHC})] + V_1[E_2(e_{RHC})],$$
$$v_2(u_j) = \sum\sum_{i<i'\in B(j)}\left(\frac{\pi_i\pi_{i'}-\pi_{ii'}}{\pi_{ii'}}\right)\left(\frac{a_i}{\pi_i}-\frac{a_{i'}}{\pi_{i'}}\right)^2 + \sum_{i\in B(j)}\left(\frac{a_i}{\pi_i}\right)^2\alpha_i$$

and, hence, $E_2[v_2(u_j)] = V_2(u_j)$, $j = 1,\cdots,M$.

With some algebra we find the Theorem: An unbiased estimator of $V(e_{RHC})$ is

$$v = \sum_m\left(\frac{Q_j}{p_j}\right)^2\left[\sum\sum_{i<i'\in B(j)}(\pi_i\pi_{i'}-\pi_{ii'})\left(\frac{a_i}{\pi_i}-\frac{a_{i'}}{\pi_{i'}}\right)^2/\pi_{ii'} + \sum_{i\in B(j)}\left(\frac{a_i}{\pi_i}\right)^2\alpha_i\right]$$
$$+ B\sum_m\sum_m Q_iQ_{i'}\left[\left(\frac{u_i}{p_i}-\frac{u_{i'}}{p_{i'}}\right)^2 - \left(\frac{v_2(u_i)}{p_i^2}+\frac{v_2(u_{i'})}{p_{i'}^2}\right)\right].$$

PROOF

Let $\quad D = B\sum_m\sum_m Q_iQ_{i'}\left(\frac{u_i}{p_i}-\frac{u_{i'}}{p_{i'}}\right)^2.$

Then, $\quad E_2(D) = B\sum_m\sum_m Q_iQ_{i'}\left[\frac{V_2(u_i)}{p_i^2}+\frac{V_2(u_{i'})}{p_{i'}^2}+\left(\frac{w_i}{p_i}-\frac{w_{i'}}{p_{i'}}\right)^2\right].$

So, $\quad B\sum_m\sum_m Q_iQ_{i'}\left[\left(\frac{u_i}{p_i}-\frac{u_{i'}}{p_{i'}}\right)^2 - \left[\left(\frac{v_2(u_i)}{p^2}+\frac{v_2(u_{i'})}{p_{i'}^2}\right)\right]\right]$

is an unbiased estimator of

$$B\sum_m\sum_m Q_iQ_{i'}\left(\frac{w_i}{p_i}-\frac{w_{i'}}{p_{i'}}\right)^2 = v_1(t_{RHC})$$

which is an unbiased estimator of $V_1(t_{RHC})$.

$$E_2\left[\sum\sum_{i<i'\in B(j)}(\pi_i\pi_{i'}-\pi_{ii'})\left(\frac{a_i}{\pi_i}-\frac{a_{i'}}{\pi_{i'}}\right)^2/\pi_{ii'} + \sum_{i\in B(j)}\left(\frac{a_i}{\pi_i}\right)^2\alpha_i\right]$$
$$= \sum\sum_{i<i'\in A(j)}(\pi_i\pi_{i'}-\pi_{ii'})\left(\frac{a_i}{\pi_i}-\frac{a_{i'}}{\pi_{ii'}}\right)^2 + \sum_{i\in A(j)}\frac{a_i^2}{\pi_i}\alpha_i\right] = V_2(u_j).$$

Network and Adaptive Procedures 191

So,
$$E_2(v) = \sum_m \left(\frac{Qj}{p_j}\right)^2 V_2(u_j) + v_1(t_{RHC})$$
$$= V_2(e_{RHC}) + v_1(t_{RHC}).$$

So,
$$E(v) = E_1 V_2(e_{RHC}) + V_1(t_{RHC}).$$

So,
$$E(v) = E_1 V_2(e_{RHC}) + V_1(t_{RHC})$$
$$= E_1 V_2(e_{RHC}) + V_1 E_2(e_{RHC})$$
$$= V(e_{RHC}).$$

Note that in v finally one should take $\pi_i = \frac{d_j}{C_j}$ and $\pi_{ii'} = \frac{d_j(d_j-1)}{C_j(C_j-1)}$ for i and $i' (\neq i)$ in case $B(j)$'s are taken as SRSWOR from $A(j)$. ∎

8.3.2 CONSTRAINING ADAPTIVE SAMPLES

Recall that in Adaptive Sampling with C_i as the cardinality of $A(i)$, the network of i including one has with

$$t_i = \frac{1}{C_i} \sum_{j \in A(i)} y_j$$

the quantity
$$T = \sum_1^N t_i \text{ equals } Y = \sum_1^N y_i$$

and that
$$t = \sum_1^N t_i b_{si} I_{si}$$

is taken as an estimator for $T = Y$.

Let us suppose that $\sum_{i \in s} C_I = C(s)$, say, is too large, creating a tough problem to effectively survey the network sample $N(s) = \sum_{i \in s} A(i)$.

So, let us take simple Random Sub-samples $B(i)$ of sizes d_i from $A(i), i \in s$ on choosing a suitably modest positive integer $L << C(s)$ for the s at hand such that $\sum_{i \in s} d_i < L$.

Let
$$u_i = \frac{1}{d_i} \sum_{j \in B(i)} y_j, \ i \in s.$$

Let E_R, V_R denote expectation, variance operators with respect to SRSWOR of $B(i)$ from $A(i)$ independently across i in s, E_p, V_p the expectation, variance operators over selection of the original sample s with probability $p(s)$ from

$U = (1, \cdots, i, \cdots, N)$ and $E = E_p E_R$ and $V = E_p V_R + V_p E_R$ the overall expectation, variance operators for the above noted sampling designs/schemes.

Now, $\quad E_R(u_i) = t_i,$

$$V_R(u_i) = \left(\frac{1}{d_i} - \frac{1}{C_i}\right) \frac{1}{(C_i - 1)} \sum_{\hat{j} \in A(i)} \left(y_{\hat{j}} - t_i\right)^2$$

with an unbiased estimator as

$$v_R(w_i) = \left(\frac{1}{d_i} - \frac{1}{c_i}\right) \frac{1}{(d_i - 1)} \sum_{\hat{j} \in B(i)} (y_j - u_i)^2$$

because $\quad E_R v_R(u_i) = V_R(u_i), \ i \in s.$

For this Constrained Adaptive Sampling with the above size restrictions our proposed unbiased estimator for $Y = T$ is

$$e = \sum_{u_i} b_{si} I_{si}$$

with b_{si} free of $\underline{Y} = (y_1, \cdots, y_i, \cdots y_N)$ subject to

$$\sum_s p(s) b_{si} = 1 \ \forall \ i \quad \text{and} \quad I_{si} = 1/0 \text{ for } i \in s | i \notin s.$$

Then, $\quad E_R(e) = \sum t_i b_{si} I_{si} = t.$

$$M(e) = E(e-T)^2 = E_p E_R \left[(e - E_R(e)) + (t - T)\right]^2$$
$$= E_p \left[\sum V_R(u_i) b_{si}^2 I_{si}\right] + M_p(t)$$
$$= E_p E_R \left[\sum v_R(u_i) b_{si}^2 I_{si}\right] + E_p m(t).$$

Here $\quad M_p(t) = -\sum\sum_{i<j} d_{ij} a_i a_j \left(\frac{t_i}{a_i} - \frac{t_j}{a_j}\right)^2 + \sum_i \frac{t_i^2}{a_i} \alpha_i$

with $a_i (\neq 0)$ as some constants,

$$\alpha_i = 1 + \frac{1}{\pi_i} \sum_{\substack{i'=1 \\ \neq i}}^N \pi_{ii'} - \sum_1^N \pi_i \quad \text{and}$$

$$d_{ij} = E_p(b_{si}I_i - 1)(b_{sj}I_{sj} - 1);$$

further $\quad m(t) = -\sum\sum_{i<j} d_{sij} I_{sij} a_i a_j \left(\frac{t_i}{a_i} - \frac{t_j}{a_j}\right)^2 + \sum \left(\frac{t_i^2}{a_i}\right) \alpha_i C_{si} I_{si}$

such that $\quad E_p(d_{sij}I_{sij}) = d_{ij}$ and $E_p(C_{si}I_{si}) = 1.$

Network and Adaptive Procedures

In order to derive an unbiased estimator of $M(e)$ let us consider further:

$$f_i = u_i^2 - v_R(u_i) \text{ so that } E_R(f_i) = t_i^2$$

$$a_{sij} = -\sum\sum_{i<j} d_{sij} I_{sij} a_i a_j \left(\frac{u_i}{a_i} - \frac{u_j}{a_j}\right)^2$$

$$b_{sij} = a_{sij} + \sum\sum_{i<j} d_{sij} I_{sij} a_i a_j \left[\frac{v_R(u_i)}{a_i^2} + \frac{v_R(u_j)}{a_j^2}\right].$$

Then, $E_R(a_{sij}) = -\sum\sum_{i<j} d_{sij} I_{sij} a_i a_j \left[\left(\frac{t_i}{a_i} - \frac{t_j}{a_j}\right)^2 + \left(\frac{v_R(u_i)}{a_i^2} + \frac{v_R(u_j)}{a_j^2}\right)\right]$

$$E_R(b_{sij}) = -\sum\sum_{i<j} d_{sij} I_{sij} a_i a_j \left[\left(\frac{t_i}{a_i} - \frac{t_j}{a_j}\right)^2\right].$$

So, an unbiased estimator for $M(e)$ is

$$m(e) = \sum v_R(u_i) b_{si}^2 I_{si} + \sum \frac{f_i}{a_i} \alpha_i C_{si} I_{si}$$
$$- \sum\sum_{i\leq j} d_{sij} I_{sij} a_i a_j \left[\left(\frac{u_i}{a_i} - \frac{u_j}{a_j}\right)^2 - \left(\frac{v_R(u_i)}{a_i^2} + \frac{v_R(u_j)}{a_j^2}\right)\right]$$

because

$$E[m(e)] = E_p E_R \left[\sum v_R(u_i) b_{si}^2 I_{si} + \sum \frac{t_i^2}{a_i} \alpha_i - \sum\sum_{i<j} d_{ij} a_i a_j \left(\frac{t_i}{a_i} - \frac{t_j}{a_j}\right)^2\right]$$
$$= M(e).$$

9 Analytical Methods

Abstract. Analytical surveys: Contingency tables.

9.0 ABSTRACT

So far we have been addressing ourselves to problems of inference making about characteristics defined as parameters as values defined on the members of a finite population of individuals. This is in the area of descriptive surveys. As opposed to this we may be interested in speculating about underlying random processes which have had their effects as the realization of the values of individuals defined on the finite population at hand. When dealing with inference issues concerning such processes, we talk about analytical surveys. Now we pass on to the issues which need certain special approaches not yet covered by us.

9.1 ANALYTICAL SURVEYS: CONTINGENCY TABLES

A descriptive survey aims at estimating parameters relating to a given finite population with fixed but unknown values to start with. An analytical survey, on the contrary, is intended to study through surveys of samples drawn from a finite population the underlying processes which may be modeled to have generated such finite population values. We shall illustrate several such situations in the sub-sections below.

9.1.1 CONTINGENCY

Suppose the individuals are cross-classified according to the values of two variables into rXc categories namely into r alternative categories like their preferences in favor of different kinds of sports and according to their economic standings into c categories. Suppose a total of 1006 individuals are thus classified into $r = 4$ categories, namely their top likings for Football, Tennis, Cricket, and Badminton and $c = 3$ categories upper, middle, and economic levels of well-being. The frequencies for these $rc = 4X3 = 12$ classes are distributed into 12 cells in Table-called a Contingency Table.

Suppose our purpose is to examine how these two traits in people, namely their financial stature and love for one of four specified types of sport, are associated. A standard approach is to statistically test the Independence of the two characteristics in question. A procedure for this is as follows.

Let p_{ij} = Probability that a person belongs to the category i according to one characteristic A and to the category j in respect of the other characteristic

TABLE 9.1
Contingency Table of 1006 People Showing Their Frequencies in 4×3=12 Cells According to Their Financial Stature and Top Priority in Likings for Specific Kinds of Sport

Love of Sports	Financial Level			Marginal Totals
	Upper	Middle	Lower	
Football	57	168	148	373
Tennis	81	132	25	238
Cricket	128	78	89	295
Badminton	21	53	26	100
Marginal Totals	287	431	288	1006

B; here $i = 1, \cdots, r+1$ and $j = 1, \cdots, c+1$, say. Suppose $\hat{P}_{ij}$'s are their respective estimators obtained from a sample s of n units chosen according to a design p with probability $p(s)$.

Let $$P_{i0} = \sum_{j=1}^{c+1} P_{ij}, P_{0j} = \sum_{i=1}^{r+1} P_{ij} \text{ and } h_{ij} = P_{ij} - P_{i0}P_{0j}.$$

We then have to test the null hypothesis $H_0 : P_{ij} = P_{i0}P_{0j}$ for every $i = 1, \cdots r$ and $j = 1, \cdots c$ against the alternative that $h_{ij} \neq 0$ at least for one pair (ij). Let H be the matrix of the $\frac{\delta h_{ij}}{\delta P_{ij}}$, $\underline{h}$; $\hat{\underline{H}}$ be the corresponding analogues replacing P_{ij}'s by $\hat{P}_{ij}$'s; and $\frac{V}{n}$ the matrix of variance-covariances corresponding to $\hat{P}_{ij}$'s. The test for independence of the two characteristics is provided by the Wald statistic

$$X_W = n\hat{\underline{h}}' \left(\hat{H}' \hat{V} \hat{H} \right)^{-1} \hat{\underline{h}}$$

with $\hat{V}$ obtained on replacing the entries in V by the corresponding estimators. Writing

$$\underline{\hat{p}}_r = \left(\hat{P}_{01}, \cdots, \hat{P}_{0c} \right)', \qquad \underline{\hat{p}}_c = \left(\hat{P}_{01}, \cdots, \hat{P}_{0c} \right)',$$
$$\hat{p}_r = Diag\left(\underline{\hat{p}}_r \right) - \underline{\hat{p}}_r \underline{\hat{p}}_r', \qquad \hat{p}_c = Diag\left(\hat{p}_c \right) - \hat{p}_c \hat{p}_c,$$

an approximate test is provided by the Pearson statistic

$$X_P = n\, \hat{\underline{h}}' \left(\underline{\hat{P}}_r^{-1} \otimes \underline{\hat{P}}_C^{-1} \right) \hat{\underline{h}}$$

Analytical Methods

writing $\otimes$ to denote Kronecker products. Denoting by $\delta_1 \geq \cdots \geq \delta_T, T = rc$ the eigenvalues of

$$\left(\underline{P}_r^{-1} \otimes \underline{P}_C^{-1}\right)(H'VH),$$

it is known that

$$X_P = \sum_{i=1}^{T} \delta_i X_1^2(i)$$

with $X_1^2(i)$ for each i as a chi-square variate with 1 degree of freedom.

If δ_1 may be guessed, $\frac{X_P}{\delta_1}$ provides a conservative test, treating X_P as a chi-square variable with T degrees of freedom.

9.1.2 CORRELATION, REGRESSION ESTIMATION

Suppose x and y are two real variables such that

$$\rho = \frac{E(y - E(y))(x - E(x))}{\sqrt{E(y - E(y))^2}\sqrt{E(x - E(x))^2}}$$

is the total correlation coefficient between y and x, and $\beta(y|x) = \rho\frac{\sigma_y}{\sigma_x}$ is the coefficient of regression of y on x. Now the question is how to estimate these ρ and β by a sample survey; here σ_y, σ_x are the standard deviations, respectively, of y and x. A simple answer is that y_i, x_i-values for i in the entire population are first presumed to be available, and census estimates of ρ and β are first written down as follows:

$$R_N = \frac{\sum_1^N (y_i - \bar{Y})(x_i - \bar{X})}{\sqrt{\sum_1^N (y_i - \bar{Y})^2}\sqrt{\sum_1^N (x_i - \bar{X})^2}}$$

and

$$B_N = B_N(y|x) = \frac{\sum_1^N (y_i - \bar{Y})(x_i - \bar{X})}{\sum_1^N (x_i - \bar{X})^2}$$

Now from a sample drawn from $U = (1, \cdots i, \cdots N)$ as s with a probability $p(s)$ according to a design p admitting positive inclusion-probability π_i, the following estimators for R_N and B_N respectively seem quite plausible, namely

$$r = \frac{\sum_{i \in s}\frac{1}{\pi_i}\left(y_i - \frac{\sum_{i \in s} y_i/\pi_i}{\sum_{i \in s}\frac{1}{\pi_i}}\right)\left(x_i - \frac{\sum_{i \in s} x_i/\pi_i}{\sum_{i \in s}\frac{1}{\pi_i}}\right)}{\sqrt{\sum_{i \in s}\frac{1}{\pi_i}\left(y_i - \frac{\sum_{i \in s} y_i/\pi_i}{\sum_{i \in s}\frac{1}{\pi_i}}\right)^2}\sqrt{\sum_{i \in s}\frac{1}{\pi_i}\left(x_i - \frac{\sum_{i \in s}\frac{x_i}{\pi_i}}{\sum_{i \in s}\frac{1}{\pi_i}}\right)^2}}$$

and
$$b = b(y/x) = \frac{\sum_{i \in s} \frac{1}{\pi_i} \left(y_i - \frac{\sum_{i \in s} \frac{y_i}{\pi_i}}{\sum_{i \in s} \frac{1}{\pi_i}} \right) \left(x_i - \frac{\sum_{i \in s} \frac{x_i}{\pi_i}}{\sum_{i \in s} \frac{1}{\pi_i}} \right)}{\sum_{i \in s} \frac{1}{\pi_i} \left(x_i - \frac{\sum_{i \in s} \frac{x_i}{\pi_i}}{\sum_{i \in s} \frac{1}{\pi_i}} \right)^2}.$$

A rationale behind these choices will be discussed in the next sub-section. How to provide estimated measures of their errors will also be discussed there in these connections.

9.1.3 LINEARIZATION

Suppose $\theta_1, \cdots, \theta_j, \cdots, \theta_K$ denote K distinct finite population totals, say,

$$\theta_j = \sum_{i=1}^{N} \xi_{ji}, j = 1, \cdots, K$$

with ξ_j as a real-valued variable with its value ξ_{ji} taken on the ith element of $U = (1, \cdots, i, \cdots, N)$. Suppose $\underline{\theta} = (\theta_1, \cdots, \theta_K)$ and our interest is to estimate a real-valued function $f(\underline{\theta}) = f(\theta_1, \cdots \theta_j, \cdots \theta_K)$ of $\underline{\theta}$ on surveying a sample s chosen from U with a selection-probability $p(s)$ according to a design P. Let based on s unbiased estimators $t_j, j = 1, \cdots, K$ be available, respectively, for $\theta_j, j = 1, \cdots K$.

Let
$$t_j = \sum_{i \in s} \xi_{ji} b_{si}$$

be an unbiased linear homogeneous estimator for θ_j so that

$$\sum_{s \ni i} p(s) b_{si} = 1 \ \forall \ i \quad \text{with} \quad b_{si}$$

free of ξ_{ji}'s, $j = 1, \cdots, K$.

As an estimator for $f(\underline{\theta})$ it seems prudent to try the estimator $f(\underline{t})$ writing $\underline{t} = (t_1, \cdots, t_j \cdots, t_K)$. Assuming the sample-size n large enough, making Taylor series expansion and omitting higher-order terms with higher-order differential coefficients, it seems plausible to approximate $f(\underline{t})$ by

$$g(\underline{t}) = f(\underline{\theta}) + \sum_{j=1}^{K} (t_j - \theta_j) \left(\frac{\delta f(\underline{t})}{\delta t_j} \right) |_{\underline{t}=\underline{\theta}}$$

neglecting all other terms. Then, we get

$$E_p g(\underline{t}) = f(\underline{\theta}) \quad \text{since} \quad E_p (t_j - \theta_j) = 0 \ \forall j.$$

Analytical Methods

Since
$$g(t) = \left[f(\underline{\theta}) - \sum_{1}^{K} \theta_j \lambda_j \right] + \sum_{j=1}^{K} t_j \lambda_j$$

writing
$$\lambda_j = \left. \frac{\delta f(t)}{\delta t_j} \right|_{\underline{t}=\underline{\theta}}$$

approximate $E_p\, f(\underline{t})$ by $f(\underline{\theta})$ and approximate the variance of $f(\underline{t})$ by

$$V_p g(\underline{t}) = V_p \left(\sum_{j=1}^{K} t_j \lambda_j \right)$$

$$= V_p \left[\sum_{j=1}^{K} \lambda_j \left(\sum_{i \in s} \xi_{ji} b_{si} \right) \right]$$

$$= V_p \left[\sum_{i \in s} b_{si} \Psi_i \right], \quad \text{writing} \quad \Psi_i = \sum_{j=1}^{K} \lambda_j \xi_{ji}.$$

Thus,
$$V_p f(\underline{t}) \simeq \sum_{i=1}^{N} \psi_i^2 C_i + \sum_{\substack{1 \\ i \neq j}}^{N} \sum_{1}^{N} \psi_i \psi_j C_{ij}.$$

Writing
$$C_i = \sum_{s \ni i} p(s) b_{si}^2 - 1 \text{ and}$$

$$C_{ij} = \sum_{s \ni i,j} p(s) b_{si} b_{sj} - 1.$$

Now choosing C_{si}, C_{sj} free of ξ_{ij}'s subject to $E_p (C_{si} I_{si}) = C_i$ and $E_p (C_{sij} I_{sij}) = C_{ij}$, recalling $I_{si} = 1$ if $s \ni i$ and $I_{si} = 0$, otherwise, and $I_{sij} = I_{si} I_{sj}$, it follows that we may employ the reasonable measure of error of $f(\underline{t})$ as an estimator for $f(\underline{\theta})$ as

$$v(f) = \sum_{i=1}^{N} \left(\hat{\Psi}_i \right)^2 C_{si} I_{si} + \sum_{i \neq j} \sum \hat{\Psi}_i \hat{\Psi}_j C_{sij} I_{sij}$$

taking
$$\hat{\Psi}_i = \sum_{j=1}^{K} \hat{\lambda}_j \xi_{ji}, \quad \text{and taking} \quad \hat{\lambda}_j \quad \text{as}$$

$$\hat{\lambda}_j = \lambda_j |_{\underline{\theta}=\underline{t}}.$$

Let us cite a few examples in application:

(i) $K = 2$, $\theta_1 = \sum_1^N y_i = Y$, $\theta_2 = \sum_1^N x_i = X$, $\xi_{1i} = y_i$, $\xi_{2i} = x_i$

$$\underline{\theta} = (\theta_1, \theta_2), f(\underline{\theta}) = \frac{\theta_1}{\theta_2} = \frac{Y}{X} = R, \text{ say},$$

$$t_1 = \sum_{i \in s} \frac{y_i}{\pi_i}, \quad t_2 = \sum_{i \in s} \frac{x_i}{\pi_i}, \quad \underline{t} = (t_1, t_2),$$

$$f(\underline{t}) = \frac{t_1}{t_2} = \sum_{i \in s} \frac{y_i}{\pi_i} \bigg/ \sum_{i \in s} \frac{x_i}{\pi_i}, \quad \text{assuming } \pi_i \text{ as also}$$

$$\pi_{ij} = \sum_{s \ni i,j} p(s), i \neq j \quad \text{as positive quantities we get}$$

$$\lambda_1 = \delta \left(\frac{t_1}{t_2} \right) \bigg|_{\substack{t_1 = Y, \\ t_2 = X}} = \frac{1}{X}, \quad \hat{\lambda}_1 = \frac{1}{t_2}$$

$$\lambda_2 = \frac{\delta}{\delta t_2} \left(\frac{t_1}{t_2} \right) \bigg|_{\substack{t_1 = Y \\ t_2 = X}} = -\frac{t_1}{t_2^2} \bigg|_{\substack{t_1 = Y \\ t_2 = X}} = -\frac{Y}{X^2} = -\frac{R}{X}.$$

So, $\hat{\lambda}_2 = -\frac{t_1}{t_2^2}$.

So, for $\hat{R} = f(t_1, t_2) = \frac{t_1}{t_2}$,

$E_p(\hat{R})$, is approximately equal to $\frac{\theta_1}{\theta_2} = R$, and $V_p(\hat{R})$ is approximately equal to

$$V_p(\lambda_1 t_1 + \lambda_2 t_2) = V_p \left(\frac{\sum_{i \in s} \frac{y_i}{\pi_i}}{X} - \frac{R}{X} \sum_{i \in s} \frac{x_i}{\pi_i} \right)$$

$$= \frac{1}{X^2} V_p \left(\sum_{i \in s} \frac{y_i - R x_i}{\pi_i} \right)$$

$$= \frac{1}{X^2} V_p \left(\sum_{i \in s} \frac{y_i}{\pi_i} \right) \bigg|_{y_i = y_i - R x_i}$$

and $v(\hat{R}) = v \left[\hat{\lambda} t_1 + \hat{\lambda} t_2 \right]$

$$= v \left[\frac{1}{t_2} \left(t_1 - \frac{t_1}{t_2} t_2 \right) \right]$$

$$= \frac{1}{\left(\sum_{i \in s} \frac{x_i}{\pi_i} \right)^2} \sum \sum_{i < j \in s} \left(\frac{\pi_i \pi_j - \pi_{ij}}{\pi_{ij}} \right) \left(\frac{y_i}{\pi_i} - \frac{y_j}{\pi_j} \right)^2 \bigg|_{y_i = \left(y_i - \frac{t_1}{t_2} x_i \right)}.$$

Analytical Methods

(ii) $$R_N = \frac{\sum_1^N (y_i - \bar{Y})(x_i - \bar{X})}{\sqrt{\sum(y_i - \bar{Y})^2}\sqrt{\sum(x_i - \bar{X})^2}}$$

$$= \frac{N\sum_1^N y_i x_i - \left(\sum_1^N y_i\right)\left(\sum_1^N x_i\right)}{\sqrt{N\sum y_i^2 - \left(\sum_1^N y_i\right)^2}\sqrt{N\sum_1^N x_i^2 - \left(\sum_1^N x_i\right)^2}}$$

$K = 6$, $\xi_{1i} = 1$, $\xi_{2i} = y_i x_i$, $\xi_{3i} = y_i$, $\xi_{4i} = x_i$, $\xi_{5i} = y_i^2$, $\xi_{6i} = x_i^2$;

$\theta_1 = N$, $\theta_2 = \sum_1^N y_i x_i$, $\theta_3 = \sum y_i$, $\theta_4 = \sum x_i$, $\theta_5 = \sum x_i^2$, $\theta_6 = \sum x_i^2$.

So, $R_N = f(\theta_1, \theta_2, \theta_3, \theta_4, \theta_5, \theta_6) = f(\underline{\theta})$.

And, $t_1 = \sum_{i \in s} \frac{1}{\pi_i}$, $t_2 = \sum_{i \in s} \frac{\xi_{2i}}{\pi_i}$, $t_3 = \sum_{i \in s} \frac{y_i}{\pi_i}$, $t_4 = \sum \frac{x_i}{\pi_i}$,

$t_5 = \sum \frac{y_i^2}{\pi_i}$, $t_6 = \sum_{i \in s} \frac{x_i^2}{\pi_i}$.

So, R_N may be estimated by

$r = f(t_1, t_2, t_3, t_4, t_5, t_6) = f(\underline{t})$

$$= \frac{\left(\sum_{i \in s} \frac{1}{\pi_i}\right)\left(\sum_{i \in s} \frac{y_i x_i}{\pi_i}\right) - \left(\sum_{i \in s} \frac{y_i}{\pi_i}\right)\left(\sum_{i \in s} \frac{x_i}{\pi_i}\right)}{\sqrt{\left(\sum_{i \in s} \frac{1}{\pi_i}\right)\left(\sum_{i \in s} \frac{y_i^2}{\pi_i}\right) - \left(\sum_{i \in s} \frac{y_i}{\pi_i}\right)^2}\sqrt{\left(\sum_{i \in s} \frac{1}{\pi_i}\right)\left(\sum_{i \in s} \frac{x_i^2}{\pi_i}\right) - \left(\sum_{i \in s} \frac{x_i}{\pi_i}\right)^2}}$$

$$= \frac{\sum \frac{1}{\pi_i}\left(y_i - \frac{\sum \frac{y_i}{\pi_i}}{\sum \frac{1}{\pi_i}}\right)\left(x_i - \frac{\sum \frac{x_i}{\pi_i}}{\sum \frac{1}{\pi_i}}\right)}{\sqrt{\sum \frac{1}{\pi_i}\left(y_i - \frac{\sum \frac{y_i}{\pi_i}}{\sum \frac{1}{\pi_i}}\right)^2}\sqrt{\sum \frac{1}{\pi_i}\left(x_i - \frac{\sum \frac{x_i}{\pi_i}}{\sum \frac{1}{\pi_i}}\right)^2}}.$$

To calculate approximately the variance of r one has to calculate

$$\lambda_j = \left.\frac{\partial f}{\partial t_j}(\underline{t})\right|_{\underline{t}=\underline{\theta}} \quad \text{for} \quad j = 1, \cdots, 6,$$

and in order to calculate a reasonable estimator for this approximate $V_p(r)$ one needs to evaluate

$$\hat{\lambda}_j = \lambda_j\big|\underline{\theta} = \underline{t}, j = 1, \cdots, 6.$$

(iii) $$B_N = \frac{\sum_1^N (y_i - \bar{Y})(x_i - \bar{X})}{\sum_1^N (x_i - \bar{X})^2} = \frac{N\sum_1^N y_i x_i - \left(\sum_1^N y_i\right)\left(\sum_1^N x_i\right)}{N\sum_1^N x_i^2 - \left(\sum_1^N x_i\right)^2}.$$

So, letting $K = 5$,

$$\xi_{1i} = 1,\ \xi_{2i} = y_i x_i,\ \xi_{3i} = y_i,\ \xi_{4i} = x_i,\ \xi_{5i} = x_i^2,$$

$$t_1 = \sum_{i \in s} \frac{1}{\pi_i},\ t_2 = \sum_{i \in s} \frac{y_i x_i}{\pi_i},\ t_3 = \sum_{i \in s} \frac{y_i}{\pi_i},\ t_4 = \sum_{i \in s} \frac{x_i}{\pi_i},\ t_5 = \sum_{i \in s} \frac{x_i^2}{\pi_i}.$$

So, B_N may be estimated by

$$b = \frac{\sum_{i \in s} \frac{1}{\pi_i}\left(y_i - \frac{\sum_{i \in s} \frac{y_i}{\pi_i}}{\sum_{i \in s} \frac{1}{\pi_i}}\right)\left(x_i - \frac{\sum_{i \in s} \frac{x_i}{\pi_i}}{\sum_{i \in s} \frac{1}{\pi_i}}\right)}{\sum_{i \in s} \frac{1}{\pi_i}\left(x_i - \frac{\sum_{i \in s} \frac{x_i}{\pi_i}}{\sum_{i \in s} \frac{1}{\pi_i}}\right)^2}.$$

Now evaluating

$$b = f(t_1, \cdots, t_5) = f(\underline{t}),\quad \underline{\theta} = \left(N, \sum_1^N y_i x_i, \sum_1^N y_i, \sum_1^N x_i, \sum_1^N x_i^2\right),$$

$$\lambda_i = \left.\frac{\partial f(\underline{t})}{\partial t_j}\right|_{\underline{t}=\underline{\theta}},\ j=1,\cdots 5\ \text{ and }\ \hat{\lambda}_j = \lambda_j\Big|\underline{\theta} = \underline{t},\ j = 1,\cdots 5,$$

evaluating $V_p(b)$ approximately and a suitable estimator for it by calculating λ_j and $\hat{\lambda}_j$, $j = 1, \cdots 5$, is now a routine affair.

9.1.4 JACK-KNIFING

This is a method of reducing the magnitude in the bias of an estimator for a real-valued parameter when the size of the sample is modest.

Let θ be a real parameter and t be an initial statistic based on a sample s of size n intended to estimate θ.

Let us write $t(n)$ to emphasize that t is based on a sample of size n.

We suppose that $t(n)$ is a biased estimator of θ and its bias $B(n)$ is expressible as

$$B_n = B_n(\theta) = E(t(n) - \theta) = Et(n) - \theta = \frac{b_1(\theta)}{n} + \frac{b_2(\theta)}{n^2} + \cdots,$$

where $b_1(\theta), b_2(\theta),\ldots$ are certain functions of θ such that $b_1(\theta) \neq 0$. Then, we may derive in the following way an alternative estimator with a reduced bias. Let the sample s of size n be divided into g disjoint parts each of size m so that $n = gm$ such that n, g, m are each a positive integer. Let us calculate

Analytical Methods

the statistic $t = t(n)$ now based on $(n - m)$ units composed of the values of the variable for the $(g - 1)$ groups, omitting successively the values for the jth group taking $j = 1 \cdots, g$. Then let us write $t_j (n - m)$ to denote the statistic t based on the $(n - m)$ values for the sample s from which the values in the jth group $(j = 1, \cdots, m)$ are dropped.

Then, we may write down the bias of $t_j (n - m)$ as

$$B_j (n - m) = \frac{b_1 (\theta)}{(n - m)} + \frac{b_2 (\theta)}{(n - m)^2} + \cdots$$

Let us construct a new estimator as

$$e_j = e_j (n) = g t(n) - (g - 1) t_j (n - m).$$

Then, we get

$$E(e_j) = g \left[\theta + \frac{b_1(\theta)}{n} + \frac{b_2(\theta)}{n^2} + \cdots \right] - (g-1) \left[\theta + \frac{b_1(\theta)}{n-m} + \frac{b_2(\theta)}{(n-m)^2} + \cdots \right]$$

$$= \theta + b_1(\theta) \left[\frac{g}{n} - \frac{(g-1)}{n\left(1 - \frac{1}{g}\right)} \right] + b_2(\theta) \left[\frac{g}{n^2} - \frac{(g-1)}{n^2 \left(1 - \frac{1}{g}\right)^2} \right] + \cdots$$

$$= \theta - \frac{b_2(\theta)}{n^2} \left(\frac{g}{g-1} \right) + \cdots +.$$

Now, we see that

$$B_n \left(t_{(n)}\right) \text{ is of } 0 \left(\frac{1}{n} \right)$$

and $B_n (e_j)$ is of $0 \left(\frac{1}{n^2} \right)$.

For a moderate sample-size, compared to $t(n)$, the revised statistic e_j has a reduced magnitude of a bias.

The average of $e_j, j = 1. \cdots, g$, namely,

$$\bar{e} = \frac{1}{g} \sum_{j=1}^{g} e_j = g\, t(n) - \left(\frac{g-1}{g} \right) \sum_{j=1}^{g} t_j (n - m)$$

is called a jack-knife statistic derived in this way from the statistic t.

Then, bias of $\bar{e}$ is of order $\left(\frac{1}{n^2} \right)$.

Interestingly, the mean-square error is then estimated by

$$m(\bar{e}) = \frac{1}{g} \left[\frac{1}{(g-1)} \sum_{j=1}^{g} (e_j - \bar{e})^2 \right]$$

$$= \left(\frac{g-1}{g} \right) \sum_{j=1}^{g} \left[t_j (n-m) - \frac{1}{g} \sum_{j=1}^{g} t_j (n-m) \right]^2.$$

Quenouille (1949) introduced the jack-knife technique, and Tukey (1954) showed how as above the Jack-Knife statistic may be used in estimating the Mean Square Error of non-linear statistics.

9.1.5 BOOTSTRAP

Efron's (1982) bootstrap technique of independent random sampling of n units from a random sample of size n at hand planned to produce an empirical cumulating distribution as a replica for the unavailable true histogram for the original distribution of a random variable. It provides rational inference procedures in the general statistical parlance. In the context of finite populations, considerable research has been reported in the literature. Chaudhuri and Stenger (2005) have given a comprehensive report on the developments. Chaudhuri and Saha (2004) and Pal (2009) have also made some contributions worthy of attention. In this context Rao and Wu's (1988) procedure bears a brief reproduction. They start taking a probability sample s from a finite survey population by a general procedure such that every sample unit is distinct from every other unit in the sample, each of which is of a common size. Inclusion-probability of each population unit as well as of every pair of units is strictly positive. As a consequence one may start with the Horvitz and Thompson (1952) unbiased estimator for the finite population total admitting the Yates and Grundy (1953) unbiased variance estimator for it which Rao and Wu (1988) need and assume to be positive. Then they give a method of drawing a sample from the initial sample in such a way that this may be independently repeated a very large number of times, each producing with respect to this probability distribution for these samples called bootstrap samples an exactly unbiased estimator for the finite population total with its variance equal to the Yates-Grundy unbiased variance estimator for the Horvitz-Thompson estimator for the total. This unbiased estimate of the population total is called a Bootstrap estimate which is obtained as the average of estimates over the numerous replications of the bootstrap sample drawing. As an average over these bootstrap sample frequencies a bootstrap variance estimate is also obtained.

A single bootstrap sample typically written as s_b^*, the bth bootstrap sample and b takes values $1, 2, \cdots, B$ with B as large enough, say, 1000 or even 10,000. A bootstrap estimate for Y is taken as $\bar{t} = \frac{1}{B}\sum_{b=1}^{B} t_b$, each t_b, based on s_b^* is suitably derived from s_b^*, and $\bar{v} = \frac{1}{B-1}\sum_{b=1}^{B}(t_b - \bar{t})^2$ is taken as the bootstrap variance of the estimate $\bar{t}$. The main importance of the bootstrap technique in the context of finite population sampling is its use in handling non-linear functions of several estimators for different finite population totals of real variables, in getting point estimates for them, variances of these estimators, and most importantly developing confidence intervals for non-linear parametric functions of real-variate totals. However, in the case of general statistical theory, an empirical distribution function has the Glivenko-Cantelli property of being almost sure uniform convergence (vide Loeve, 1977) to the true distribution function, but no such elegant theoretical property is yet established for a bootstrap technique applicable to finite populations.

Let $t_1, \cdots, t_K$ be respective bootstrap estimates for finite population totals $\theta_1, \cdots, \theta_K$, respectively. Then, Rao and Wu (1988) recommend the use of $f(t_1, \cdots, t_K)$ as a bootstrap estimate of a linear or non-linear function $f(\theta_1, \cdots, \theta_K)$. Writing $\underline{t} = (t_1, \cdots, t_K)$, $\underline{\theta} = (\theta_1, \cdots, \theta_K)$, and $\underline{t}_b$ as $\underline{t}$ calculated from the bth bootstrap sample, the bootstrap variance estimator for $f(\underline{t})$ is taken as

$$\frac{1}{(B-1)} \sum_{b=1}^{B} \left[f(\underline{t}_b) - \frac{1}{B} \sum_{b=1}^{B} f(\underline{t}_b) \right]^2.$$

The most notable use of the bootstrap technique in the Finite Population context is in the construction of Confidence Interval (CI) with a given Confidence Coefficient (CC) of say $100(1-\alpha)\%$ in the following two ways.

(I) Percentile Method:

Calculating $f(\underline{t}_b)$ for $b = 1, \cdots, B$ with B taken very large, say, 1000 or 10,000, find two values t_L and t_U, say, such that $100\left(\frac{\alpha}{2}\right)\%$ of the values $f(\underline{t}_b)$ lie below t_L and $100\left(\frac{\alpha}{2}\right)\%$ of the values $f(\underline{t}_b)$ lie above t_U so that $100(1-\alpha)\%$ of these $f(\underline{t}_b)$ values within the interval (t_L, t_U) which then provides a $100(1-\alpha)\%$ CI for $f(\underline{\theta})$ with a CC of $100(1-\alpha)\%$.

(II) Double Bootstrap Confidence Interval:

First a bootstrap variance estimate

$$\frac{1}{(B-1)} \sum_{b=1}^{B} \left(f(\underline{t}_b) - \frac{1}{B} \sum_{b=1}^{B} f(\underline{t}_b) \right)^2 = v(f)$$

is calculated. Then,

$$\frac{f(\underline{t}_b) - \frac{1}{B}\sum_{b=1}^{B} f(\underline{t}_b)}{\sqrt{v(f)}} = \frac{f(\underline{t}_b) - f(\underline{t})}{\sqrt{v(f)}}, \text{ say,}$$

is calculated, and the value

$$\delta_b = \frac{f(\underline{t}_b) - f(\underline{t})}{\sqrt{v(f)}}$$

is calculated for $b = 1, \cdots, B$. Then are calculated two real values δ_L and δ_U such that $100\left(\frac{\alpha}{2}\right)\%$ of the B values of δ_b, $b = 1, \cdots, B$ lie below δ_L and $100\left(\frac{\alpha}{2}\right)\%$ of the B values of δ_b, $b = 1, \cdots, B$ lie above δ_U. Then, (δ_L, δ_U) gives the $100(1-\alpha)\%$ CI for $(\underline{\theta})$ of $100(1-\alpha)\%$ CC. Since $v(f)$ is calculated from a first set of B bootstrap samples and (δ_L, δ_U) is calculated from a second set of B bootstrap samples, this method of construction of CI is thus named.

9.1.6 PERMANENT RANDOM NUMBERS: BUSINESS SURVEYS

In a Business Survey which is to be executed on successive occasions over several years, we cannot have a formulation with a finite population fixed at $U = (1, \cdots, i, \cdots, N)$ as we have noted so far. This is because as we proceed chronologically forward some business units may disappear, some new entries may be born, some may break away from the parent bodies, some may be amalgamated, and some may get split and carry on business as two entities with at least one having a new credential. Consequently, revising a population on every occasion when sampling is needed for fresh estimation of totals or changes in values of totals may be a cumbrous exercise. A simpler device is practicable using Permanent Random Numbers (PRN) in the following way.

Suppose we start with an initial population $U = (1, \cdots, i, \cdots, N)$ of N units identifiably labeled as $i = 1, 2, \cdots, N$. Let with each i a random variable X_i be generated as independently uniformly distributed over the Left-closed, Right-open unit interval $[0, 1)$. Let these be arranged increasingly ordered as $X_{(i)}, i = 1, \cdots, N$ such that $X_{(1)} \leq X_{(2)} \leq \cdots \leq X_{(N)}$, and let the labels $(1), (2), \cdots (i) \leq \cdots \leq N$ be identified with the corresponding $X_{(i)}$'s, $i = 1, \cdots, N$. Then, as shown by Ohlsson (1992), the first $n (< N)$ labels constitute an SRSWOR of size n from U. In order to draw a sample on a nearby subsequent occasion, we should first find out the M new units which have introduced themselves and the R units which out of the existing N units have dropped out. So, retaining the selected labels $(1), \cdots, (n)$ with the values $X_{(1)}, \cdots X_{(n)}$ called the Permanent Random Numbers (PRN), now with the new labels $N+1, N+2, \cdots, N+M$ generate independently uniformly distributed random numbers over $[0, 1)$ as $X_{N+1}, \cdots, X_{N+M}$. Now considering the earlier $X_1, \cdots, X_N$ and omitting the X_j's corresponding to the labels j which have vanished from the scene, find the labels $(1), \cdots (N), (N+1), \cdots, (N+M)$ corresponding to $X_{(1)} \quad X_{(2)}, \quad X_{(N+M)}$ with $X_{(1)} \leq \cdots \leq X_{(N)} \leq \cdots \leq X_{(N+M)}$. Notice that for the first occasion like $(1), \cdots, (n)$ the last labels $(N-n+1), \cdots, (N)$ also give us an SRSWOR of size n. If n is too large relative to N, then these numbers should be reduced modulo (N). On the second occasion, then, if an SRSWOR of size n' is needed, then either the first n' labels out of $(1), \cdots (N), \cdots (N+M-R)$ or the last n' labels out of $(1), \cdots (N), \cdots (N+M-R)$ modulo $(N+M-R)$ will be an SRSWOR of the size n'. If the fitst sample is the first n labels out of $(1) \cdots, (N)$ defined above and the second sample is the first n' labels out of $(1) \cdots, (N+M-R)$, then several members of the two samples may be common. When such a phenomenon of a maximum overlapping of the two samples occurs, we say we have a positive co-ordination in the sampling on two consecutive occasions. In order to reduce the overlap a simple device to adopt is to shift the selection process sufficiently away to the right of points $(1), \cdots, (n)$ when taking the second sample of the first n' points $(1), \cdots, (N+M-R)$. If we have the minimal overlap we say we have the negative co-ordination between the two samples for the two occasions.

Similarly if the last n labels out of $(N-n+1), \cdots, (N+M-R)$ modulo $(N+M-R)$ is the sample on the first occasion and the last n' labels out of $(N+M-R-n'+1), \cdots, (N+M-R)$ modulo $(N+M-R)$ then a positive co-ordination may result. In order to avoid too much overlap suitable shifting to the right is desirable. A great advantage in using the Permanent random numbers is thus choosing the shifting point to the right so as to achieve desirable extent of an overlap in the successive samples. Estimation is no problem because the above sequential random sampling is theoretically equivalent to SRSWOR.

In unequal probability sampling, so far the concept PRN has been applied only to Ha'jek's (1964, 1971) Poisson's Scheme of sampling. A Poisson sampling scheme demands choosing an integer n and associating with each i in $U = (1, \cdots, i, \cdots, N)$ a number π_i such that $0 < \pi_i < 1 \ \forall \ i \in U$ and $\sum_1^N \pi_i = n$. As a next step N independent Bernoullian trials are performed for the N units of U with probability of success π_i for the respective i in U. Each unit for which a success is achieved is taken into the sample, the other units achieving failures or no successes are kept outside the sample. Equivalently, with each i is associated a random number X_i distributed uniformly and independently over the Left-closed, Right-open unit interval $[0, 1)$ and a unit i is taken in the sample if $X_i \leq \pi_i$ and kept outside the sample, otherwise. Writing

$$I_{si} = 1 \quad \text{if} \quad i \in s$$
$$= 0 \quad \text{if} \quad i \notin s$$

the sample size for s is

$$v(s) = \sum_1^N I_{si}$$

which is a random number. Then,

$$E_p v(s) = E_p \left(\sum_1^N I_{si} \right) = \sum_1^N \pi_i = n.$$

If normed size-measures

$$p_i \left(0 < p_i < 1, \sum_1^N p_i = 1 \right)$$

are available such that $n\, p_i \leq 1$, then conventionally one takes $\pi_i = np_i \ \forall \ i \in U$. Choosing an arbitrary point A in $[0, 1)$, i is selected in the sample if $A < X_i \leq (A + n\, p_i) \bmod (1)$, $i = 1, \cdots, N$. If no such p_i is available, then Poisson's sampling is specialized into the Bernoulli sampling on taking

$$p_i = \frac{1}{N} \quad \forall \quad i \in U$$

throughout in the above Poisson sampling procedure and so on taking $\pi_i = \frac{n}{N} \; \forall \, i \in U$.

For the Poisson scheme of sampling the inclusion-probability of i is obviously π_i and that of every pair of distinct units $(i,j), i \neq j$ is $\pi_{ij} = \pi_i \pi_j$. Consequently, the Horvitz-Thompson estimator for

$$Y = \sum_{1}^{N} y_1 \quad \text{is} \quad t_{HT} = \sum_{i \in s} \frac{y_i}{\pi_i}$$

with a variance

$$V(t_{HT}) = \sum_{i=1}^{N} \frac{y_i^2}{\pi_i}(1 - \pi_i)$$

which has the obvious unbiased estimator as

$$v(t_{HT}) = \sum_{i \in s} y_i^2 \frac{(1 - \pi_i)}{\pi_i^2}.$$

These are pleasant features, of course, but a glaring shortcoming of the Poisson sampling scheme is that the sample-size $v(s)$ is a random variable including the possibility of an empty sample with $v(s) = 0$. So, a provision to guard against this eventuality is to employ the "ratio estimator" for Y as

$$t_R = \frac{\sum_{i \in s} \frac{y_i}{\pi_i}}{v(s)} E_p(v(s)) = \frac{n \sum_{i \in s} \frac{y_i}{\pi_i}}{v(s)}, \qquad \text{if } v(s) > 0$$

and $\quad t_R = 0, \qquad \text{if } v(s) = 0.$

Another alternative is to employ Modified Poisson Sampling which is a Poisson scheme in which an empty sample is to be discarded and to be continued so with a stopping rule to discontinue until a first non-empty sample is realized.

Letting P_0 be the probability of drawing an empty Poisson sample, it is to be calculated from the equation

$$P_0 = \Pi_{i=1}^{N}\left[1 - \pi_i(1 - P_0)\right].$$

Then for this modified scheme the inclusion probability of i becomes

$$\Pi_i' = (1 - P_0)\pi_i$$

and that of $(i,j), i \neq j$ is

$$\Pi_{ij}' = \Pi_i \Pi_j(1 - P_o).$$

Another alternative to the Poisson scheme is the Sequential Poisson Scheme to be implemented as follows.

With X_i's generated independently and identically distributed (iid) uniformly over the Left-closed, Right-open interval $[0,1)$, take with P_i's as normed size-measures,

$$\xi_i = \frac{X_i}{np_i}, i=1,\cdots,N.$$

Then, arranging them in order as

$$\xi_{(1)} \leq \xi_{(2)} \leq \cdots \leq \xi_{(1)}$$

identify the corresponding labels $(1), \cdots (N)$. The first n labels $(1), \cdots, (n)$ constitute the sequential Poisson sample. These X_i's are the PRN's, not ξ_i's because the latter undergo changes with the p_i's which vary with time.

In case of Poisson sampling where $v(s)$ not only may be zero, but also may be too high as N itself or thereabout, another sampling method called Collocated sampling was devised by Ken Brewer. In it the labels $1, \cdots i, \cdots N$ in U are randomly re-arranged and their new locations $L_1, \cdots, L_i \cdots, L_N$ are noted. For example, if $N=4$ and $1,2,3,4$ are randomly re-arranged as $2,4,1,3$, then we take $L_1 = 3, L_2 = 1, L_3 = 4$, and $L_4 = 2$. Now a random variable $\in$ distributed uniformly over $[0,1)$ is generated. Then, taking

$$R_i = \frac{L_i - \in}{N}, \ i = 1, \cdots, N$$

as replacements for X_i in the Poisson Sampling scheme, the Collocated Sampling scheme is executed. With this, variability in $v(s)$ is reduced as well as its probability of taking the zero-value.

9.1.7 BALANCED REPEATED REPLICATION

In stratified sampling the replicated sampling, rather Mahalanobis's half-sampling technique may be conveniently employed in estimating a finite population mean and in unbiasedly estimating the variance of the estimator of the population mean. Let there be H strata of respective sizes

$$N_h, h = 1, \cdots, H, \sum_{h=1}^{H} W_h = 1, W_h = \frac{N_h}{N}$$

and let from the respective strata samples of sizes $n_h = 2, h = 1, \cdots, H$ be independently selected by the SRSWOR method. For the population mean

$$\bar{Y} = \sum W_h \bar{Y}_h, \ \bar{Y}_h = \frac{1}{N_h} \sum_{1}^{N_h} y_{hi}$$

let

$$\bar{y}_{st} = \sum_{h=1}^{N} W_h \bar{y}_h, \bar{y}_h = \frac{1}{2}(y_{h_1} + y_{h_2})$$

be the customary unbiased estimator, neglecting $\frac{n_h}{W_h} = \frac{2}{W_h}$, the variance of $\bar{y}_{st}$ is

$$V(\bar{y}_{st}) = \frac{1}{4} \sum W_h^2 d_h^2 = V, \quad \text{writing}$$
$$d_h = (y_{h_1} - y_{h_2}).$$

Let the sample (y_{h_1}, y_{h_2}) be divided into two half-samples on taking

$$\delta_h = 1 \quad \text{if } y_{h_1} \text{ is taken in the first half-sample}$$
$$= 0 \quad \text{if } y_{h_2} \text{ is taken in the first half-sample.}$$

Let δ_h^j be the jth $(j = 1, 2, \cdots)$ choice of δ_h and

$$t_{1j} = \sum W_h \left[\delta_h^j \, y_{h_1} + \left(1 - \delta_h^j\right) y_{h_2} \right]$$
$$t_{2j} = \sum W_h \left[\left(1 - \delta_h^j\right) y_{h_1} + \delta_h^j y_{h_2} \right].$$

Then, $\quad t_j = \frac{1}{2} (t_{1j} + t_{2j}) = \bar{y}_{st}.$

Let $\quad \bar{t} = \frac{1}{2^H} \sum_{j=1}^{2^H} t_j = \bar{y}_{st}.$

Then, $\quad V(\bar{t}_j) = \frac{1}{4} (t_{1j} - t_{2j})^2 .$

Also, writing $\Psi_{hj} = \left(2\delta_h^j - 1\right),$ we get

$$\frac{1}{4} (t_{1j} - t_{2j})^2 = \frac{1}{4} \left[\sum_h W_h^2 d_h^2 + \sum \sum_{h \neq K} W_h W_K \Psi_{hj} \Psi_{Kj} d_h d_K \right]$$

and $\quad \frac{1}{2^H} \left[\frac{1}{4} \sum_{j=1}^{2^H} (t_{1j} - t_{2j})^2 \right] = \frac{1}{4} \sum W_h^2 d_h^2 = V, \quad \text{say,}$

because $\quad \sum_{j=1}^{2^H} \psi_{hj} \psi_{Kj} = 0 \quad \forall \quad h \neq K.$

But calculation of

$$\sum_{j=1}^{2^H} (t_{1j} - t_{2j})^2$$

will be difficult even for H as small as 10 because $2^{10} = 1024$. A device is needed to make the sum

$$\sum_{j=1}^{T} W_{hj} \, W_{Kj} = 0$$

Analytical Methods

for a modestly valued T. This is easily achieved keeping T as small as one of the numbers $H+1, H+2$, or $H+3$, because one may choose ψ_{hj}'s as the entries 1 or 0 in the cells of a Hadamard matrix which is a square matrix of an order which is a multiple of 4 with all its entries as 1 or 0. For such a moderate-valued T one achieves for the entries ψ_{hj}'s in such a Hadamard matrix the value zero for

$$\sum_{j=1}^{T} \psi_{hj}\psi_{Kj} = 0 \quad \forall \quad h \neq K.$$

Consequently, it follows that

$$\frac{1}{T}\sum_{j=1}^{T} \frac{1}{4}\left[\sum_{h=1}^{H} W_h d_h \psi_{hj}\right]^2 = \frac{1}{4}\sum W_h^2 d_h^2 = V.$$

This is balanced repeated replication. The main advantage of this BRR is in evaluating estimated unbiased variance estimator for non-linear estimators for non-linear parameters.

For example, to estimate the finite population correlation coefficient between y and x in stratified sampling, we may write it as

$$R_N = \frac{\left(\sum_h W_h \sum_{i=1}^{N_h} y_{hi} x_{hi}\right) - \left(\sum W_h \sum_1^{N_h} y_{hi}\right)\left(\sum W_h \sum x_{hi}\right)}{\sqrt{\sum W_h \sum y_{hi}^2 - \left(\sum W_h \sum y_{hi}\right)^2}\sqrt{\sum W_h \sum x_{hi}^2 - \left(\sum W_h \sum x_{hi}\right)^2}}.$$

Clearly, this involves five mean parameters. Now for the strata observed values in the samples of size two each per stratum are (y_{h1}, x_{h1}) and (y_{h2}, x_{h2}). Defining δ_h^j, ψ_h^j as before for $j = 1, \cdots, T$ by BRR an estimator r for this R_N and variance estimator v for r may be easily worked out by the BRR technique.

A.1 Reviews and Further Openings

Chaudhuri and Stenger (2005) in the Preface, pp XV–XX gave quite a detailed summary of their understanding about how the Survey Sampling as a subject should be growing giving wide citations of the authors committed to the tasks they have taken up. Interested readers are invited hereby to its perusal. More modern developments noticed by us including a few aspects in our hands are also covered as they should be in this text which is not a research monograph. But a few more may be briefly cited here.

Bose, Chaudhuri, Dihidar, and Das (2010) have employed an unbiased method for estimating the prevalence rate of a disease in a locality based on a spatial smoothening facilitation of a transformed Hartley-Ross ratio-type estimator from a two-stage SRSWOR sample. They have derived a highly complex variance estimator as well.

Theirs is a model-cum-design-based approach. A simpler variance estimator will be presently provided in Appendix 3 below.

Though Stenger (1977) showed an exception to Godambe's (1955) fundamental non-existence result about a UMV Homogeneous Linear Unbiased Estimator (HLUE) for a finite population total by his Sequential Sampling approach, no further work around it has been found. Only Basu (1969) has expressed his positive ideas about sequential sampling in finite populations with his Bayesian methods in Survey Sampling. Except for research in Small Area Estimation (SAE), other applications of Bayesian and sequential methods in Survey Sampling are quite few and far between.

Another age-old inferential approach in survey sampling receiving no current attention is the "game theoretic" approach started by Chaudhuri (1969). He supposed that in suitably adopting a sampling strategy of choosing a sampling design and an estimator for a finite population total one may imagine that Nature as the player I is to choose certain values as an adversary of player II who is the Statistician who is to choose his values so that the Pay-off is to be chosen to protect each other's interests. In order to formalize this Chaudhuri (1969) introduced the following.

Lemma A.1

Given the real numbers a_i and b_i such that

(i) $\quad\quad\quad a_1 \geq a_2 \geq \cdots \geq a_N\quad$ and
(ii) $\quad\quad\quad b_1 \geq b_2 \geq \cdots \geq b_N,\quad$ one has

$$\sum_1^N a_i b_i \geq \frac{1}{N}\left(\sum_1^N a_i\right)\left(\sum_1^N b_i\right) \quad\quad\text{(iii)}$$

and
$$\sum_1^N a_i b_{(N-i+1)} \leq \sum_1^N a_i b_{j_i} \leq \sum_1^N a_i b_i \quad\quad\text{(iv)}$$

taking $(j_1, \cdots, j_i \cdots, j_N)$ as a permutation of $(1, 2, \cdots, N)$.

This Lemma is used by him to

(I) prove Minimaxity of Equal Probability Sampling and also to
(II) proceed to find a varying probability sampling design as an optimal one if an HLUE for a population total is pre-assigned.

Works of Blackwell and Girshick (1954) and Ghosh (1958) in this context are worthy of attention.

Brewer's (1979) asymptotic approach in finite population also does not seem to attract many takers. Exceptionally, however, Chaudhuri and Roy (1997) have demonstrated its utility in deriving asymptotically optimal survey sampling strategies in the context of randomized response surveys (details omitted here). ∎

Kish's (1965) concept of Design Effect which means the ratio of the variance of an estimator $\hat{\theta}$ of a finite population parameter θ calculated with respect to the true sampling design employed to produce $\hat{\theta}$ to the variance of $\hat{\theta}$ calculated under simple random sampling (SRSWR) is revised as follows.

Skinner (1989) instead considers an estimator v_0 of the variance of $\hat{\theta}$ calculated under SRSWR and calls the ratio of variance of $\hat{\theta}$ under the true design to the expected value of v_0 under the true design as a summary measure called the Mis-specification effect, meff $(\hat{\theta}, v_0)$, for this v_0 is taken as a summary measure of how far v_0 underestimates the true variance of $\hat{\theta}$.

A.2 Case Studies

1. Estimation of infant and maternal mortality in a district.

In December 1998, sponsored by UNICEF in a district near Kolkata, India, a field-based empirical study was implemented by the Indian Statistical Institute with the present author's active participation to produce some of the findings inter alia as reported below.

By Maternal mortality "rate" is meant the number of mothers dying of puerperal causes out of 10^5 eligible mothers aged 15–49 years last birthday (lbd) in a specified time and in a specified time and place. The corresponding number out of 10^5 live births is called the maternal "ratio." Mortality of under-5 year-old children is suitably defined as usual. In developed countries a maternal mortality incidence is known, vide World Health Organization (WHO) reports, to be about 1 out of 1800 women. In developing countries it is found often to be about 1 out of 48 women only. In India an ideal is to achieve maternal mortality below 200 per 10^5 live births and mortality of under-5-year children under 70 per 1000 such children in a year.

In India the Sample Registration System (SRS) is the primary source of data on infant as well as maternal mortality. Throughout the country such registration is not quite adequate. So, our endeavor had a social need.

The district surveyed was North 24 Parganas in the state of West Bengal in India with 4094 square kilometers of area and 7.3 million of people as per Census of India, 1991.

The district contained 1 district hospital, 3 sub-divisional hospitals, 5 rural hospitals, 9 state general and 1 ESI hospitals. In addition it had 22 Block Primary Health Centers (BPHC), 49 Primary Health Centers (PHC), plus 678 sub-centers, each serving 5000 and each PHC serving 30,000 people, besides 40 other clinics and health centers to serve the urban and rural people.

For sampling our selection units (SU) were the 22 Blocks of 8 accessible ones called A and 14 accessible with difficulties called D, each with 1 BPHC but with 16 PHC's for A and 33 PHC's for the D-type Blocks. From the A-Blocks we randomly chose 3 PHC-areas and from the D-Blocks independently selected 3 random PHC-areas. The 3+3=6 BPHC's were also selected as close to the 6 selected PHC's each. Also 12 sub-centers within chosen BPHC's, 12 of them within the selected PHC areas and 25 more selected outside the selected BPHC and PHC areas. Next we selected 101 villages which are within 49 sub-centers, 25 PHC's, 12 BPHC's, and 4 half samples—1 each from A- and D-blocks.

From the selected villages 20% households were selected systematically at random giving 4661 households from 101 sampled villages.

Since independent half-samples were selected, unbiased estimation of totals and variance estimation were accomplished easily. To estimate maternal mortality, rate/ratio adaptive sampling was employed, covering related adult women of sampled households with reported incidences of maternal mortalities. Questionnaires were prepared and canvassed following guidelines of WHO, UNICEF, and NFHS (Indian National Family Health Surveys). Also Network and Sisterhood methods were both tried. We considered the following definitions in particular:

(i) Neonatal Mortality is the probability of dying in the first month of life.
(ii) Post neonatal mortality is the probability of dying in 1 month-before 12 months of life.
(iii) Infant mortality is the probability of dying before the first birthday.
(iv) Child mortality is the probability of dying after the first but before the fifth birthday.
(v) Maternal mortality is the death of a woman while pregnant or within 42 days of termination of pregnancy, irrespective of the duration and the site of pregnancy from any cause related to or aggravated by the pregnancy or its management but not from accidental or incidental causes (cf WHO, 1992).

Findings are not considered interesting enough to be disclosed but may be found in the Indian Statistical Institute publication in December 1998 of this UNICEF-sponsored investigation.

---

2. An ILO-sponsored project on the investigation of prevalence of Child Labor implemented by Advanced Survey Research Centre in West Tripura District in a state of India in October 2006 to March, 2007.

ILO, Geneva, 2002, in its report "A future without child labour" gave an initial concept of "Unconditional worst form of child labour" as a follow-up of certain resolutions adopted in the Child Labour Convention, ILO, 1999. In October, 2006 the Supreme Court, the apex court in India banned Child Labor in any form in India with an immediate effect. In India a child has the upper limit of the age as 14 years. The ILO allows a child to be at most 18 year, old. Consequently, our team with me as the chairman of ASRC, a society registered in West Bengal on 15 June, 2004, could by law fully execute this sample-survey-based empirical investigation. Incidentally, it was experienced that every child encountered during the survey announced his/her age as above 14 to obey the Indian law remaining ignorant about the ILO dispensation.

To present the salient features of our investigation results, the following particulars are appended below.

By y we denote a child laborer (CL)–related real variable. In the district we identify cities/towns and villages, each supposed to be composed of a number of strata of households (HH) and establishments (EST). The households

Case Studies

(HH)/establishments (EST) are our selection units (SU). The individual child laborers (CL) are our final observation units (OU) and we are to find to which HH/EST these CL's are "linked." First turning to the villages we let

i denote an OU
m_i = Number of SU's linked to ith OU

A_{lhu} = Set of OU's linked to the uth HH/EST of hth stratum of lth village, ($u = 1, \cdots, U_{lh}$; $h = 1, \cdots, H_l$; $l = 1, \cdots, M$, supposing there are M villages in the district).

$$w_{lhu} = \sum_{i \in A_{lhu}} \frac{y_i}{m_i}$$

t_{lhu} = The network of uth HH/EST in hth stratum of lth village, a cluster being formed of HH/EST along with H closest neighboring HH/EST extending to them if a linked CL is encountered in any but with no extension in case no CL is found to be linked and the network being constructed omitting the HH/EST with no linked CL. An HH/EDT with no linked CL is by courtesy taken as a singleton network.

$N(l, h, u)$ = Set of OU's linked to T_{lhu}.
C_{lhu} = Number of OU's linked to $N(l, h, u)$.
s_{lh} = Sample of HH/EST's chosen from hth stratum of lth village.
u_{lh} = Number of HH/EST chosen from U_{lh}.
Let m = Number of villages to choose from M villages.
So, $N_l = \left[\frac{M}{m}\right]$ for some l and $\left[\frac{M}{m}\right] + 1$ for the other villages in the district, so that $\sum_m N_l = M$ the sum $\sum_m$ is over the groups formed in applying the RHC scheme.
Let $p_l \equiv$ a normed size-measure of lth village, say, formed using the census population figure for the lth village.
Q_l = seem of the p_l's within the respective groups of N_l villages.
Our task is to estimate

$$T = \sum_l \sum_h \sum_u t_{lhu}$$

which equals $Y = \sum_l \sum_h \sum_u y_{lhu}$, writing y_{lhu} for the value of y for the uth OU of hth stratum in lth village.
So, our estimate for $Y = T$ is taken as

$$\hat{T} = \sum_m \frac{Q_l}{p_l} \left[\sum_{h=1}^{H_l} \left(\frac{U_{lh}}{u_{lh}}\right) \sum_{u \in s} t_{lhu} \right],$$

writing s as the SRSWOR of HH/EST's chosen from the hth stratum of the lth village chosen;

$$= \sum_m \frac{Q_l}{p_l}\alpha_{lh}, \quad \text{writing} \quad \alpha_{lh} = \frac{U_{lh}}{u_{lh}}\sum_{h=1}^{l}\alpha_{lh}.$$

Restricting to $u_{lh} \geq 2$, let

$$Q_l = \sum_{h=1}^{H_l} U_{lh}^2 \frac{\left(\frac{1}{u_{lh}} - \frac{1}{U_{lh}}\right)}{(u_{lh}-1)} \left[\sum_{u \in s}\left(t_{lhu} - \frac{\sum_{u \in s} t_{lhu}}{u_{lh}}\right)^2\right].$$

An estimate of $V\left(\hat{T}\right)$ is

$$v = \left(\frac{\sum_m N_l^2 - M}{M^2 - \sum_m N_l^2}\right)\left[\sum_m \theta_l \left(\frac{\alpha_l}{p_l}\right)^2 - \left(\hat{T}\right)^2\right].$$

A similar approach was made to cover the cities/towns, choosing them by the SRSWOR instead of by the RHC method. The rest is similar. From the West Tripura district we planned to address in all 400 urban and 200 rural CL's. In the actual survey we could survey 400 urban and 242 rural CL's meeting the resource provisions. The following structured questionnaire was addressed.

<u>Part of the Questionnaire Addressed to a Sampled Child Laborer.</u>

(i) Who escorted you here?
[Code: Parents-1, Relative(s)-2, Friend(s)-3, Unknown person(s)-4.]
(ii) Mode of transport used for first arrival.
(Code: Train-1, Bus/Lorry-2, Autorickshaw-3, Boat-4, Long Walk-5, Others-6).
(iii) Number of times place of work changed since first arrival.
Last week: , Last month: , Last year:
(iv) Whether exploitation perceived.
Yes-1, No-2.
Type: Code: Commercial-1, sexual-1, Both-3.
(v) Contraband drug handled/transported.
Code: No -1, Handled-2, Transported-3, Both-4.
(vi) How transported
Code: To an organization-1, to a gang-2, to a male person-3, to a female person-4.

Further details are available in

(I) Chaudhuri (2010) and

Case Studies

(II) "Children in the unconditional worst forms of child labor," by Arijit Chaudhuri, Indian Statistical Institute, Kolkata, and Advanced Survey Research Centre (web:www:asrc.net.in), 1 May, 2007.

---

3. The author, as the (Council of Scientific and Industrial Research of India) CSRI, Emiritus Scientist, Honorary Visiting Professor of Applied Statistics Unit, Indian Statistical Institute (ISI) and the Chairman, Advanced Survey Research Centre (ASRC), was approached by the Ministry of Statistics & Programme Implementation (MOSPI), Government of India, to undertake, in April 2005, an "Application of Small Area Techniques for Estimation of District Level Parameters" covering Rural India. (401)

This was executed during 27 April, 2005, through 26 December, 2005, and the finaly revised report submitted in July 2006 was accepted in July 2006.

The report gave estimates plus their estimated coefficients of variation (CV) for all the districts of all the 28 states and union territories (UT) of India based on the NSSO survey results for its 55th round on the following 20 target parameters.

1. Total male population
2. Total female population
3. Total employed people
4. Total unemployed people
5. Total number of children under 14 years
6. Total number below poverty line
7. Total number of households with monthly per capita expenses below Rs.250/-
8. Total number not in labor force
9. Average Monthly per capita expenses
10. Monthly average per capita expense on food
11. Monthly average per capita expense on non-food
12. Literacy rate
13. Educational expenses
14. Percent below poverty line
15. Total enrolled in schools
16. Percent enrolled in schools
17. Gini's coefficient
18. Percentiles of expenditure distribution
19. Food intake and calorie consumption
20. Poverty line

First the district population figures, by Projection methods using Census and NSSO figures over the years were determined to be used as size-measures in Small Area Estimation.

In NSSO surveys roughly every district is taken as a stratum and hence is surveyed with no omussion. But the villages in each district are sampled and households in small numbers are sampled from the selected villages; thus, district-level estimates are obtained on applying two-stage sampling procedures. But the total number of second-stage units, i.e., the households sampled and surveyed per district, are so small in numbers that the district estimates turn out too short in occurences. So, borrowing strength from neighboring district-wide sample data seems to be a need for attempting higher accuracy levels in the district-wide estimates. Clearly it is a deviation from standard small area estimation principles, because here the districts are sampled in conscientious planned manners and they are our "Small Areas."

In order to effectively enhance the sample-size in estimation we borrow strength in employing Synthetic Generalized Regression (Syn Greg) method.

Let y be a variable of interest and d denote a typical district out of all the D districts in a state/UT. Let for Y_d, the district total of y an NSSO-data-based estimate be $t_y(d)$. Let for a size-measure variable x, the district estimate be X_d. As the NSSO estimation involves systematic sampling, the half-sampling technique will be used in estimation and variance-estimation.

Let y_{ij} denote the y-value for the jth household of the ith village in the district and x_{ij} the number of members of this household. By z_i we mean the 1991 Census total population of the ith village, Z is the total of z_i's for the entire district, and $p_i = \frac{z_i}{Z}$ the normed size-measure used to select the ith village from the district. Also M_i is the total number of households in the ith village, as determined by a listing survey conducted by NSSO within each selected village. Also, n is the number of villages selected m_i the number of households sampled out of M_i in the selected ith village. Writing $\sum_z$ to denote summing over the m_i households let

$$Q_i = \frac{1 - np_i}{np_i \left(\frac{M_i}{m_i} \sum_z x_{ij}\right)}.$$

The device to borrow strength we postulate a linear regression of y on x through the origin and a common slope for all the districts in the state concerned. Denoting by $\sum_1$ the summing over the selected n villages and by $\sum_d$ the sum over the districts in the state, this regression slope is estimated by

$$b_Q = \frac{\sum_d \sum_1 Q_i \sum_2 y_{ij} x_{ij}}{\sum_d \sum_1 Q_i \sum_2 x_{ij}^2}.$$

Then a synthetic greg estimator for Y_d's

$$t_g(d) = t_y(d) + l_Q(X_d - t_x(d)).$$

Case Studies

For the two independent half-samples let

$$t_g^{(r)}(d) = t_g^{(r)}(d) + b_Q^{(r)}\left(X_d - t_x^{(r)}(d)\right), r = 1, 2$$

be the synthetic greg estimators with obvious notations. So, let the final estimator for Y_d be

$$t_{gd} = \frac{1}{2}\left(t_g^{(1)}(d) + t_g^{(2)}(d)\right)$$

and the estimated standard error (se) for it is

$$s_{gd} = \frac{1}{2}\left(\left|t_g^{(1)}(d) - t_g^{(2)}(d)\right|\right)$$

and the estimates coefficient of variation is

$$CV = CV(t_{gd}) = \frac{\left|t_g^{(1)}(d) - t_g^{(2)}(d)\right|}{\left|t_g^{(1)}(d) + t_g^{(2)}(d)\right|} X100.$$

In addition, we calculated as follows the Empirical Bayes Estimators (EBE) for Y_d also with the following additional postulations:

Let $\quad v_d = s_{gd}^2 = \frac{1}{4}\left(t_{gd}^{(1)} - t_{gd}^{(2)}\right)^2.$

Let $\quad t_{gd}|Y_d \stackrel{iid}{\cap} N(Y_d, v_d)$ .... (i)

$\quad Y_d \stackrel{iid}{\cap} N(\beta X_d, A)$ .... (ii)

with β and A as two unknown constants. Thus, combining (i) and (ii) the Bayes Estimator for Y_d is

$$t_{\beta d} = \left(\frac{A}{A + v_d}\right)t_{gd} + \left(\frac{v_d}{A + v_d}\right)(\beta X_d). \quad\quad\quad (1)$$

This cannot be used as A, β are unknown.

Let $\quad \tilde{\beta} = \frac{\sum_d t_{gd} X_d/(A + v_d)}{\sum_d X_d^2/(A + v_d)}$ .... (2)

and $\quad \sum_d \left(t_{gd} - \tilde{\beta} X_d\right)^2/(A + v_d) = (D - 1)$ .... (3)

Then, solving (2) and (3) by iteration using Newton and Raphson's technique starting with an initial choice of A as zero estimates $\hat{A}, \hat{B}$ are derived to produce

$$t_{EB_d} = \left(\frac{\hat{A}}{\hat{A}+v_d}\right)t_{gd} + \left(\frac{v_d}{\hat{A}+v_d}\right)(\hat{\beta}X_d)$$

the EBE for Y_d.

To work out a CV for t_EBd, let

$$r_d = \frac{\hat{A}}{\hat{A}+v_d}, \quad m_{1d} = \frac{\hat{A}v_d}{\hat{A}+v_d} = r_d v_d$$

$$m_{2d} = \frac{(1-r_d)^2 X_d^2}{\sum_d X_d^2/\left(\hat{A}+v_d\right)},$$

$$m_{3d} = \frac{v_d^2}{\left(\hat{A}+v_d\right)^3}\left[\frac{2}{D}\sum_d \left(\hat{A}+v_d\right)^2\right].$$

Then, from Prasad and Rao's (1990) results, $m_d = m_{1d} + m_{2d} + 2\, m_{3d}$ is taken as the estimated mean square error (MSE) of t_{EBd} about Y_d and hence its CV as

$$CV(t_{EBd}) = 100\frac{+\sqrt{m_d}}{t_{EBd}}.$$

Usual procedures are followed to estimate ratio parameters and their standard errors.

Detailed numerical findings, to save space, are not reproduced here from the Report already referenced.

---

4. Report on a project by Indian Statistical Institute (ISI) to implement audit by sampling methods.

The Finance (Internal Audit) Department, Government of West Bengal approached ISI in 2001 to help them carry out audit of their account relating to (1) Public Works Department (PWD), (2) Public Health Engineering (PHE) and (3) Irrigation & Water Works (IW) department of the Govt. of West Bengal (WB), by employing sophisticated statistical procedures. The present author undertook and executed the task as the principal investigator when he was a professor in ISI.

At that time there were in all 17 districts in West Bengal. Also in the state there were 232, 46, and 85 district offices, respectively, under PWD, PHE, and IW with a total of 363 offices altogether in the state. The approved budget allocated for the previous year to the respective was taken as the size-measure in selecting the sampled offices applying the RHC scheme while choosing the

Case Studies

samples of offices out of 7 districts suitably chosen with no probability attached. Totals 43, 14, and 22 and in all 79 PWD, PHE, and IW offices were sampled.

From the selected offices out of several types as books some were selected at random. Then, from the sampled books were randomly selected a few pages then a few lines from each selected page. Thus, taking as y a variable for an entry we needed to estimate $Y = \sum_{i=1}^{N} y_i$ for all the n offices in the district.

Again, $y_i = \sum_{j=1}^{M_i} \sum_{k=1}^{B_{ij}} \sum_{l=1}^{P_{ijk}} \sum_{m=1}^{L_{ijkl}} y_{ijkl}$.

Only the offices are sampled by RHC scheme using normed-size-measures p_i with their group-based sums as Q_i's. But the (i) office departments (M_i in number) of which m_i are sampled, (ii) books (B_{ij} in number) of which b_{ij} are sampled, (iii) the pages (P_{ijk} in number) of which p_{ijk} are sampled, (iv) the lines (L_{ijkl} in number) of which l_{ijkl} are sampled and the entries along the columns of the lines are recorded, and the sampling in each case is implemented by the SRSWOR method. So, y_i is unbiasedly estimated by

$$y_i^* = \frac{M_i}{m_i} \sum_{j=1}^{m_i} \frac{B_{ij}}{b_{ij}} \sum_{k=1}^{b_{ij}} \frac{P_{ijk}}{p_{ijk}} \sum_{l=1}^{p_{ijk}} \frac{L_{ijkl}}{l_{ijkl}} \sum_{m=1}^{l_{ijk}} y_{ijklm}.$$

Here y_{ijklm} is the value of the entry on the mth column of the lth line of the kth page of the jth book of the ith office-department. Finally the district total is unbiasedly estimated by

$$\hat{Y} = \sum_{n} y_i^* \frac{Q_i}{p_i}.$$

This is calculated from an RHC sample of n offices chosen from the district. We obtained only the district totals. So, we did not present figures for the state of West Bengal as a whole because the districts were purposively taken so as to convey the most important of the district-wise figures.

In order to show how the standard error of Y is to be estimated, we introduce the following additional notations:

$$e_{ijkl} = \frac{L_{ijkl}}{l_{ijkl}} \sum_{m=1}^{l_{ijkl}} y_{ijklm}$$

$$C_{ijk} = \frac{P_{ijk}}{p_{ijk}} \sum_{l=1}^{p_{ijk}} l_{ijkl}$$

$$a_{ij} = \frac{B_{ij}}{b_{ij}} \sum_{k=1}^{b_{ij}} c_{ijk}$$

$$y_i^* = \frac{M_i}{m_i} \sum_{j=1}^{m_i} a_{ij}$$

These respectively yield the variance components as

$$\hat{V}(e_{ijkl}) = \frac{L_{ijkl}^2}{(l_{ijkl}-1)}\left(\frac{1}{l_{ijkl}} - \frac{1}{L_{ijkl}}\right)\sum_1^{l_{ijkl}}\left(y_{ijklm} - \frac{1}{i_{ijkl}}\sum_1^{l_{ijkl}}(y_{ijklm})\right)^2$$

$$\hat{V}(c_{ijk}) = \frac{P_{ijk}^2}{(p_{ijk}-1)}\left(\frac{1}{p_{ijk}} - \frac{1}{P_{ijk}}\right)\sum_1^{p_{ijk}}\left(l_{ijkl} - \frac{\sum l_{ijkl}}{p_{ijk}}\right)^2 + \frac{P_{ijk}}{p_{ijk}}\sum_1^{p_{ijk}}\hat{V}(l_{ijk})$$

$$\hat{V}(a_{ij}) = \frac{B_{ij}^2}{(b_{ij}-1)}\sum_1^{b_{ij}}\left(C_{ijk} - \frac{\sum C_{ijk}}{b_{ij}}\right)^2 + \frac{B_{ij}}{b_{ij}}\sum_1^{b_{ij}}\hat{V}(C_{ijk})$$

$$\hat{V}(y_i^*) = \frac{M_i^2}{m_i-1}\left(\frac{1}{m_i} - \frac{1}{M_i}\right)\sum_1^{m_i}\left(a_{ij} - \frac{\sum a_{ij}}{m_i}\right)^2 + \frac{M_i}{m_i}\sum_1^{m_i}\hat{V}(a_{ij})$$

$$\hat{V}(\hat{Y}) = \frac{\sum_n N_i^2 - N}{N^2 - \sum_n N_i^2}\sum_n\sum_n Q_i Q_{i'}\left(\frac{y_i^*}{p_i} - \frac{y_{i'}^*}{p_{i'}}\right)^2 + \sum_n \frac{Q_i}{p_i}\hat{V}(y_i^*).$$

In $\hat{V}\left(\hat{Y}\right)$, obviously N_i denotes the number of district offices falling in the ith group when the districts are selected by RHC scheme, $\sum_n$ the sum over the offices falling across the groups, and $\sum_n \sum_n$ the pairs with no repetitions falling in the groups.

We present only the method we applied. The numerical findings are omitted, saving the space.

---

5. District-wise survey on literacy in West Bengal, 2008-2009.

The Ministry of Mass Education Extension Govt. of West Bengal in 2008 assigned to the Advanced Survey Research Centre (website: www.asrc.net.in) under the present author's chairmanship to undertake a comprehensive field survey in all the then 17 districts excluding only Kolkata of West Bengal to examine how well the training was being imparted to the learners in the district urban and rural centers for "Continuing Education Programme" (CEP) and "Literacy Extension Programme" (LEP) in West Bengal.

It is well-known that in some Indian states including West Bengal all people are not literate even after attaining the age of 6 years. That is why some efforts are made at the government initiative to assess and promote the percentage of literacy among people aged at least 6 years. But after some such drive is made, it is also considered important to examine from time to time how the people found to be literate at a time as to whether and how they are proceeding with their studies.

For our study during October 2008 through January 2009 we selected 4 municipalities and 4 Panchayat Samities, equivalently blocks in each of the

17 districts and 8 CEP/LEP centers in each. We designed test questions on language (L) and arithmetic (A) at random. Further randomly choosing from the registers put 100 urban and 100 rural students to these test questions. Our findings in brief turned out as follows:

(i) Meeting the criteria set by us on the basis of the test-score-related performances, 10 of the 17 districts show more than 50% of the overall learners, male, female, urban, and rural combined are adjudged "Literate" enough. In some districts the situation seemed dismal with a mere 18%–33% as literate while 2 particular districts showed as high percentages as 67% and 69%.
(ii) No general conclusions could be reached to discriminate among the males versus females or rural versus urban.
(iii) The learners interviewed were, especially the females and the rural ones, found to be quite keen to learn.
(iv) About 50% of the learners were found present on an average in the centers when interviewed.
(v) Because of inadequate lighting, facilities, and day-time occupations in earning livelihood, attendance was found to be lower than expected.
(vi) Because of irregular arrival of supporting funds, functioning of the centers was handicapped.
(vii) No discriminatory political factors could be identified to hamper smooth functioning of the centers.

Statistical aspects were found to be rather routine in nature. So, the details are suppressed. Chaudhuri (2010) has given some details. We refrain from reproducing them here.

A.3 Exercises and Solutions Supplementaries

Relating to Chapters 1 & 2

1. Question:

From a finite survey population of 24 households an SRSWOR is needed to produce a sample mean of the household expenses on purchase of warm clothing. How many samples will you take to keep its value less than Rs. 500/- in error vis-à-vis the population mean?

Solution:

$$\text{Prob}[|\bar{y} - Y| < 500] \geq 0.95$$

$$\text{Prob}\left[|\bar{y} - Y| \leq \lambda \sigma(\bar{y})\right] \geq 1 - \frac{1}{\lambda^2}$$

or

$$\text{Prob}\left[|\bar{y} - Y| \leq \lambda S_y \sqrt{\frac{1}{n} - \frac{1}{N}}\right] \geq 1 - \frac{1}{\lambda^2}$$

writing

$$S_y^2 = \frac{1}{N-1} \sum_{1}^{N} (y_i - \bar{Y})^2.$$

Let

$$1 - \frac{1}{\lambda^2} = 0.95 \quad \text{or} \quad \lambda = \sqrt{20}.$$

Then,

$$500 = \sqrt{20}\, S_y \sqrt{\left(\frac{1}{n} - \frac{1}{N}\right)}$$

or

$$\frac{1}{n} = \frac{1}{N} + \frac{250000}{20 S_y^2}$$

or

$$n = \frac{N}{1 + \frac{N}{S_y^2} 12500}$$

$$= \frac{24}{1 + \frac{1}{S_y^2} 300000}.$$

Now, anticipating different possible values of S_y^2 and tabulating them one may work out a value of n in the range $[2, 24]$.

Alternatively, assuming $\frac{\bar{y}-\bar{Y}}{\sigma(\bar{y})}$ as approximately a standard normal deviate we may observe

$$\text{Prob}[|\bar{y} - \bar{Y}| \leq 500] \geq 0.95$$

giving,
$$\text{Prob}\left[\left|\frac{\bar{y} - \bar{Y}}{\sigma(\bar{y})}\right| \leq \frac{500}{\sigma(\bar{y})}\right] \geq 0.95.$$

From this it follows that

$$\frac{500}{S_y\sqrt{\left(\frac{1}{n} - \frac{1}{N}\right)}} = 1.96 \simeq 2$$

or
$$\frac{1}{n} = \frac{1}{N} + \frac{250 \times 250}{S_y^2}$$

or
$$n = \frac{N}{1 + \frac{N}{S_y^2}625000} = \frac{24}{1 + \frac{1500000}{S_y^2}}.$$

Tabulating anticipated S_y^2-values possible choice of n may turn up within the range [2,24).

2. Question:

Work out a reasonable solution for the following problem posed by J.E. Freund (1994, p 369).

"A country's military intelligence knows that an enemy built certain new tanks serially numbered from 1 to K. If three of these tanks are captured and their serial numbers are 210, 38 and 155, find a suitable value of this K, giving your rationale for it."

Solution:

Using the theory on pp 24–25, an unbiased estimator of this K is $\hat{K} = \frac{210}{3} \times 4 - 1 = 279$.

Now the variance of $\hat{K}$ works out, following the contents on pp 24–25, as below:

$$V\left(\hat{K}\right) = \left(\frac{4}{3}\right)^2 V(X),$$

writing X as the largest sample serial number.

Now,
$$V(X) = E(X+1)X - E(X) - E^2(X)$$
$$= \frac{(n+1)n}{(n+2)}(K+2)(K+1)$$
$$- \frac{n(K+1)}{(n+1)} - \frac{n^2(K+1)^2}{(n+1)^2}.$$

Replacing K by $\hat{K}$ an estimate of $V\left(\hat{K}\right)$ may be taken as

$$v = \left(\frac{4}{3}\right)^2 \left[\frac{4 \times 3}{5}(281)(280) - \frac{3(280)}{4} - \frac{9(280)^2}{16}\right]$$

$$= \frac{16}{9} \times \frac{280}{16 \times 5} [4 \times 3 \times 16 \times 281 - 5 \times 4 \times 3 - 9 \times 5 \times 280]$$

$$\simeq \frac{280}{9 \times 5} \times 280 \times 67 - \frac{280}{3} \times 4$$

$$\simeq \left(\frac{280 \times 8}{3}\right)^2 \frac{1}{5} - \frac{280}{3} \times 4$$

$$\simeq \left(\frac{280 \times 8}{3 \times 2}\right)^2.$$

So, $\sqrt{v} = 280$.

So, $\hat{K} \pm 3\sqrt{v} = 279 \pm 840$ contains the unknown value K with a very high probability.

Relating to Chapter 3

3. Question:
Given N numbers $\pi_i(2)$ such that $0 < \pi_i(2) < 1$ $\forall i = 1, \cdots, N$ and $\binom{N}{2}$ numbers $\pi_{ij}(2)$ such that $0 < \pi_{ij}(2) < 1$, for every i, j $(i \neq j)$ equal to $i = 1, \cdots, N$ and $j = 1, \cdots, N$, show that it is possible to choose a sample of 2 distinct units of $U = (1, \cdots i, \cdots, N)$ with $\pi_i(2)$ as the inclusion-probability of i of U in the sample if $\sum_{j \neq i} \pi_{ij}(2) = \pi_i(2)$ $\forall$ $i \in U$, noting $\pi_{ij}(2) = \pi_{ji}(2)$ $\forall i \neq j$.

Solution:
Draw the unit i of U in the sample s on the first draw with probability $p_i = \frac{\pi_i(2)}{2}$; leaving this i aside take on the second draw from the remaining $(N-1)$ units of U with the conditional probability $p_{ij} = \frac{\pi_{ij}(2)}{\pi_i(2)}$ for the units j of U other than i. Then start further sampling.

Then, the inclusion-probability of i in the sample is

$$\pi_i = p_i + \sum_{j \neq i} p_j p_{ji}$$

$$= \frac{\pi_i(2)}{2} + \sum_{j \neq i} \frac{\pi_j(2)}{2} \frac{\pi_{ji}}{\pi_j(2)}$$

$$= \frac{\pi_i(2)}{2} + \frac{1}{2} \sum_{j \neq i} \pi_{ji} = \frac{\pi_i(2)}{2} + \frac{\pi_i(2)}{2} = \pi_i(2).$$

Also, the inclusion-probability in the sample of i and j according to this scheme is

$$\pi_{ij} = p_i\, p_{ij} + p_j\, p_{ji}$$

$$= \frac{\pi_i(2)}{2} \frac{\pi_{ij}(2)}{\pi_i} + \frac{\pi_j(2)}{2} \frac{\pi_{ji}(2)}{\pi_j(2)}$$

$$= \pi_{ij}(2).$$

4. Question:

Suppose with $\pi_i(2)$ and $\pi_{ij}(2)$ as the inclusion-probability of the units i and $j(i \neq j)$ of $U = (1, \cdots, i, \cdots, N)$ are drawn in a sample of size 2 from U. Let with probability α $(0 < \alpha < 1)$ and with probability $(1 - \alpha)$, respectively, additionally $(n - 2)$ and $(n - 3)$ units be taken from the remaining population $U = (1, \cdots i, \cdots N)$ excluding i and j both by SRSWOR. Find the inclusion-probabilities π_i and π_{ij} $(i \neq j)$ in this finally implemented sample selection. Also find the expectation and variance of the final sample-size.

Solution:

$$\pi_i = \pi_i(2) + \left(1 - \pi_i(2)\right)\left[\alpha \frac{n-2}{N-2} + (1-\alpha)\frac{n-3}{N-2}\right]$$

$$= \pi_i(2) + \left(1 - \pi_i(2)\right)\left[\frac{n-3}{N-2} + \frac{\alpha}{N-2}\right].$$

The expected sample-size is

$$\gamma = \sum_{i=1}^{N} \pi_i = 2 + (N-2)\left[\frac{n-3}{N-2} + \frac{\alpha}{N-2}\right]$$

$$= n - 1 + \alpha = n - (1 - \alpha)$$

$$\pi_{ij} = \pi_{ij}(2) + \pi_i(2)\left[\alpha\frac{n-2}{N-2} + (1-\alpha)\frac{n-3}{N-2}\right]$$
$$+ \pi_j(2)\left[\alpha\frac{n-2}{N-2} + (1-\alpha)\frac{n-3}{N-2}\right]$$
$$+ \left[1 - \pi_i(2) - \pi_j(2)\right]\left[\alpha\frac{(n-2)(n-3)}{(N-2)(N-3)}\right.$$
$$\left. + (1-\alpha)\frac{(n-3)(n-4)}{(N-2)(N-3)}\right].$$

Variance of the sample-size is

$$V(\gamma(s)) = \sum\sum_{i\neq j}\pi_{ij} - \gamma(\gamma - 1)$$
$$= 2 + \frac{4(N-1)}{(N-2)}(n-3+\alpha)$$
$$+ \frac{(N-1)(N-4)}{(N-2)(N-3)}(n-3)(2\alpha + n - 4)$$
$$- (n-1+\alpha)(n-2+\alpha).$$

5. Question:

Find an unbiased estimator for the variance of the Hartley-Ross unbiased ratio-estimator for a finite population total.

Solution:

The Hartley-Ross estimator for total is

$$t_{HR} = X\bar{r} + \frac{n(N-1)}{(n-1)}(\bar{y} - \bar{r}\bar{x}).$$

This is based on SRSWOR in n draws.
Hence, $\pi_i = \frac{n}{N}$ and $\pi_{ij} = \frac{n(n-1)}{N(N-1)}, i, (i,j)\,(i \neq j)$

$$V(t_{HR}) = E(t_{HR}^2) - Y^2$$
$$= E(t_{HR}^2) - \left(\sum_1^N y_i^2 + \sum\sum_{i\neq j}y_iy_j\right).$$

So, $\hat{V}(t_{HR})$ is

$$v_{HR} - t_{HR}^2 - \left(\frac{N}{n}\sum_{i\in s}y_i^2 + \frac{N(N-1)}{n(n-1)}\sum\sum_{i\neq j}y_iy_j\right).$$

Alternatively, collecting the coefficients of y_i for i in s, one may write $t_{HR} = \sum_{i\in s}y_i b_{si}$. Since $E(t_{HR}) = Y$, b_{si}'s are subject to $\sum_{s\ni i}b_{si} = \binom{N}{n}\,\forall i$.

A usual formula for $V(t_{HR})$ and $\hat{V}(t_{HR})$ may be derived. But the formula, vide Bose et al. (2010) is quite complicated.

6. Question:

How will you randomly choose 1 out of 7 using a table of random number?

Solution:

Let A denote one of the digits $0, 1, \cdots, 8, 9$ and B denote a specific one out of the set of $1, 2, \cdots, 7$, say. Then, the probability that 7 of the set B is found is

$$P(A|B) = \frac{P(A \cap B)}{P(B)} = \frac{\frac{1}{10}}{\frac{7}{10}} = \frac{1}{7}.$$

7. Question:

(a) How will you randomly choose a residential household in a big city, say, Kolkata?

(b) How will you choose a residential household in a big city like Kolkata with a probability proportional to the number of members it contains?

Solution:

Let N be the total number of residential buildings in Kolkata, labeled $i = 1, \cdots, N$. Let X_i be the number of households the ith building contains and $X = \sum_{i=1}^{N} X_i$. Further, let l_{ij} denote the number of members the jth household contains in the ith residential building in the city, $L_i = \sum_{j=1}^{X_i} l_{ij}$ which is the number of members among the ith building containing X_i households and $L = \sum_{1}^{N} L_i$ be the total number of members living in all the residential buildings in the city.

Let one, say, the ith building be chosen randomly, i.e., with probability $\frac{1}{N}$.

Let $T \geq \underset{1 \leq i \leq N}{max}(L_i)$. Then draw a random number of r, say, such that $1 \leq r \leq T$. If $r \leq L_i$, then retain ith building; else reject (i, r) and repeat until the building i happens to be selected. Then the probability p_i that the ith building is selected is

$$p_i = \frac{1}{N} \frac{L_I}{T} (1 - Q)^{-1} \text{ with}$$

$$Q = 1 - \frac{1}{N} \frac{1}{T} \sum_{1}^{N} L_i = 1 - \frac{L}{NT}.$$

So, $p_i = \frac{L_i}{L}, i = 1, \cdots, N$

From the ith building containing X_i households choose one at random, i.e., with probability $\frac{1}{X_i}$.

Let $M \geq \underset{1 \leq j \leq X}{max}(L_{ij})$.

Now take a random number R such that $1 \leq R \leq M$. If $R \leq l_{ij}$, then take the jth household of the ith building in the sample; else reject (R, j) and repeat until eventually the jth household of the ith building is selected. Then, the probability that the jth household of the ith building is selected is

$$q_{ij} = \frac{1}{X_i} \frac{l_{ij}}{M} \left(1 - \frac{1}{MX_i} \sum_{j=1}^{X_i} l_{ij}\right)^{-1}$$

$$= \frac{1}{X_i} \frac{l_{ij}}{M} \frac{MX_i}{L_i} = \frac{l_{ij}}{L_i}.$$

Hence, the probability that the jth household of the ith building is selected at random from all the households in the city is

$$p_i \, q_{ij} = \frac{L_i}{L} \frac{l_{ij}}{L_i} = \frac{l_{ij}}{L}.$$

This answers the question (b). Here L need not be known.

To answer (a) first choose a building with probability $\frac{1}{N}$. For this chosen building, say, the ith find X_i, the number of households it contains.

Then, take a number $H \geq \underset{1 \leq i \leq N}{max}(X_i)$ and choose a random number A between 1 and H. Then, if $A \leq X_i$, choose the ith building in the sample or else reject the choice (i, A). Then, the probability that the ith building is eventually selected is

$$R_i = \frac{1}{N} \frac{X_i}{H} \left[1 - \frac{1}{NH} \sum_1^N X_i\right]^{-1} = \frac{X_i}{X}.$$

Choose one of the X_i households, say, the jth in the ith building at random, i.e., with probability $\frac{1}{X_i}$. Then, the final probability that any jth households in the city selected is

$$\theta = \frac{1}{X_i} \frac{X_i}{X} = \frac{1}{X}.$$

Most importantly this X is not required to be known.
This answers (a).
The answers (a) and (b) are reached following the Lahiri's (1951) method as we have already described.

Relating to Chapter 4

8. Question:

Let N clusters be formed with the ith cluster containing M_i elements. Examine how cluster sampling may be justified from the consideration of efficiency.

Solution:
The population mean is

$$\overline{Y} = \frac{\sum_{i=1}^{N} \sum_{j=1}^{M_i} y_{ij}}{\sum_{1}^{N} M_i} = \frac{\sum_{1}^{N} M_i \overline{Y}_i}{\sum_{1}^{N} M_i},$$

$$\overline{Y}_i = \frac{1}{M_i} \sum_{j=1}^{M_i} y_{ij}.$$

Each M_i is known, $i = 1, \cdots, N$.
Let s_1 be an SRSWOR of m elements taken out of $\sum_{i=1}^{N} M_i = M$ elements in the population. As opposed to this s_1 let an SRSWOR s_2 of n clusters be chosen and the ith sample cluster of M_i elements be completely surveyed such that $m = \sum_{i \in s_2} M_i$.

Let
$$t_1 = \frac{1}{m} \sum_{i,j \in s_1} \sum y_{ij} \text{ and}$$

$$t_2 = \frac{N}{n} \sum_{i \in s_2} \bar{y}_i \left(\frac{M_i}{M} \right) = \frac{N}{nM} \sum_{i \in s_2} M_i \bar{y}_i$$

writing
$$\bar{y}_i = \frac{1}{M_i} \sum_{j=1}^{M_i} y_{ij} = \overline{Y}_i.$$

Then,
$$E(t_1) = \frac{1}{M} \sum_{i=1}^{N} \sum_{j=1}^{M_i} y_{ij} = \overline{Y}$$

$$E(t_2) = \frac{1}{M} \sum_{i=1}^{N} M_i \overline{Y}_i = \overline{Y}$$

$$V(t_1) = \frac{\left(\sum_{1}^{N} M_i \right) - m}{m \sum_{1}^{N} M_i} S^2$$

$$= \frac{M - m}{Mm} S^2$$

$$S^2 = \frac{1}{M-1} \sum_{i=1}^{N} \sum_{j=1}^{M_i} \left(y_{ij} - \overline{Y} \right)^2$$

Exercises and Solutions Supplementaries

$$V(t_2) = \frac{N^2}{(N-1)} \left(\frac{N-n}{Nn}\right) \frac{1}{M^2} \sum_{i=1}^{N} \left[M_i \bar{y}_i - \bar{Y}\right]^2$$

$$= \frac{N^2}{M^2} \left(\frac{N-n}{Nn}\right) \frac{1}{(N-1)} \sum_{i=1}^{N} \left[\sum_{j=1}^{M_i} y_{ij} - \bar{Y}_i\right]^2.$$

There seems to be little to infer about the relative magnitudes of $V(t_1)$ versus $V(t_2)$.

Relating to Chapter 5

9. Question:

Show that Brewer's model assisted predictor for a finite population total is also a generalized regression (greg) predictor.

Solution:

Brewer's (1979) predictor for Y is of the form

$$t_B = \left(\sum_{i \in s} y_i\right) + \frac{\sum_{i \in s} y_i \left(\frac{1-\pi_i}{\pi_i}\right)}{\sum_{i \in s} x_i \left(\frac{1-\pi_i}{\pi_i}\right)} \left(X - \sum_{i \in s} x_i\right)$$

while a greg predictor of Y is of the form

$$t_g = \left(\sum_{i \in s} \frac{y_i}{\pi_i}\right) + \left(X - \sum_{i \in s} \frac{x_i}{\pi_i}\right) \frac{\sum_{i \in s} y_i x_i Q_i}{\sum_{i \in s} x_i^2 Q_i}$$

with Q_i as any positive real number.

Let us take $Q_i = \frac{1-\pi_i}{\pi_i x_i}, i \in U.$

Then, t_g reduces to

$$t_g = \left(\sum_{i \in s} \frac{y_i}{\pi_i}\right) + \left(X - \sum_{i \in s} \frac{x_i}{\pi_i}\right) \frac{\sum_{i \in s} y_i \frac{1-\pi_i}{\pi_i}}{\sum_{i \in s} x_i \frac{1-\pi_i}{\pi_i}}$$

$$= \sum_{i \in s} \frac{y_i}{\pi_i} + \left(X - \sum_s x_i - \sum \frac{x_i}{\pi_i}(1-\pi_i)\right) \frac{\sum_{i \in s} y_i \left(\frac{1-\pi_i}{\pi_i}\right)}{\sum x_i \left(\frac{1-\pi_i}{\pi_i}\right)}$$

$$= \sum_{i \in s} y_i + \frac{\sum_{i \in s} y_i \frac{1-\pi_i}{\pi_i}}{\sum x_i \frac{1-\pi_i}{\pi_i}} \left(X - \sum_{i \in s} x_i\right)$$

$$= t_B.$$

10. Question:

Give a suitable estimated measure of error of Brewer's predictor for a finite population total.

Solution:

Since the Brewer's predictor

$$t_B = \sum_{i \in s} y_i + \frac{\sum_{i \in s} y_i \left(\frac{1-\pi_i}{\pi_i}\right)}{\sum_{i \in s} x_i \left(\frac{1-\pi_i}{\pi_i}\right)} \left(X - \sum_{i \in s} x_i\right)$$

is also the generalized regression (greg) predictor

$$t_g = \sum_{i \in s} \frac{y_i}{\pi_i} + \left(X - \sum_{i \in s} \frac{x_i}{\pi_i}\right) \frac{\sum_{i \in s} y_i x_i Q_i}{\sum_{i \in s} x_i^2 Q_i}$$

in the special case when

$$Q_i = \frac{1 - \pi_i}{\pi_i - x_i}$$

it is appropriate to estimate

$$M = E_p \left(t_g - Y\right)^2$$
$$= \sum\sum_{i<j} (\pi_i \pi_j - \pi_{ij}) \left(\frac{E_i}{\pi_i} - \frac{E_j}{\pi_j}\right)^2 + \sum_1^N \frac{E_i^2}{\pi_i} \beta_i$$

writing $E_i = y_i - B_Q x_i$, $\beta_i = 1 + \frac{1}{\pi_i} \sum_{j+i} \pi_{ij} - \sum_1^N \pi_i$

$$B_Q = \frac{\sum_1^N y_i x_i Q_i \pi_i}{\sum_1^N x_i^2 Q_i \pi_i} \text{ by}$$

$$m = \sum\sum_{i<j} \left(\frac{\pi_i \pi_j - \pi_{ij}}{\pi_{ij}}\right) \left(\frac{e_i}{\pi_i} - \frac{e_j}{\pi_j}\right)^2 + \sum_{i \in s} \frac{e_i^2}{\pi_i^2} \beta_i$$

writing
$$e_i = y_i - b_Q x_i,$$
$$b_Q = \frac{\sum_{i \in s} y_i x_i Q_1}{\sum_{i \in s} x_i^2 Q_i}$$

and take $Q_i = \frac{1-\pi_i}{\pi_i x_i}$ in m and finally take that value of m as an estimated measure of error of t_B.

Relating to Chapter 6

11. Question:

Suppose y is a real-valued variable relating to a stigmatizing characteristic like expenses on treatment of AIDS, gain or loss last month through gambling, money earned or spent in dubious means with y_i as values relevant to an ith person $i = 1, \cdots, N$. To gather a response from a sampled person suppose an investigator approaches with a box of cards marked either (i) genuine with C $(0 < C < 1)$ as their proportion or (ii) marked $x_i, ...x_j.....x_M$ with respective proportions $q_1, ...q_j...q_M$ $(0 < q_j < 1, j = 1, \cdots, M$ such that $C + \sum_{j=1}^{M} q_j = 1)$. The device produces the RR from the ith person as

$z_i = y_i$ if "genuine" card appears
$ = x_j$ if x_j "marked" card appears

Then, $$E_R(z_i) = C y_i + \sum_{j=1}^{M} q_j x_j$$

So, $r_i = \dfrac{1}{C}\left(z_i - \sum_{1}^{M} q_j x_j\right)$ satisfies $E_R(r_i) = y_i$ and

$$V_i = V_R(r_i) = \frac{1}{C^2} V_R(z_i)$$
$$= \frac{1}{C^2}\left[E_R(z_i^2) - E_R^2(z_i)\right]$$
$$= \frac{1}{C^2}\left[\left(C y_i^2 + \sum_{j=1}^{M} q_j x_j^2\right) - \left(C^2 y_i^2 + \left(\sum_{1}^{M} q_j x_j\right)^2\right.\right.$$
$$\left.\left. + 2 C y_i \left(\sum_{1}^{M} q_j x_j\right)\right)\right]$$
$$= \frac{1}{C^2}\left[C(1-C) y_i^2 - 2 C y_i \left(\sum_{1}^{M} q_j x_j\right)\right.$$
$$\left. + \sum q_j x_j^2 - \left(\sum q_j x_j\right)^2\right]$$
$$= \alpha\, y_i^2 + \beta\, y_i + \theta, \text{ say, with } \alpha, \beta \text{ and } \theta$$

as known quantities.

Solution:
An unbiased estimator for V_i is
$$v_i = \frac{\alpha\, r_i^2 + \beta\, r_i + \theta}{(1 + \alpha)}.$$

For a proof Chaudhuri (1992) may be seen.

12. Question:

With the setup in Question 11 find an RR-based estimator for y_i with a variance less than V_i.

Solution:

Let from the ith sample person the RR be

$z_i = y_i$ if "genuine" card comes

$ = x_j + fy_i$ if x_j "marked" card comes.

Then,

$$E_R(z_i) = Cy_i + \sum_1^M (x_j + fy_i)$$
$$= [C + f(1-C)]y_i + \sum q_j x_j = Ay_i + B, \text{ say.}$$

Then,
$$r_i = (z_i - B)/A \quad \text{and} \quad E_R(r_i) = y_i,$$
$$A = C + f(1-C)$$
$$E_R(z_i^2) = Cy_i^2 + \sum_j q_j(x_j + fy_i)^2.$$

So, $V_R(z_i) = Cy_i^2 + f^2 y_i^2 \sum q_j$
$$+ 2fy_i \sum q_j x_j - (Ay_i + B)^2 + \sum q_j x_j^2$$
$$= Ty_i^2 + Fy_i + G,$$

writing
$$T = C + f^2(1-C) - A^2,$$
$$F = 2f \sum q_j x_j - 2AB,$$
$$G = \sum q_j x_j^2 - B^2.$$

Now compare $\alpha y_i^2 + \beta y_i + \theta$ versus $Ty_i^2 + Fy_i + G$ to check their comparative values given $\alpha, \beta, \theta, T, F, G$.

13. Question:

Corresponding to

$$t_g = \sum_{i \in s} \frac{y_i}{\pi_i} + \left(X - \sum_{i \in s} \frac{x_i}{\pi_i} \right) b_Q$$

writing
$$b_Q = \frac{\sum_{i \in s} y_i x_i Q_i}{\sum_{i \in s} x_i^2 Q_i}, \quad Q_i > 0,$$

based on direct responses y_i and fixed and known x_i's we shall take

$$e_g = t_g\big|_{y_i = r_i}$$

for RR's as r_i satisfying

$$E_R(r_i) = y_i, \; V_R(r_i) = \alpha y_i^2 + \beta y_i + \theta.$$

The following property is given by Chaudhuri and Roy (1997) when the DR's y_i are subject to the following model specifications:

$$E_m(y_i) = \beta x_i, \; V_m(y_i) = \sigma_i^2, \; C_m(y_i, y_j) = \sigma_{ij}, \text{ namely,}$$

(i) $\lim E_p(t) = Y$, $\lim E_p(e) = R$ with $R = \sum_1^N r_i$
(ii) $\lim E_p E_n E_R(e - y)^2 \geq \lim E_p E_m E_R (e_g - Y)^2$
and also
(iii) $\lim E_p E_m (t - Y)^2 \geq \lim E_p E_m (t_g - y)^2$
$$= \sum \sigma_i^2 \left(\frac{1}{\pi_i} - 1\right) + \sum\sum \sigma_{ij} \left(\frac{\pi_{ij}}{\pi_i \pi_j} - 1\right).$$

Solution:
See Chaudhuri and Roy (1997).

14. Question:
As an alternative to Warner's (1965) RR procedure, suppose the respondent is approached with a box of identical cards differing only in that a proportion p $(0 < p < 1, p \neq \frac{1}{2})$ of them are marked A and the others marked A^C. An ith respondent is to honestly respond the number g_i which is the first trial on which a Match is obtained between the person's trait and the card type. How is the proportion θ of persons bearing A estimated, and how do we measure the accuracy in estimation?
Solution:
See Chaudhuri and Christofides (2013) to find

$$E_R(g_i) = \frac{y_i}{p} + \frac{1 - y_i}{1 - p},$$

$$V_R(g_i) = y_i \frac{(1-p)}{p^2} + (1 - y_i) \frac{p}{(1-p)^2}$$

$$\Rightarrow r_i = \frac{p(1-p)g_i - 1}{1 - 2p} \; . \; \ni \; . \; E_R(r_i) = y_i,$$

$$V_i = V_R(r_i) = \frac{p^3}{(1-2p)^2} + \frac{1 - p + p^2}{(1-2p)} y_i,$$

$$\Rightarrow v_i = V_i\big|_{y_i = r_i}, \; i \in s \; . \; \ni \; . \; E_R(v_i) = V_i.$$

Relating to Chapter 7

15. Question:

Suppose in an office there are $N=1186$ employees attached to $D=39$ departments. By a suitably designed survey utilizing Basic Pays (z_i's) in choosing the sample by Lahiri's (1951) scheme of PPAS (Probability proportional to aggregate size, i.e., $p(s) \propto \left(\frac{\sum_{i \in s} z_i}{\sum_1^N z_i}\right)$) a sample is supposed to be drawn. You are required to estimate the dearness allowances (y_i) in the monthly pays for all office workers and also department-wide and also using the "take-home" pays of all the workers with the known totals department-wide. Suggest suitable estimators giving justifications.

Solution:
First postulate the model M_d to write

$$y_i = \beta_d x_i + \epsilon_i, \; i \in U_d, \; d = 1, \cdots, D$$
$$E_m(\epsilon_i) = 0, \; V_m(\epsilon_i) = \sigma^2, \; (\sigma_i > 0 \; \forall \; i)$$
$$\epsilon_i\text{'s independent } \forall \; i$$
$$I_{si} = 1/0 \text{ if } i \in s / i \notin s$$
$$I_{di} = 1/0 \text{ if } i \in U_d / i \notin U_d, \; d = 1, \cdots, D.$$

For $Y_d = \sum_{i \in U_d} y_i$, one estimator is

$$t_1 = \sum_{i \in U_d} \frac{Y_i I_{si}}{\pi_i}$$

wtiting $s_d = s \cap U_d$, $\hat{\beta}_{Qd} = \frac{\sum_{i \in s} y_i x_i Q_i I_{di}}{\sum_{i \in s} x_i^2 Q_i I_{di}}$,

$Q_i (>0)$, $e_{di} = y_i - \hat{\beta}_{Qd} x_i$, $X_d = \sum_1^N x_i - I_{di}$;

a second estimator is

$$t_2 = X_d \hat{\beta}_{Qd} + \sum_{i \in s} e_{di} \frac{I_{di}}{\pi_i}$$
$$= \sum_{i \in s} y_i \frac{I_{di}}{\pi_i} \hat{\beta}_{Qd} \left(X_d - \sum_{i \in s} x_i \frac{I_{di}}{\pi_i} \right)$$
$$= \sum_{i \in s} \frac{y_i}{\pi_i} I_{di} g_{sdi},$$

writing $\quad g_{sdi} = 1 + \left(X_d - \sum_{i \in s} x_i \frac{I_{di}}{\pi_i} \right) \frac{x_i Q_i \pi_i}{\sum_{i \in s} x_i^2 Q_i \pi_i}$

Then,

$$_1\hat{V}(t_2) = \sum\sum_{i<j\in s} \frac{\pi_i\pi_j - \pi_{ij}}{\pi_{ij}} \left(e_{di}\frac{I_{di}}{\pi_i} - e_{dj}\frac{I_{dj}}{\pi_j} \right)^2$$

and

$$_2\hat{V}(t_2) = \sum\sum_{i<j\in s} \left(\frac{\pi_i\pi_j - \pi_{ij}}{\pi_{ij}} \right) \left(\frac{e_{di}I_{di}g_{sdi}}{\pi_i} - \frac{e_{di}I_{di}g_{sdi}}{\pi_j} \right)^2.$$

Thirdly, revising the model M_d by M so that β_d is taken as β for every $d = 1, \cdots, D$, let

$$\hat{\beta}_Q = \frac{\sum_{i\in s} y_i x_i Q_i}{\sum_{i\in s} x_i^2 Q_i},$$

the third estimator is

$$t_3 = X_d\hat{\beta}_Q + \sum_{i\in s} e_i \frac{I_{di}}{\pi_i}, \; e_i = y_i - \hat{\beta}_Q x_i.$$

Writing

$$g'_{sdi} = I_{di} + \left(X_d - \sum_{iins} \frac{x_i I_{di}}{\pi_i} \right) \frac{x_i Q_i \pi_i}{\sum_{i\in s} x_i^2 Q_i},$$

$$t_3 = \sum_{i\in s} \frac{y_i}{\pi_i} g'_{sdi}.$$

Its two variance estimators are

$$\nu_{s1} = \sum\sum_{i<j<\in s} \Delta_{ij} \left(\frac{e_i I_{di}}{\pi_i} - \frac{e_j I_{dj}}{\pi_j} \right)^2, \; \Delta_{ij} = \frac{\pi_i\pi_j - \pi_{ij}}{\pi_{ij}}$$

$$\nu_{s2} = \sum\sum_{i<j<\in s} \Delta_{ij} \left(\frac{e_i g'_{sdi}}{\pi_i} - \frac{e_j g'_{sdj}}{\pi_j} \right)^2.$$

For rational interpretations one may consult Chaudhuri and Adhikary (1998).

16. Question:

How will you modify the estimation procedures if the figures are available for four consecutive years and you need to estimate for the current year?

Solution:
Apply Kalman Filtering in tandem over territorial borrowing as is done across the departments. Chaudhuri and Maiti (1994, 1997) and Chaudhuri, Adhikary, and Seal (1997) may be consulted.

Relating to Chapter 8

17. Question:

In Chapter 8 constrained Network and constrained Adaptive sampling have been discussed. How will you constrain in a multi-stage sampling with the RHC scheme followed in the first stage but SRSWOR in the next four stages?

Solution:

With usual notations $t = \sum_n y_i \frac{Q_i}{p_i}$ in a single-stage RHC sample unbiasedly estimates $Y = \sum_1^N y_i$. Suppose all the n groups cannot be covered though the survey is already in progress. Take an SRSWOR of $m\,(2 < m < n)$ of the n initial groups.

Let
$$e = \frac{n}{m} \sum_m y_i \frac{Q_i}{p_i}.$$

Then, writing E_n, E_m, V_n, V_m, E, V for the obvious operators one gets

$$E_m(e) = t = \sum_n z_i, \text{ say, } z_i = y_i \frac{Q_i}{p_i}$$

$$V_m(e) = n^2 \left(\frac{1}{m} - \frac{1}{n}\right) \frac{1}{(n-1)} \sum_n \left(z_i - \frac{\sum_n z_i}{n}\right)^2$$

$$= E_m(e^2) - t^2.$$

Then,
$$\nu_m(e) = n^2 \left(\frac{1}{m} - \frac{1}{n}\right) \frac{1}{(m-1)} \sum_n \left(z_i - \frac{\sum_m z_i}{m}\right)^2$$

has $E_m \nu_m(e) = V_m(e)$.

Writing
$$B = \frac{\sum_n N_i^2 - N}{N^2 - \sum_n N_i^2},$$

$$w = B\left[\frac{n}{m}\sum_m Q_i \frac{y_i^2}{p_i^2} - (e^2 - \nu_m(e))\right] \text{ it follows}$$

$$\nu(y) = v = \nu_m(e) + w$$

$$= (1+B)\nu_m(e) + B\left[\frac{n}{m}\sum_m Q_i \frac{y_i^2}{p_i e} - e^2\right]$$

has $E\nu(y) = V(e).$

Exercises and Solutions Supplementaries

Let
$$y_i = \sum_1^{M_i} y_{ij} = \sum_1^{M_i}\sum_1^{T_{ij}} y_{ijk} = \sum_1^{M_i}\sum_1^{T_{ij}}\sum_1^{P_{ijk}} y_{ijkr}$$
$$= \sum_1^{M_i}\sum_1^{T_{ij}}\sum_1^{P_{ijk}}\sum_1^{ijkru} y_{ijkru}.$$

Let
$$d_{ijkr} = \frac{L_{ijkr}}{l_{ijkr}} \sum_{l_{ijkr}} y_{ijkru},$$
$$c_{ijk} = \frac{P_{ijk}}{p_{ijk}} \sum_{p_{ijk}} d_{ijkr},$$
$$b_{ij} = \frac{T_{ij}}{t_{ij}} \sum_{t_{ij}} c_{ijk},$$
$$r_i = \frac{M_i}{m_i} \sum_{m_i} b_{ij}.$$

Then
$$g = \frac{n}{m} \sum_m r_i \frac{Q_i}{p_i}$$

is an unbiased estimator for Y with obvious notations above. Letting

$$a_{ijkr} = L_{ijkr}\left(\frac{1}{l_{ijkr}} - \frac{1}{L_{ijkr}}\right) \sum_{l_{ijkr}} \left(y_{ijkru} - \frac{\sum_u y_{ijkru}}{l_{ijkr}}\right)^2$$

$$w_{ijk} = P_{ijk}^2\left(\frac{1}{p_{ijk}} - \frac{1}{P_{ijk}}\right) \frac{\sum_{kr}\left(d_{ijkr} - \frac{\sum_r d_{ijkr}}{p_{ijk}}\right)^2}{p_{ijk} - 1}$$
$$+ \frac{P_{ijk}}{p_{ijk}} \sum_r Q_{ijkr}$$

$$\nu_{ij} = T_{ij}^2\left(\frac{1}{t_{ij}} - \frac{1}{T_{ij}}\right) \frac{\sum_k\left(c_{ijk} - \frac{\sum_k c_{ijk}}{t_{ij}}\right)^2}{(t_{ij} - 1)} + \frac{T_{ij}}{t_{ij}} \sum_k w_{ijk}$$

$$\nu_i = M_i^2\left(\frac{1}{m_i} - \frac{1}{M_i}\right) \frac{\sum_i\left(b_{ij} - \frac{\sum_j b_{ij}}{m_i}\right)^2}{(m_i - 1)} + \frac{M_i}{m_i} \sum_j \nu_{ij}.$$

Finally,
$$\nu(g) = \nu(r) + \frac{n}{m}\sum_m \nu_i \frac{Q_i}{p_i}$$

writing
$$\nu(r) = \phi\nu(y)|_{y_i = r_i}$$
and
$$E\nu(g) = V(e).$$

Chaudhuri (2003) is a relevant reference.

References

Abraham, T.P., Khosla, R.K. and Kathuria, O.P. (1969). Some investigations on the use of successive sampling in pest and disease surveys. JISAS, 21(2), 43-57.

Anderson, T.W. (1958). Introduction to Multivariate Statistical Analysis. John Willey, NY.

Arnab, R. (1981). Sampling on two occasions with varying probabilities. AJS, 23, 360-364.

Asok, C. (1980). A note on the comparison between simple mean and mean based on distinct units in sampling with replacement. Amer. Stat. 34, 158.

Avadhani, M.S. and Sukhatme, B.V. (1965). Controlled random sampling. JISAS, 17, 1, 34-42.

Avadhani, M.S. and Sukhatme, B.V. (1966). A note on ratio and regression methods of estimation under controlled simple random sampling. JISAS, 18, 2, 17-20.

Avadhani, M.S. and Sukhatme, B. V. (1967). Controlled random sampling with equal and unequal probabilities. AJS, 9, 8-15.

Avadhani, M.S. and Sukhatme, B.V. (1968). Simplified procedure for designing controlled simple random sampling. AJS, 10, 1-7.

Avadhani, M.S. and Sukhatme, B.V. (1970). A comparison of two sampling procedures with an application to successive sampling. Applied Statistics, 19, 231-259.

Avadhani, M.S. and Sukhatme, B.V. (1972). Sampling on several successive occasions with equal and unequal probabilities and without replacement. AJS, 14, 109-119.

Avadhani, M.S. and Sukhatme, B.V. (1973). Controlled sampling with equal probabilities and without replacement. Int. Stat. Rev. 41, 175-182.

Avadhani, M.S. and Sukhatme, B.V. (1979). Controlled sampling with equal probabilities and without replacement. ISR, 41, 175-182.

Barabesi, L, Franceschi, S. and Marcheselli, M. (2012). A randomised response procedure for multiple sensitive questions. Statistical Papers, 53, 703-718.

Basu, D. (1958). On Sampling with and without replacement. Sankhyā. 20, 287-294.

Basu, D. (1969). Role of the suffiency and likelihood principles in sample survey theory. Sankhyā, Ser A. 31, 441-454.

Basu, D. (1971). An essay on the logical foundations of survey sampling. Part I. In Foundations of statistical inference. ed. Godambe, VP. and Sprott, DA. Holt, Rinehart & Winston. Toronto, Canada. 203-242.

Binder, D.A. and Dick, J.P. (1989). Modeling and estimation for repeated surveys. Survey Methodology. 15, 29-45.

Binder, D.A. and Hidiroglou, M.A. (1988). Sampling in time. Handbook of Statistics Vol. 6. ed. Krishnaiah, North Holland, Amsterdam.

Blackwell, D. and Girshick, M.A. (1954). Theory of games and statistical decisions. John Wiley, NY.

Blight, B.J.N. and Scott, A.J. (1973). A stochastic model for repetitive surveys. JRSS, Ser B.,35, 61-66.

Bose, M., Chaudhuri, A., Dihidar, K. and Das, S. (2010). Model-cum-design-based estimation of the prevalence rate of a disease in a locality using spatial smoothing. Statistics, 45, 3, 293-305.

Box, G.E.S. and Jenkins, G.M. (1970). Time series analysis: Forecasting and control. Holden-Day, Amsterdam, NL.

Brewer, K.R.W. (1963). Ratio estimation and finite populations: some results deducible from the assumption of an underlying stochastic process. AJS, 5, 93-105.

Brewer, K.R.W. (1979). A class of robust sampling designs for large-scale surveys. JASA, 74, 911-915.

Cassel, C.M., Sarndal, C.E. and Wretman, J.H. (1976). Some results on generalized difference estimation and generalized regression estimation for finite populations. Biometrika, 63, 615-620.

Chakrabarti, M.C. (1963). On the use of incidence matrices in sampling from finite populations. JISA, 1, 78-85.

Chaudhuri, A. (1969). Minimax solutions of some problems in sampling from a finite population. Cal. Stat, Assoc. Bull. 18, 1-24.

Chaudhuri, A. (1977). On some problems of choosing the sample size in estimating finite population totals. BISI, 47(4), 116-119.

Chaudhuri, A. (1977). Some applications of the principles of Bayesian sufficiency and invariance to inference problems with finite populations. Sankhyā, Ser. C. 39, 140-149.

Chaudhuri, A. (1978). On the choice of sample size for the Horvitz-Thompson estimator. JISAS, 30(1), 35-42.

Chaudhuri, A. (1985). An optimal and related strategies for sampling on two occasions with varying probabilities. JISAS, 37(1), 45-53.

Chaudhuri, A. (1987). Randomized response surveys of finite populations: A unified approach with quantitative data. JSPI, 15, 157-165.

Chaudhuri, A. (1988). Optimality of sampling strategies. In HBS, 6, 47-96.

Chaudhuri, A. (1992). Randomized response: Estimating mean square errors of linear estimators and finding optimal unbiased strategies. Metrika, 39, 341-357.

Chaudhuri, A. (2000). Network and adaptive sampling with unequal probabilities. Cal. Stat. Assoc. Bull, 50, 237-253.

Chaudhuri, A. (2001). Using randomized response from a complex survey to estimate a sensitive proportion in a dichotomous finite population. JSPI, 94, 37-42.

Chaudhuri, A. (2003). Estimation from an under-covered sample in a complex survey for auditing. CSAB, 54, 115-120.

Chaudhuri, A. (2010). Essentials of Survey Sampling. Prentice Hall of India.

Chaudhuri, A. (2011a). Unbiased estimation of a sensitive proportion in general sampling by three non-ramdomized schemes. Jour. Stat. Theo. Prac., 6(2), 376-381.

Chaudhuri, A. (2011b). Randomized Response and Indirect Questioning Techniques in Surveys. Chapman and Hall, CRC Press, Taylor & Francis Group, Boca Raton, FL.

Chaudhuri, A. (2012a). Developing small domain statistics-modeling in survey sampling (e-Book, Lambert Academic Publishing, Saarbrucken, Germany.)

Chaudhuri, A. (2012b). Panel Rotation sampling in unequal probability sampling. Invited paper presented at the 8th Triennial of Calcutta Statistical Association.

Chaudhuri, A. (2013). Panel rotation with general sampling schemes. Jour. Ind. Soc. Agri. Stat. 67(3), 301-304.

Chaudhuri, A. (2013). Panel rotation with general sampling schemes. - to appear in JISAS.

Chaudhuri, A. (2013). Panel rotation with general sampling schemes. Jour. Ind. Soc. Agri. Stat. 67(3), 301-304.

Chaudhuri, A. (2014). Network and Adaptive Sampling. To be published by Science Publishers.

Chaudhuri, A. and Arnab, R. (1977). On the relative efficiencies of a few strategies of sampling with varying probabilities on two occasions. Cal. Stat. Assoc. Bull., 26 (101-104), 25-38.

Chaudhuri, A. and Arnab R. (1978). On the role of sample size in determining efficiency of Horvitz-Thompson estimators. Sankhya, Ser C., 40, 104-109.

Chaudhuri, A. and Arnab, R. (1979). On estimating the mean of a finite population sampled on two occasions with varying probabilities. AJS, 21(2), 162-165.

Chaudhuri, A. and Arnab, R. (1982). On unbiased variance-estimation with various multi-stage sampling strategies. Sankhya, Ser B., 44(1), 92-101.

Chaudhuri, A. and Christofides, T.C. (2007). Item count technique in estimating the proportion of people with a sensitive feature. JSPI, 137, 589-593.

Chaudhuri, A. and Christofides, T.C. (2008). Indirect questioning: How to rival randomized response techniques. Int. J. Pure and Appl. Math., 43(2), 283-294.

Chaudhuri, A. and Christofides, T.C. (2013). Indirect Questioning in Sample Surveys. Springer-Verlag, Berlin, Heidelberg, Germany.

Chaudhuri, A., Christofides, T.C. and Saha, A. (2009). Protection of privacy in efficient application of randomized response techniques. Stat. Math. a PPLI, 18, 389-418.

Chaudhuri, A. and Dihidar, K. (2009). Estimating means of stigmatizing qualitative and quantitative variables from discretionary responses randomized or direct. Sankhya, Ser B., 71, 123-136.

Chaudhuri, A. and Maiti, T. (1994). Borrowing strength from past data in small domain prediction by Kalman filtering—a case study. Comm. Stat. Theo. Meth., 23(12), 3507-3514.

Chaudhuri, A. and Maiti, T. (1997). Small domain estimation by borrowing strength across time and domain—a case study. Comm. Stat. Simul. Comp., 26(4), 1547-1557.

Chaudhuri, A. and Mukerjee, R. (1985). Optionally randomized responses techniques. CSAB, 34, 225-229.

Chaudhuri, A. and Mukerjee, R. (1987). Randomized responses techniques: A review. Stat. Neerlandica, 41, 27-44.

Chaudhuri, A. and Mukerjee, R. (1988). Randomized Response: Theory and Techniques. Marcel Dekker, New York.

Chaudhuri, A. and Mukhopadhyay, P. (1978). A note on how to choose the sample size for Horvitz-Thompson estimation. Bull. Cal. Stat. Assoc., 27, 149-154.

Chaudhuri, A. and Pal, S. (2002). On certain alternative mean square error estimators in complex survey sampling. JSPI, 104, 363-375.

Chaudhuri, A. and Pal, S. (2003). Estimating from under-covered sample in a complex survey for auditing. Cal. Stat. Assoc. Bull., 54, 37-42.

Chaudhuri, A. and Roy, D. (1997). Model assisted survey sampling strategies with randomized response. JSPI, 60, 61-68.

Chaudhuri, A. and Saha, A. (2004). Extending Sitter's mirror-match bookstrap to cover Rao-Hartley-Cochran sampling in two-stages with simulated illustrations. Sankhya, 66(4), 791-802.

Chaudhuri, A. and Saha, A (2005). Optional versus compulsory randomized response techniques in complex surveys. JSPI, 135, 516-527.

Chaudhuri, A. and Stenger, H. (2005). Survey Sampling: Theory and Methods. 2nd Ed., Chapman and Hall, CRC Press, Taylor & Francis Group, New York.

Chaudhuri, A. and Vos, J.W.E. (1988). Unified Theory and Strategies of Survey Sampling. North Holland. Amsterdam.

Chaudhuri, A., Adhikary, A.K. and Dihidar, S. (2000). Mean square error estimator in multi-stage sampling. Metrika, 52, 115-131.

Chaudhuri, A., Adhikary, A.K. and Seal, A.K. (1997). Small domain estimation by empirical Bayes and Kalman Filtering procedures—A case study. Comm. Stat. - Theo Meth., 26(7), 1613-1621.

Chikkagoudar, M.S. (1966). A note on inverse sampling with equal probabilities. sankhya, Ser A., 28, 93-96.

Chotai, J. (1974). A note on Rao-Hartley-Cochran method for PPS over two occasions. Sankhya, Ser C., 36, 173-180.

Choudhry, G.H. and Rao, J.N.K. (1989). Evaluation of small area estimators: An empirical study. Pre-print seen through the courtesy of the second author.

Christofides, T.C. (2009). Randomized response without a randomized device. Advances and Applications in Statistics, 11, 15-28.

Chua, T.C. and Tsui, A.K. (2000). Procuring honest responses indirectly. JSPI, 90, 107-116.

Cochran, W.G. (1977). Sampling Techniques. Wiley Eastern Limited, Delhi, India.

Cramer, H. (1946). Mathematical methods of statistics. Princeton University Press.

Dalenius, T. (1957). Sampling in Sweden—Contributions to the methods and theories of sample survey practice. Almqvist and Wicksell, Sweden.

Dalenius, T. and Gurney, M. (1951). The problem of optimum stratification. II. Skand. Akt. 34, 133-134.

Dalenius, T. and Hodges, J.L., Jr. (1959). Minimum variance stratification. JASA, 54, 88-101.

Das, M.N. (1982). Systematic sampling without drawback. Tech. Rep. 8206. ISI, Delhi.

David, H.A. (1978). Contributions to survey sampling and applied statistics papers in honor of H.O. Hartley Academic Press Inc. Cochran, W.G. Laplace's Ratio Estimator pp 3-10.

Doss, D.C., Hartley, H.O. and Somayajulu, G.R. (1979). An exact small sample theory for post-stratification. JSPI, 3, 235-247.

Droitcour, E.M., Larson, E.J. and Scheuren, F.J. (2001). The three card method: Estimating sensitive items with permanent anonymity of response. Proc. SOC. Stat. Sec. ASA, Alexandria, VA.

Eckler, A.R. (1955). Rotation Sampling. Ann Math. Stat., 26, 664-685.

Efron, B. (1982). The jackknije , the resampling plans, Soc. Ind. Appl. Math. CBMS. Nat. Sc. Found. Monograph 38.

References

Eichorn, B. and Hayre, L.S. (1983). Scrambled randomized response methods for obtaining sensitive data. JSPI, 7, 307-316.

Ericson, W.A. (1969). Subjective Bayesian models in sampling finite populations (with discussion). JRSS Ser. B, 31, 195-233.

Ericson, W.A. (1970). On a class of uniformly admissible estimators of a finite population total. AMS 41, 1369-1372.

Eriksson, SC. (1973). A new model for RR. ISR, 41, 101-113.

Fay, R.E. and Herriot, R.A. (1979). Estimation of income from small places: An application of James-Stein procedures to census data. JASA, 74, 269-277.

Foody, W. and Hedayat, A. (1977). On theory and applicability of BIB designs with repeated blocks. AS, 5, 932-945.

Fox, J.A. and Tracy, P.E. (1986). Randomized Response: A Method for Sensitive Surveys. Sage, London.

Franklin, L.A. (1989a). Randomized response sampling from dichotomous population with continuous randomization, Survey Methodology, 15(2), 225-235.

Franklin, L.A. (1989b). A comparison of estimators for randomized response sampling with continuous distributions from dichotomous population. Comm. Stat. Theo. Math., 18(2), 489-505.

Freund, J.E. (1994). Mathematical Statistics, 5th Ed. Prentice Hall.

Ghangurde, P.D. and Rao, J.N.K. (1969). Some results on sampling over two occasions. Sankhya, Ser A., 31, 463-472.

Ghosh, J.K. (1963). A game theory approach to the problem of optimum allocation in stratified sampling with multiple characters. Cal. Stat. Assoc. Bull., 12, 4-12.

Godambe, V.P. (1955). A unified theory of sampling from finite populations. JRSS, 17, 269-278.

Godambe, V.P. (1966). A new approach to sampling from finite populations, I, II. JRSS, Ser B, 310-319, 320-328.

Godambe, V.P. (1968). Bayesian sufficiency in survey sampling. AISM, 20, 363-373.

Godambe, V.P. (1969). Some aspects of the theoretical developments in survey sampling. NDSS, 27-58.

Godambe, V.P. and Thompson, M.E. (1977). Robust near optimal estimation in survey practice. BISI, 47, 3, 129-146.

Goodman, L.A. and Kish, L. (1950). Controlled selection—A technique in probability sampling. JASA, 55, 350-372.

Goswamy, J.N. and Sukhatme, B.V. (1965). Ratio method of estimation in multiphase sampling with several auxiliary variables. JISAS, 17(1), 83-103.

Greenberg, B.G, Abul-Ela, A.L, Simmons, W.P. and Horvitz, D.G. (1969). The unrelated question RR model: Theoretical framework. JASA, 64, 520-539.

Gupta, V.K., Mandal, B.N. and Parsad, R. (2012). Combinatorics in sample surveys vis-a-vis controlled selection. Lambert Academic Publication.

Gupta, V.K., Nigam, A.K. and Kumar, P. (1982). On a family of sampling schemes with inclusion probability proportional to size. Biometrika, 69, 191-196.

Gurney, M. and Daly, J.F. (1965). A multi-variate approach to estimation in periodic sample surveys. Proc. Amer. Stat. Assoc. Soc. Stat. Sec. 245-257.

Ha'jek, H. (1964). Asymptotic theory of rejective sampling with varying probabilities from a finite population. AMS, 35, 1491-1523.

Ha'jek, H. (1971). Comment on a paper by Basu, D. In Foundations of Statistical Inference pp 203-242. Ed. Godambe, V.P. and Sprott, D.A. Holt, Rinehart & Winston, Toronto, Canada.

Ha'jek, J. (1949). Representative sampling by the method of two phases in Czeck-Statistics Obzor. 29, 384-394.

Ha'jek, J. (1958). Some contributions to the theory of probability sampling. BISI, 36(3), 127-133.

Ha'jek, J. (1959). Optimum strategy and other problems in probability sampling. Cas. pet. mat., 84, 387-475.

Hall, W.J., Wijsman, R.A. and Ghosh, J.K. (1965). The relationship between sufficiency and invariance with sequential analysis. AMS, 36, 575-614.

Hansen, M.H. and Hurwitz, W.N. (1943). On the theory of sampling from finite populations. Ann. Math. Stat., 14, 333-362.

Hansen, M.H. and Hurwitz, W.N. (1946). The problem of non-response in sample-surveys. JASA, 41, 517-529.

Hanurav, T.V. (1962). Some sampling schemes in probability sampling. Sankhya, Ser. A, 24, 421-428.

Hanurav, T.V. (1966). Some aspects of unified sampling theory. Sankhya, Ser. A, 28, 175-204.

Hartley, H.O. and Ross, A. (1954). Unbiased ratio estimators. Nature, 174, 270-271.

Hedayat, A., Lin, B. and Stufken, J. (1989). The construction of IIPS sampling through a method of employing boxes. Ann. Stat. 17(4), 1886-1905.

Hege, V.S. (1965). Sampling designs which admit uniformly minimum variance unbiased estimators. Cal. Stat. Assoc. Bult., 14, 160-162.

Heiden, P.G.M., Van der, Gills, V.G., Bout, J. and Hox, J.J. (2000). A comparison of randomized response, computer-assisted self-interview and face-to-face direct questioning. Eliciting sensitive information in the context of welfare and unemployment benefit. Soc. Math. & Res., 28, 505-537.

Hendriks, W.A. (1944). The relative efficiencies of groups of farms as sampling units. JASA, 39, 366-376.

Horvitz, D.G. and Thompson, D.J. (1952). A generalization of sampling without replacement from a finite universe. JASA, 77, 89-96.

Horvitz, D.G., Shah, B.V. and Simmons, W.R. (1967). The unrelated question RR model Proc. Soc. Stat. Sec., ASA, 65-72.

Huang, K.C. (2004). A survey technique for estimating the proportion and sensitivity in a dichotomous finite population. Stat. Neerlandica, 58, 75-82.

James, W. and Stein, C. (1961). Estimation with quadratic loss. Proc. of Fourth Berkeley symposium on Math. Stat. California Press. 361-379.

Jessen, R.J. (1942). Statistical investigation of a sample survey for obtaining farm facts; Iowa Agri-Exper. Station Research Bill. No. 304.

Jones, V.M. (1965). The Jackknife Method. Proc. IBM Sci. Comp. Symp. Statist. 185-201.

Jones, R.G. (1980). Best linear unbiased estimators for repeated surveys. J.R.S.S, Ser B, 42, 221-226.

Kathuria, O.P. (1975). Some estimators in two-stage sampling on successive occasions with partial matching at both stages. Sankhya, 37(C), 146-162.

Kerkvliet, J. (1994). Estimating a logic model with randomized data: The case of cocaine use. AJS, 36(1), 9-20.

Kish, L. (1965). Survey Sampling, Wiley, NY.

References

Korwar, R.M. and Serfling, R.J. (1970). On averaging over distinct units in sampling with replacement. AMS, 41(6), 2132-2134.

Kulldorff, G. (1963). Some problems of optimum allocation for sampling on two occasions. ISR, 31, 24-57.

Lahiri, D.B. (1951). A method of sample selection providing unbiased ratio estimators. BISI, 33(2), 133-140.

Lanke, J. (1975). Some contributions to the theory of survey-sampling. Unpublished Ph.D. thesis. Univ. of Lund, Sweden.

Ljungqvist, L. (1993). A unified approach to measures of privacy in randomized response models: A utilitarian perspective. JASA, 88, 97-103.

Loeve, M. (1977). Probability Theory 4th Ed. Springer-Verlag.

Mahalanobis, P.C. (1938). Statistical report on the experimental crop census 1937. Indian Central Jute Committee.

Mahalanobis, P.C. (1940). A sample survey of the acreage under jute in Bengal. Sankhya, 4, 511-530.

Mahalanobis, P.C. (1942). Sample Surveys; 1942. Presidential Address, Proc. Ind. Sc. Congress. 25-46.

Mahalanobis, P.C. (1944). On large-scale sample surveys. Philosophical Transactions of Royal Society, London, 231, (B), 329-451.

Mahalanobis, P.C. (1946). Recent experiments in statistical sampling in the Indian Statistical Institute. JRSS. Ser A, 109, 325-378.

Mangat, N.S. (1994). An improved randomized response peocedure. Biometrika, 77(2), 439-442.

Mangat, N.S., Singh, R., Singh, S., Bellhouse, D.R. and Kashani, H.P. (1995). On efficiency of estimator using distinct respondents in randomized response survey. Survey Methodology, 21, 21-23.

Meinhold, R.J. and Singlpurwalla, N.D. (1983). Understanding the Kalman filter. AS, 37, 123-127.

Midzuno, H. (1950). An outline of the theory of sampling systems. AISM, 1, 149-156

Midzuno, H. (1952). An outline of the theory of sampling systems. AISM, 1, 149-156.

Miller, J.D., Cisin, I.H. and Harrel, A.V. (1986). A new technique for surveying deviant behaviour: Item count estimates of marijuana, cocaine and heroin. ASA paper. St. Petersburg, FL.

Mukhopadhyay, P. (1998). Theory and Methods of Survey Sampling. Prentice Hall of India.

Murthy, M.N. (1957). Ordered and unordered estimators in sampling without replacement. Sankhya, 18, 379-390.

Murthy, M.N. (1977). Sampling Theory and Methods. Stat. Pub. Soc. Cal. Ind.

Neyman, J. (1934). On the two different aspects of the representative method: The method of stratified sampling and the method of purposive selection. JRSS, 97, 558-625.

Neyman, J. (1938). Contributions to the theory of sampling human populations. JASA, 33, 101-116.

Nigam, A.K., Kumar, P. and Gupta, V.K. (1984). Some methods of inclusion probability proportional to size sampling. JRSS, Ser. B, 46, 564-571.

O'Hagan, A. (1987). Bayes linear estimator for randomized response models. JASA, 82, 580-585.

Ohlsson, E. (1992). SAMU—The system for co-ordination of samples from the Business Register at Statistics Sweden—A methodological description, R & D Report 1992:18 Stockholm: Statistics Sweden.

Ohlsson, E. (1995). Co-ordination samples using permanent random numbers. In Business Survey Methods. Ed. Cox, B.G., Binder, D.A., Chinnappa, B. Nanjamma, Christianson, A., Cokedge, M.J. & Kott, P.S., John Wiley & Sons, New York.

Olkin, I. (1958). Multi-variate ratio estimation for finite populations. Biometrika, 45, 154-165.

Pal, S. (2009). Rao and Wu's re-scaling modified to achieve coverages. JSPI, 139, 3552-3558.

Park, Y.S., Choi, J.W. and Kim, K.W. (2007). A balanceed multi-level rotation sampling design and its efficient composite estimators. JSPI, 137(2), 594-610.

Pathak, P.K. (1961). Use of "order statistics" in sampling without replacement. Sankhya, Ser. A . 23, 409-414.

Pathak, P.K. (1962). On simple random sampling with replacement. Sankhya, Ser. A., 24, 287-302.

Pathak, P.K. and Rao, T.J. (1967). Inadmissibility of customary estimators in sampling over two occasions. Sankhya, Ser. A., 29, 49-56.

Patterson, H.D. (1950). Sampling on successive occasions with partial replacement of units. JRSS, Ser B, 12, 241-255.

Pfeffermann, D. (1989). Robust small area estimation combining time series and cross-sectional data. Unpublished MS.

Politz, A.N. and Simmons, W.R. (1949). An attempt to get the "not at home" into the sample without callbacks. JASA, 44, 9-31.

Prasad, N.G.N and Rao, J.N.K. (1990). The estimation of the mean squared errors of small area estimators. JASA, 85, 163-171.

Quelnouille, M.H. (1949). Approximate tests of correlation in time-series. JRSS, 11, 68-84.

Raj, Des. (1956). Some estimators in sampling with varying probabilities without replacement. JASA, 51, 269-284.

Raj, Des. (1965). On sampling over two occasions with probability proportionate to size. AMS, 36, 327-330.

Raj, Des and Khamis, S.H. (1958). Some remarks on sampling with replacement. JASA, 51, 269-284.

Rao, J.N.K. (1975). Unbiased variance-estimation for multi-stage designs. Sankhya, 37, 133-139.

Rao, J.N.K. (1979). On deriving mean square errors and other non-negative unbiased estimators in finite population sampling. JISA, 17, 125-136.

Rao, J.N.K. (2003). Small Area Estimation. Wiley Interscience, NY, USA.

Rao, J.N.K. and Bellhouse, D.R. (1978). Optimal estimation of a finite population mean under generalized random permutation models. JSPI, 2, 125-141.

Rao, J.N.K., Hartley, H.O. and Cochran, W.G. (1962). On a simple procedure of unequal probability sampling without replacement . Jour. Roy. Stat. Soc. Ser. B, 24, 482-491.

Rao, J.N.K and Graham, J.E. (1964). Rotation designs for sampling on repeated occasions. JASA, 59, 492-509.

References

Rao, J.N.K. and Nigam, A.K. (1990). Optimal controlled sampling designs. Biometrika, 77(4), 807-814.

Rao, J.N.K and Wu, D.F.J. (1988). Resampling inference with complex survey data. JASA, 80, 620-630.

Royall, R.M. (1970). Finite population sampling theory under certain linear regression models. Biometrika, 57, 377-387.

Roychoudhury, D.K. (1957). Unbiased sampling design using information provided by linear function of auxiliary variate. Chapter 5, Thesis for Associateship of Indian Statistical Institute, Kolkata.

Sampford, M.R. (1967). Methods of cluster sampling with and without replacement for clusters of unequal sizes Biometrika, 49, 27-40.

Sarndal, C.E. (1982). Implications of survey design for generalized regression estimation of linear functions. JSPI, 7, 155-170.

Sarndal, C.E., Swensson, D.E. and Wretman, J.H. (1992). Model assisted survey sampling. Springer-Verlag, New York.

Scheers, N.J. (1992). A review of randomized response techniques in measurement and evaluation in counseling and development. Meas. Eval. Couns. & Dev., 25, 27-41.

Scott, A.J. (1977). On the problem of randomization in survey sampling. Sankhya, Ser. C, 39, 1-9.

Scott, A.J. and Smith, T.M.F. (1974). Linear super-population model in survey sampling. Sankhya, Ser. C, 36, 143-146.

Seber, G.A.F. and Salehi, M.M. (2012). Adaptive Sampling Designs. Springerbriefs. New York.

Sen, A.R. (1953). On the estimator of the variance in sampling with varying probabilities. JISAS, 5(2), 119-127.

Sen, A.R. (1973). Some theory of sampling on successive occasions. AJS, 15, 105-110.

Singh, D. (1968). Estimating in successive sampling using a multi-stage design. JASA., 63, 99-112.

Singh, D. and Chaudhary, F.S. (1986). Theory and Analysis of Sample Survey Designs. Wiley Eastern Limited, New Delhi, India.

Singh, D. and Kathuria, O.P. (1969). On two-stage sampling. AJS., 11, 59-66.

Singh, D. and Singh, B.D. (1965). Double sampling for stratification on successive occasions. JASA, 60, 784-792.

Singh, D. and Singh, R. (1973). Multi-purpose surveys on successive occasions. JISAS, 25, 81-90.

Singh, D., Singh, P. and Kumar, P. (1994). Handbook on Sampling Methods. Ind. Agri. Stat. Res. Inst., New Delhi,. India.

Singh, R. (1972). On Pathak and Rao's estimate in pps with replacement sampling over two occasions. Sankhya, Ser A., (34), 301-303.

Singh, R. (1980). A modified PPSWR scheme for sampling over two occasions. Sankhya, Ser C., 42, 124-127.

Skinner, C.J. (1989). Aggregated analysis: Standard errors and significant tests. In Analysis of Complex Surveys. ed. Skinner, C.J., Holt, D. and Smith, T.M.F. Wiley & Sons, New York p 24.

Smith, H.F. (1938). An empirical law describing heterogeneity in the yields of agricultural crops J. Agri. Sci., 28, 1-23.

Sudhakar, Kunte. (1978). A note on circular systematic sampling design. Sankhya, Ser. C. 40(1), 72-73.

Sukhatme, B.V. and Avadhani, M.S. (1965). Controlled selection: A technique in random sampling. AISM, $\underline{17}$, 25-28.

Sukhatme, P.V. and Sukhatme, B.V. (1954). Sampling Theory of Surveys with Applications. Ind. Soc. Agri. Stat., New Delhi, India.

Tan, M.T., Tian, G.L. and Tang, M.L. (2009). Sample surveys with sensitive questions: A non-randomized response approach. Amer. Stat., $\underline{63}$, 9-16.

Thompson, S.K. (1990). Adaptive cluster sampling. JASA, $\underline{85}$, 1050-1059.

Thompson, S.K. (1992). Sampling. John Wiley & Sons, New York.

Thompson, S.K. and Seber, G.A.F. (1996). Adaptive sampling. Wiley, New York. USA.

Tian, G.L., Yu, J.W, Tang, M.L. and Gang, Z (2007). A new non-randomized model for analyzing sensitive questions with binary outcomes. Statistics in Medicine, $\underline{26}$, 4238-4252.

Tikkiwal, B.D. (1951). Theory of successive sampling. Unpublished thesis, Ind. Council of Agri. Res., New Delhi.

Tracy, D and Mangat, N.S. (1996). Some developments in randomized response sampling during the last decades. A follow up of a revision by Chaudhuri and Mukerjee. JASS, 4(2/3), 147-158.

Tripathi, T.P. and Srivastava, O.P. (1979). Estimation on successive occasions using PPSWR sampling. Sankhya, Ser C. $\underline{41}$, 84-91.

Tukey, J.W. (1954). Bias and confidence in not quite large samples. AMS, $\underline{29}$, 614.

Warner, S.L. (1965). RR: A survey technique for eliminating evasive answer bias. JASA, $\underline{60}$, 63-69.

Wynn, H.P. (1977). Minimax purposive survey sampling design. JASA, $\underline{72}$, 655-657.

Yates, F. (1949). Sampling methods for censuses and surveys. Charles Griffin & Co., London.

Yates, F. and Grundy, P.M. (1953). Selection without replacement from within strata with probability proportional to size. JRSS, Ser. B, $\underline{15}$, 253-261.

Yu, J.W, Tian, G.L. and Tang, M.L. (2008). Two new models for survey sampling with sensitive characteristic: Design and analysis. Metrika, $\underline{67}$, 251-263.

Author Index

Abraham, TP., 166
Abul-Ela, AL., 284
Adhikary, AK., 119
Anderson, TW., 216
Arnab, R., 166, 168
Asok, C., 27
Avadhani, MS., 137, 138, 139, 165, 166

Barabesi, L., 184
Basu, D., 26, 27, 30, 42, 44, 128, 159, 261
Bellhouse, DR., 166, 184
Binder, DA., 181, 222
Blackwell, D., 262
Blight, BJN., 180
Bose, M., 261
Bouts, J., 184
Brewer, KRW., 148, 149, 155, 157, 262

Cassel, CM., 169, 211-12
Chakrabarti, MC., 138, 141
Chaudhary, FS., 106
Chaudhuri, A., 13, 51, 61, 113, 114, 115, 125, 127, 133, 134, 135, 161, 166, 167, 168, 169, 177, 183, 184, 185, 192, 193, 194, 197, 199, 200, 204, 205, 207, 222, 227, 229, 230, 231, 250, 261, 262
Chikkagoudar, MS., 30
Choi, JW., 171
Chotai, J., 165, 166
Choudhry, GH., 222
Christofides, TC., 183, 184, 194, 197, 199, 204, 205
Chua TC., 184
Cisin, IH., 199
Cochran, WG., 62, 106, 115, 118, 119, 127, 132, 133, 161, 165, 168, 176, 230, 231
Cramer, H., 155

Dalenius, T., 95
Daly, JF., 165, 166
Das, MN., 61
Das, S., 261

David, HA., 46
DeMorgan, A., 38
Dick, JP., 220
Dihidar, K., 157, 245, 247
Dihidar, S., 113, 115, 193, 261
Droitcowe, EM., 184

Eckler, AR., 165
Efron, B., 250
Eichorn, BH., 184
Eriksson, SC., 160, 184

Fay, RE., 215
Foody, W., 139
Fox, JA., 184
Francishi, S., 184
Franklin, LA., 184
Freund, JE., 24

Gang, Z., 204
Ghangurde, PD., 165, 166
Ghosh, JK., 161, 262
Gills, Van G., 184
Girshick, MA., 262
Godambe, VP., 44, 46, 148, 159, 161
Goodman, LA., 136
Goswamy, JN., 165
Graham, JE., 165
Greenberg, BG., 184, 185
Grundy, PM., 35, 51, 139, 250
Gupta, VK., 139
Gurney, M., 95, 165, 166

Ha'jek, H., 51, 58, 73, 74, 78, 79, 113, 251
Hall, WJ., 161
Hansen, MH., 53, 55, 68, 173, 174
Hanurav, TV., 10, 46
Harrel, AV., 199
Hartley, HO., 51, 58, 60, 62, 112, 115, 119, 165, 168, 230, 231
Hayre, LS., 184
Hedayat, A.S., 139
Hege, VS., 46
Heiden, V., 184

Author Index

Hendriks, WA., 103
Herriot, RA., 215
Hidiroglou, MA., 181, 222
Hodges, JL., Jr., 95
Horvitz, DG., 29, 31, 37, 71, 80, 125, 128, 184, 185, 209, 250
Hox, JJ., 184
Huang, KC., 184
Hurwitz, W.N., 53, 55, 68, 173, 175

Ikeda, J., 57

James, W., 218
Jessen, RJ., 103, 104, 164, 165
Jones, VM., 180

Kashani, HP., 184
Kathutia, OP., 166
Kerkvliet, J., 184
Khamis, SH., 27, 30
Khosla, RK., 166
Kiron, KW., 171
Kish, L., 136, 262
Korwar, RM., 27
Kulldorff, G., 165
Kumar, P., 106, 139

Lahiri, DB., 51, 54, 57, 58, 141, 166
Lanke, J., 46, 168
Larson, EJ., 184
Ljungqvist, L., 184
Loeve, M., 251

Mahalanobis, PC., 53, 84, 103
Mandal, BN., 139
Mangat, NS., 184
Marcheselli, M., 184
Meinhold, RJ., 184, 220
Midzuno, H., 141, 166
Miller, JD., 199
Mukerjee, R., 184, 192
Mukhopadhyay, P., 106, 168, 207
Murthy, MN., 72, 73, 79, 80, 106, 118

Neyman, J., 89, 125
Nigam, AK., 139, 140, 141

O'Hagan, A., 184
Ohlsson, E., 254
Olkin, I., 165

Pal, S., 5, 61, 113, 114, 125, 200, 229, 230, 250
Park, YS., 171
Parsad, R., 139
Pathak, PK., 27, 28, 29, 30, 165
Patterson, HD., 165
Politz, AN., 175

Quenouille, MH., 249

Raj, D., 30, 68, 72, 118, 165, 166
Rao, JNK., 51, 58, 62, 63, 73, 74, 78, 79, 113, 115, 116, 119, 120, 125, 139, 140, 141, 165, 166, 168, 207, 217, 218, 222, 229, 230, 232, 250, 251, 275.
Rao, TJ., 165
Ross, A., 51, 58, 60
Roy, D., 262
Royall, RM., 148, 149
Roychoudhury, DK., 69

Saha, A., 184, 192, 194, 250
Sampford, MR., 114
Sarndall, CE., 169, 211-12
Scheers, NJ., 184
Scheuren, FJ., 184
Scott, A., 160, 180
Seber, GAF., 226
Sen, AR., 57, 141, 165, 166
Serfling, RJ., 27
Shah, BV., 184
Simmons, WP., 184
Simmons, WR., 175, 184
Singh, BD., 166
Singh, D., 106, 166
Singh, P., 106
Singh, R, 165, 166, 184
Singh, S., 184
Singhpurwalla, ND., 184, 220

Author Index

Skinner, CJ., 263
Smith HF., 103
Smith, TMF., 180
Srivastava, OP., 166
Stein, C., 218
Stenger, H., 125, 229, 230, 250, 261
Sudhakar, Kunte, 62
Sukhatme, BV., 105, 106, 130, 137
Sukhatme, PV., 105, 106
Swensson, DE., 169

Tan, MT., 204
Tang, ML., 204
Thompson, DJ., 29, 31, 35, 37, 71, 80, 125, 128, 148, 161, 209, 250
Thompson, SK., 163, 226
Tian, GL, 204
Tikkiwal, BD., 165
Tracy, D., 184
Tracy, PE., 184
Tripathi, TP., 166
Tsui, AK., 184
Tukey, JW., 249

Warner, SL., 183, 184, 185, 186, 197
Wijsman, RA., 161
Wretman, JH., 169, 211-12
Wu, CFJ., 250, 251
Wynn, HP., 139, 140

Yates, F., 35, 51, 139, 165, 250
Yu, JW., 204

Subject Index

A Admissibility 57
Adaptive Sampling 225
Analytical Surveys 239
B Bias 5
Bayesian 143
BLUP 215
Bayes Estimation 216
Bootstrap 239
Business Surveys 239
BRR 239, 256
C Confidence interval 7
Complete Class 48
Cluster sampling 85
Controlled sampling 85
Contingency 239
D Double Sampling 131
Double Bootstrap Method 251
Design effects 262
E Empirical Bayes Methods 208, 215
EBLUP 215
F Fay-Herriot Methods 215
Frame 97
G GDE 128
Greg 152, 208
H Horvitz-Thompson Estimator 26
I Interval Estimation 7
Imputation 143
Indirect surveys 183
Indirect Questioning 199
ICT 199
J Jack-knifing 239, 248
K Kalman Filtering 219
L Likelihood 26, 159
Linearization 239
M MSE 5
Multi-stage sampling 85
Multiphase sampling 85
Minimaxity 262
N Non-Response 143

Not at home 143
Nominative Techniques 200
Network 225
O ORR 192
P Point Estimator 6
PPS 53
PPS Systematic 51
Post stratification 85
Panel Rotation 143
Permanent Random Number 239
Percentile Methods 251
Q Quenouille's Method 249
R Random number 12
Ratio type 60
Regression 85
Randomized response 183
RRT 184
S Standard error 5
Sampling design 4
Sufficiency 26
Stratified sampling 85
Super-population 143
Spatial Smoothing 143
Small Area Estimation 207
Synthetic greg 213
Size control 225
T Tukey's method 249